SULPHUR-CONTAINING DRUGS AND RELATED ORGANIC COMPOUNDS
Chemistry, Biochemistry and Toxicology

Volume 1: Part B
Metabolism of Sulphur Functional Groups

Ellis Horwood Series in
BIOCHEMICAL PHARMACOLOGY
Series Editor: Dr L. A. DAMANI, King's College London, University of London.

SULPLHUR-CONTAINING DRUGS AND RELATED ORGANIC COMPOUNDS
Chemistry, Biochemistry and Toxicology
Volume One: Metabolism of Sulphur Functional Groups. Parts A and B
Editor: L. A. DAMANI, King's College London, University of London.

SULPLHUR-CONTAINING DRUGS AND RELATED ORGANIC COMPOUNDS
Chemistry, Biochemistry and Toxicology
Volume Two: Analytical, Biochemical and Toxicological Aspects of Metabolic Reactions at Sulphur.
Parts A and B
Editor: L. A. DAMANI, King's College London, University of London.

SULPLHUR-CONTAINING DRUGS AND RELATED ORGANIC COMPOUNDS
Chemistry, Biochemistry and Toxicology
Volume Three: Metabolism and Pharmacokinetics of Sulphur Drugs. Parts A and B
Editor: L. A. DAMANI, King's College London, University of London.

SULPHUR-CONTAINING DRUGS AND RELATED ORGANIC COMPOUNDS
Chemistry, Biochemistry and Toxicology

Volume 1: Part B
Metabolism of Sulphur Functional Groups

Editor:

L. A. DAMANI, B.Pharm., M.Sc., Ph.D., M.R.Pharm.S.
Lecturer in Pharmacy
Chelsea Department of Pharmacy
King's College London, University of London

ELLIS HORWOOD LIMITED
Publishers · Chichester

Halsted Press: a division of
JOHN WILEY & SONS
New York · Chichester · Brisbane · Toronto

First published in 1989 by
ELLIS HORWOOD LIMITED
Market Cross House, Cooper Street,
Chichester, West Sussex, PO19 1EB, England
The publisher's colophon is reproduced from James Gillison's drawing of the ancient Market Cross, Chichester.

Distributors:

Australia and New Zealand:
JACARANDA WILEY LIMITED
GPO Box 859, Brisbane, Queensland 4001, Australia

Canada:
JOHN WILEY & SONS CANADA LIMITED
22 Worcester Road, Rexdale, Ontario, Canada

Europe and Africa:
JOHN WILEY & SONS LIMITED
Baffins Lane, Chichester, West Sussex, England

North and South America and the rest of the world:
Halsted Press: a division of
JOHN WILEY & SONS
605 Third Avenue, New York, NY 10158, USA

South-East Asia
JOHN WILEY & SONS (SEA) PTE LIMITED
37 Jalan Pemimpin # 05–04
Block B, Union Industrial Building, Singapore 2057

Indian Subcontinent
WILEY EASTERN LIMITED
4835/24 Ansari Road
Daryaganj, New Delhi 110002, India

© 1989 L. A. Damani/Ellis Horwood Limited

British Library Cataloguing in Publication Data
Damani, L. A. (Lyaguatali Abdulrasnl), *1949–*
Sulphur-containing drugs and related organic compounds.
Vol. 1
Pt. B: Metabolism of sulphur functional groups
1. Sulphur compounds
I. Title
546'.7232

Library of Congress data available

ISBN 0–7458–0216–8 (Ellis Horwood Limited)
ISBN 0–470–21258–6 (Halsted Press)

Typeset in Times by Ellis Horwood Limited
Printed in Great Britain by Hartnolls, Bodmin

Table of contents

SULPHUR-CONTAINING DRUGS AND RELATED ORGANIC COMPOUNDS:
Chemistry, Biochemistry and Toxicology

Volume 1: Metabolism of sulphur functional groups
Part A
1. Aspects of sulphur chemistry, biochemistry and xenobiochemistry, *L. A. Damani*: 2. Naturally occurring sulphur compounds, *A. B. Hanley and G. R. Fenwick*: 3. Agricultural importance of sulphur xenobiotics, *G. T. Brooks*: 4. Sulphur compounds as industrial and medicinal agents, *L. A. Damani and M. Mitchard*: 5. Inorganic sulphur compounds, *J. Westley*: 6. Thioethers, thiols, dithioic acids and disulphides: phase I reactions, *L. A. Damani*: 7. Thioethers, thiols, dithioic acids and disulphides: phase II reactions, *J. Caldwell and H. M. Given*.
Part B
1. Phosphorothionates, *F. De Matteis*: 2. Thioamides, *J. R. Cashman*: 3. Thiocarbamides, *G. G. Skellern*: 4. Carbamothioates and carbamodithioates, *Y. Segall*: 5. Sulphoxides and sulphones, *A. G. Renwick*: 6. Sulphonium salts, *P. A. Crooks*: 7. Sulphonamides, *P. A. Crooks*: 8. Sulphamates, sulphonates and sulphate esters, *A. G. Renwick*: 9. Sulphur heterocycles, *D. J. Rance*: 10. Glucosinolates, alliins and cyclic disulphides: sulphur-containing secondary metabolites, *G. R. Renwick and A. B. Hanley*

Volume 2: Analytical, biochemical and toxicological aspects of sulphur xenobiochemistry
Part A
1. Use of [^{35}S]-labelled drugs in research, *J. C. T. Lang*: 2. Analysis of thiols and disulphides, *I. D. Wilson and J. K. Nicholson*: 3. S-Oxygenases, I: chemistry and biochemistry, *D. M. Ziegler*: 4. S-Oxygenases, II: chirality of sulphoxidation reactions, *D. R. Boyd, C. T. Walsh and Y.-C. J. Chen*: 5. S-Oxygenases, III: human pharmacogenetics, *S. C. Mitchell and R. H. Waring*: 6. Thiol S-methyltransferases, I: biochemistry, *R. Weinshilboum*: 7. Thiol S-methyltransferases, II: pharmacogenetics, *R. Weinshilboum*
Part B
1. Interactions of sulphur-containing xenobiotics with cytochrome(s) P-450 and glucuronyl transferases, *H. G. Oldham*: 2. Interactions of sulphur-containing xenobiotics with S-adenosyl-L-methionine-dependent methyltransferases, *P. A. Crooks*: 3. Depletion of the sulphane pool: toxicological implications, *J. Westley*: 4. Thiol-disulphide exchange: physiological and toxicological aspects, *I. A. Cotgreave, L. Atzori and P. Moldéus*: 5. Cysteine conjugate β-lyase, I: toxic thiol production, *H. Tateishi and H. Tomisawa*: 6. Cysteine conjugate β-lyase, II: isolation, properties and structure–activity relationships, *P. N. Shaw and I. S. Blagbrough*: 7. Toxicological implications of the metabolism of carbon disulphide by microsomal monoxygenases, *R. R. Dalvi*

Volume 3: Metabolism and pharmacokinetics of sulphur-containing drugs
Part A
1. Elemental sulphur: a purgative, keratolytic and paraciticidal agent, *B. W. Burt*: 2. Phenothiazine antipsychotics, *S. C. Mitchell*: 3. H$_2$-Receptor antagonists, *M. Mitchard, R. L. McIsaac and J. A. Bell*: 4. Mucolytic agents: cysteine derivatives, *R. H. Waring*: 5. Anthelmintics — phenothiazine, thiabendazole and related compounds, *P. Galtier*: 6. Chlormethiazole: anticonvulsant, hypnotic and anxiolytic, *K. Wilson*: 7. Uricosurics and antirheumatics: sulphinpyrazone and sulindae, *A. G. Renwick, H. A. Strong and C. F. George*
Part B
1. Cytotoxic agents: 6-mercaptopurine, 6-thioguanine and related compounds, *L. Lennard*: 2. Antirheumatics: penicillamine, *J. W. Coleman and B. K. Park*: 3. Angiotensin-converting enzyme inhibitor: captopril, *B. K. Park and J. W. Coleman*: 4. Intravenous anaesthetics: thiopentone and related compounds, *D. J. Morgan*: 5. Antithyroid agents; methimazole, carbimazole and propylthiouracil, *G. G. Skellern*: 6. Antituberculous drugs: ethionamide and related compounds, *R. H. Waring*: 7. Antiinfective skin preparations: malathion, *R. C. Wester and J. R. Cashman*: 8. Miscellaneous sulphur drugs, *P. N. Shaw, I. S. Blagbrough and L. A. Damani*

Preface

Sulphur occurs widely in Nature in the elemental state, as H_2S and SO_2, in various sulphide ores of metals, and in the form of numerous sulphates; its average abundance in the biosphere has been estimated to be around 600 ppm. Sulphur is essential to the life and growth of all organisms — from microbes to man. Most microorganisms and plants can reduce oxidized forms of inorganic sulphur (e.g. sulphate), incorporating the sulphur into the sulphur amino acids (cysteine and methionine). Mammals in general are incapable of using inorganic sulphur, and their need is met by a supply of sulphur amino acids from plant sources.

Sulphur is a common element in many endogenous materials such as amino acids, enzymic and structural proteins, vitamins, co-enzymes and plant secondary metabolites. The biochemistry of endogenous organosulphur compounds has therefore been the subject of much scientific interest (Young, L. and Maw, G. A., *The Metabolism of Sulphur Compounds*, Methuen, London, 1958; Greenberg, D. M., editor, *Metabolism of Sulphur Compounds*, Academic Press, New York, 1975; Anderson, J. W., *Sulphur in Biology*, Edward Arnold, London, 1978; Jakoby, W. B. and Griffith, O. W., editors, *Sulphur and Sulphur Amino Acids*, in Methods in Enzymology, Volume 143, Academic Press, New York, 1987).

In the last two decades, the emphasis in sulphur biochemistry research has shifted towards exogenous synthetic compounds, since sulphur is a common element in numerous industrial, agricultural and medicinal compounds. In the latter case, almost all pharmacological classes are represented, e.g. H_2-receptor antagonist (cimetidine), gout prophylactic and antiplatelet (sulphinpyrazone), mucolytic (acetylcysteine), antirheumatic (sulindac, penicillamine), anaesthetic (thiopentone), cytotoxic (thioguanine), etc. Such drugs, and other sulphur xenobiotics, represent a spectrum of chemical classes, e.g. thioether (cimetidine — a drug; dimethylsulphide — a beer component), sulphoxide (sulindac — a drug; alkenyl cysteine sulphoxides — constituents of onion and garlic), sulphone (dapsone — a drug), thiol (thioguanine — a drug; propane thiol — a constituent of onion), thione (thiopentone — a drug; parathion — an insecticide), etc. In some instances the sulphur functional group may be present as a relatively inconsequential structural feature. In many other cases metabolic reactions at the appropriate sulphur functionalities play an important role in disposition and clearance. These biotransformations may also profoundly alter the pharmacological and toxicological properties of the sulphur xenobiotics. Despite this increased interest in recent years in this area of drug metabolism, i.e. in 'sulphur xenobiochemistry', until now a comprehensive compilation of metabolic and toxicological data on sulphur compounds has not been published.

Sulphur-containing Drugs and Related Organic Compounds: Chemistry, Biochemistry and Toxicology is an attempt at collecting and organizing material on sulphur xenobiochemistry into a 'library' of books, which will serve as a useful reference source on this subject. Authors of individual chapters, who are all active investigators, were asked to ensure that the coverage of material was comprehensive. At the same time it was stressed that the editor did not wish to edit a collection of annotated references. Despite the fact that the three volumes cover vastly different types of topics, at varying levels of existing knowledge, the invited contributors have tried to attain a thorough and critical exposition of their subjects.

The first of these volumes, *Metabolism of Sulphur Functional Groups*, is subdivided into Parts A and B which together cover the chemical and biochemical reactivity of organic compounds having different types of sulphur functionalities. Volume 2, *Analytical, Biochemical and Toxicological Aspects of Sulphur Xenobiochemistry* (Parts A and B) covers problems and pitfalls in analysis of sulphur xenobiotics, and describes in detail the chemistry and biochemistry of enzymes that mediate various metabolic reactions at sulphur, with emphasis on how such biotransformations can often affect the pharmacololgy and toxicology of these compounds. Chapters in Volume 3, *Metabolism and Pharmacokinetics of Sulphur-containing Drugs* (Parts A and B), are in the style of monographs, with reference to many different, but structurally related, compounds in each therapeutic class. An attempt is made in these chapters to comment on the relationship between chemical structure and metabolism/pharmacokinetics within each class of compounds.

In a multi-authored work of this nature, overlap is almost inevitable. Although this has been kept to a minimum, in many cases it was felt that repetition was desirable to present the same material from a different perspective, and, through this duplication, to make the chapters self-contained and more readable. Nonetheless an attempt has been made at extensive cross-referencing, between chapters in the same volume, and between the three volumes. In addition to the individual Part indexes, a combined index for all three volumes appears at the end of Volume 3. It is hoped that this collection of detailed, well-referenced reviews from authors actively involved in studying the metabolism of sulphur xenobiotics will lead to conceptualizations from the vast amount of compound specific data that has accumulated and furthermore to identification of fruitful areas for future research.

June, 1988 L. A. Damani

1

Phosphorothionates

F. De Matteis
Toxicology Unit, Medical Research Council, Woodmansterne Road, Carshalton,
Surrey SM5 4EF, UK

SUMMARY

1. Phosphorothionates undergo a number of metabolic reactions within the body and since the various pathways of metabolism either convert phosphorothionates into toxic metabolites or, alternatively, afford protection against their toxicity, metabolism and toxicity will be discussed together in this chapter.

2. Cytochrome P-450 converts phosphorothionates to the corresponding oxygen analogues (or oxons), an oxidative desulphuration reaction during which sulphur is liberated in a reactive form. This leads to 'suicidal' inactivation of cytochrome P-450, a response also seen with thiocarbonyl compounds (such as CS_2) which are similarly metabolized. Intermediary reactive S-oxides are also formed, which may be responsible for a more general cellular damage, for example in the liver.

3. Once the phosphorothionates have been desulphurated to the corresponding oxons, these can interact with two distinct esterases in the nervous system, a reaction of great importance in the mechanism of production of two types of neurotoxicity. (a) Inhibition of acetylcholinesterase is responsible for the symptoms of acute toxicity in mammals (as well as for the insecticidal properties) of these compounds. (b) Phosphorylation of another esterase in the nervous system (the neuropathy target esterase), followed by transfer of an alkyl group from the bound phosphate ester to the enzyme, initiates another toxic response, that of the delayed neuropathy.

4. Phosphorothionates undergo degradation by oxidative, hydrolytic and glutathione-dependent mechanisms, and these reactions inactivate the insecticides and contribute significantly to the resistance of certain insect strains which are capable of accelerating breakdown of these compounds. These degradation pathways also afford protection against the mammalian toxicity of phosphorothionates and a clear example of the protective role of hydrolytic enzymes in man is represented by the toxic epidemic observed in Pakistan among workers spraying certain formulations of parathion.

1.1 INTRODUCTION

Organophosphorus compounds, originally developed as chemical warfare agents in the 1930s, have acquired and still retain an important use as insecticides for crop protection and also for control of vector-borne diseases like malaria (see Chapter 3, Part A of this volume). Of these the phosphorothionates, parathion and malathion, are probably the most widely employed. These organic triesters possess a thiono-sulphur functional group, that is a sulphur atom double bonded to the phosphorus, and their structure is shown in Fig. 1 together with structures of some other sulphur-containing organophosphorus compounds which will be referred to in this chapter. The mechanism of action, selective toxicity and metabolism of these compounds have been very extensively investigated and are now largely understood. Their insecticidal action and acute toxicity in mammals have both been shown to be due to inhibition of acetylcholinesterase in the nervous system, leading to accumulation of toxic levels of the neurotransmitter, acetylcholine. Whereas the phosphorothionates are not directly toxic, as they are not inhibitors of acethylcholinesterase, they can be converted by mixed function monooxygenases to the corresponding phosphate triesters or oxons, which are potent inhibitors of the enzyme. Both insects and mammals are capable of this metabolic conversion of the thionates to the toxic oxons, but most mammalian species (including man) also possess more effective hydrolytic enzymes than the insects, thus inactivating the inhibitor and escaping

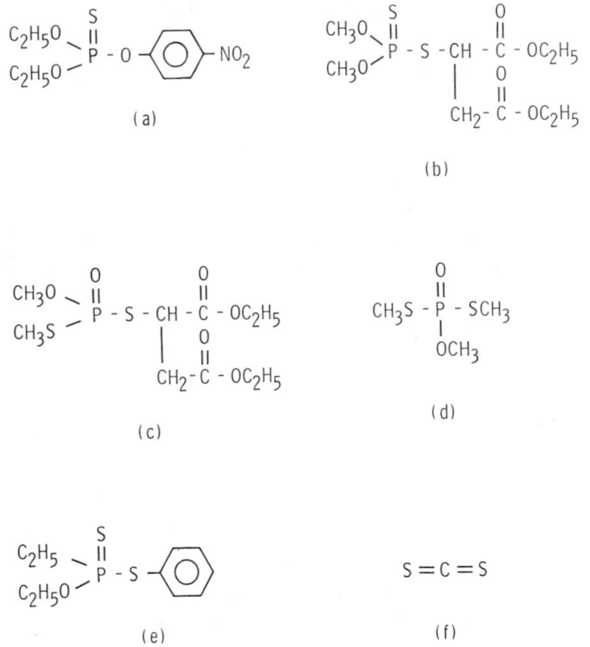

Fig. 1 — Chemical structure of several sulphur-containing chemicals discussed in this chapter: (a) parathion; (b) malathion; (c) isomalathion; (d) O,S,S-trimethyl phosphorodithiolate; (e) dyfonate; (f) carbon disulphide.

toxicity. The balance between these two, activating and inactivating, pathways of metabolism is therefore often the critical factor in determining whether these compounds will be acutely toxic or not.

Two additional types of toxic responses to the phosphorothionates have been identified and these too have been shown to be due not to the phosphorothionates themselves but to products of their metabolism. First it has been shown that, on metabolism by cytochrome P-450, a phosphorothionate will produce reactive derivatives capable of inactivating cytochrome P-450 or, in some cases, of damaging the cell where such an oxidative metabolism has taken place. The second type of toxicity involves interaction of the corresponding oxons with a second enzyme in the nervous system, the neurotoxicity target esterase (NTE), followed by transfer of an alkyl group from the bound phosphate ester to the enzyme (the so-called aging reaction), which initiate a toxic neuropathic response.

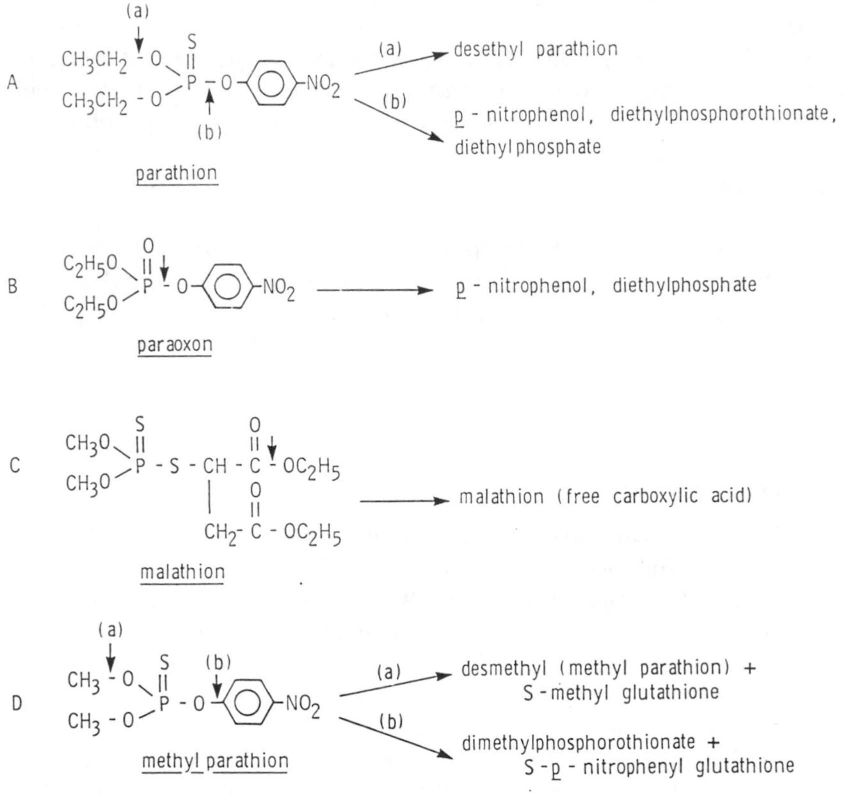

Fig. 2 — Different pathways of breakdown of phosphorothionates and related compounds, showing bonds cleaved (indicated by arrows) and products obtained in different reactions. (A) Oxidative breakdown catalysed by cytochrome P-450 involving (a) O-dealkylation or (b) a combined oxidative–hydrolytic mechanism (see also Fig. 3). (B) Purely hydrolytic pathway involving A esterases. (C) Hydrolysis of carboxyl ester function(s) of malathion by carboxyl esterases. (D) Glutathione S-transferase catalysed breakdown of methyl parathion with transfer to glutathione of either (a) a methyl group of (b) the aryl group.

Therefore it is quite clear that the various pathways of metabolism of the phosphorothionates are all related to their mechanism of toxicity and, for this reason, metabolism and toxicity will be considered together in this chapter. The main emphasis will be placed on the role of cytochrome P-450 in these reactions, as metabolism by monooxygenation appears to be a prerequisite for all types of phosphorothionate toxicity. However, the interaction of these compounds with esterases and with glutathione S-transferases will also be briefly considered, as well as the purely chemical isomerization reaction of phosphorothionates, thought to be responsible for an extensive ourbreak of acute toxicity in workers exposed to certain commercial preparations of malathion. The metabolism and toxicity of certain other compounds which also possess the thione sulphur functional group will also be discussed, as they strongly resemble the phosphorothionates in the pathway of oxidation of their sulphur atom.

The phosphoro*thionates* are sometimes referred to as phosphoro*thioates* or thiophosphates; the nomenclature followed in this chapter will be that of phosphorothionates throughout, to emphasize the presence of the characteristic thionosulphur function (P=S); the term thiolate (or S-alkyl compound) will be used to indicate the presence of a thiol-type sulphur linking an alkyl group to the phosphorus, as in certain impurities of commercial preparations of malathion which will be briefly considered as potentiators of malathion toxicity. Finally, some phosphorothionates also possess a sulphur atom linking two carbons, and reference to this thioether sulphur atom will be made when discussing its pathway of oxidative metabolism.

Phosphorothionates and related organophosphorus compounds undergo degradation by oxidative, hydrolytic and glutathione-dependent mechanisms. Fig. 2 illustrates the bonds which are cleaved in each of these different pathways and reference to this figure will be made when discussing the various degradation mechanisms.

1.2   METABOLISM BY MONOOXYGENATION

1.2.1   Oxidative desulphuration of phosphorothionates and breakdown by oxidative–hydrolytic mechanism

Organophosphorus insecticides are often administered as phosphorothionates as they are more stable in this form, but they ultimately act by inhibiting acetylcholinesterase in the nervous system, and for this effect they must first be desulphurated to the corresponding oxygen analogues (or oxons), which are the actual inhibitors of the enzyme. An advantage of administering these compounds as inactive precursors is also that they may be broken down by mammalian degrading enzymes before they are activated and made toxic, thus affording a better safety margin in their use as insecticides.

The mechanism of this critical step of metabolic activation has been elucidated during the last 35 years and the role of cytochrome P-450 in this reaction has been clarified more recently. Diggle and Gage (1951) were the first to show that paraoxon, the inhibitor of acetylcholinesterase, originates in mammalian systems through an enzyme-catalysed desulphuration of parathion. The enzymic system was then shown to be associated with the microsomal fraction from both mammalian liver and the cockroach fat body and to exhibit properties characteristic of a monooxygenase reaction, such as the dependence on molecular oxygen and NADPH (NADH being also utilized but less effectively) (Davidson, 1955; Murphy and DuBois, 1957; Nakatsugawa and Dahm, 1965, 1967).

During the reaction sulphur was shown to be liberated as an unknown metabolite that became bound to the microsomal membranes (Nakatsugawa and Dahm, 1967; Poore and Neal, 1972): the toxicological significance of this finding will be discussed later.

A complicating feature of the reaction of oxidative desulphuration was the appearance of products of degradation of the phosphorothionate (p-nitrophenol, diethyl phosphorothionate and diethyl phosphate in the case of parathion), suggesting a hydrolytic mechanism of breakdown with cleavage of the O-aryl bond (see Fig. 2). However, molecular oxygen and NADPH were again required for this degradation to take place, somehow implicating a monooxygenase mechanism (Nakatsugawa and Dahm, 1967; Neal, 1967a). To explain this apparent contradiction the hypothesis was put forward of a common intermediate resulting from the reaction of the parent phosphorothionate with an activated form of oxygen, which could then either be desulphurated (to produce the oxon) or undergo ester bond cleavage to produce the p-nitrophenol and diethyl phosphorothionate (Knaak et al., 1962; Neal, 1967a).

Support for this common oxygenate intermediate has been obtained in both enzymatic and chemical systems. In order to establish the source of the oxygen in the oxon and degradation product (whether from moleculer oxygen or water), mechanistic studies involving ^{18}O were carried out by McBain et al. (1971a) and by Ptashne et al. (1971), who incubated phosphorothionates with microsomes and NADPH, either in an $^{18}O_2$-enriched atmosphere or alternatively in ^{18}O-enriched water. The results of these studies, summarized in Fig. 3, indicate clearly that molecular oxygen is the only source of the oxygen incorporated into the oxon, while that of dialkyl phosphorothionate originates entirely from water; dialkyl phosphate incorporates two oxygen atoms, one from molecular oxygen, the other from water.

A similar mechanism was proposed by both groups of workers involving addition of oxygen to the thiono sulphur of the phosphorothionate forming an S-oxide (or sulphine) intermediate which may exist as several possible resonating structures, one of them drawn in Fig. 3. The S-oxide may then undergo one of two different reactions. (1) It may react internally to form a three-membered ring of phosphorus, sulphur and oxygen (i.e. a phosphaoxathiirane) which would then lose sulphur and produce the corresponding oxon (or P=O compound) (Ptashne et al., 1971; Herriott, 1971). This is analogous to the pathway of conversion of thiocarbonyl S-oxides to carbonyl compounds via an intermediary oxathiirane (Snyder, 1974). (2) Alternatively, the intermediary S-oxide may suffer nucleophilic attack by water and break down (in the case of parathion) to p-nitrophenol plus diethyl phosphorothionate or diethyl phosphate, sulphur or oxygen respectively also being liberated (Ptashne et al., 1971). The pathway of metabolism shown to be involved with Dyfonate (a phosphonothionate, see Fig. 1 for structure) was essentially identical (McBain et al., 1971a) except that the appropriate oxon and phosphorus containing breakdown products were isolated, as well as thiophenol.

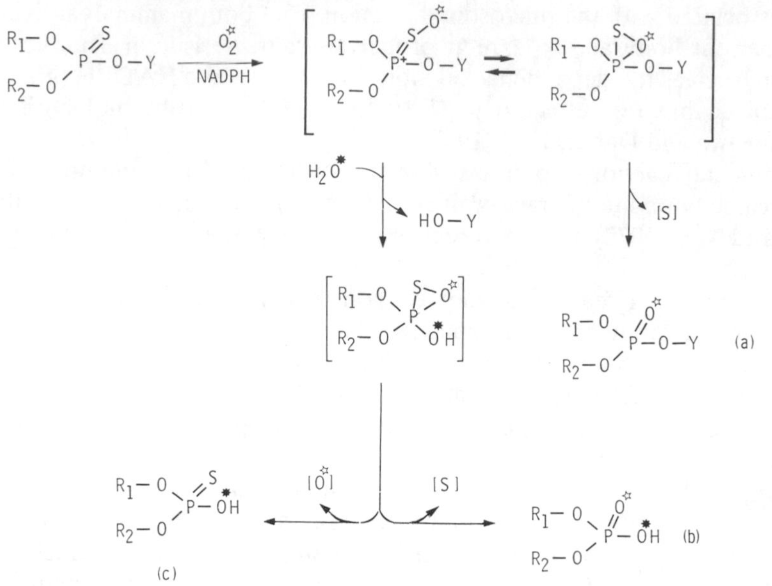

Fig. 3 — Mechanism of oxidative desulphuration and oxidative hydrolytic breakdown of a phosphorothionate, showing postulated S-oxide and phosphaoxathiirane intermediates and labelling pattern for oxygen incorporated from either O_2 or H_2O. (Adapted from Kamataki *et al.*, (1976) and see McBain *et al.* (1971a) and Ptashne *et al.* (1971) for ^{18}O labelling). The oxygen derived from molecular oxygen as indicated thus (☆) and that from water thus (★). In the case of parathion, (a) is paraoxon, (b) diethyl phosphate and (c) diethyl phosphorothionate.

Studies on the oxidation of phosphorothionates in purely chemical systems (employing either m-chloroperoxybenzoic acid or peroxytrifluoroacetic acid as the oxidants) also supported the concept of an oxygenated intermediate common to both desulphuration and hydrolytic breakdown (McBain *et al.*, 1971b; Ptashne and Neal, 1972). The former authors also isolated an unstable oxidation product of the thionate, with chemical properties compatible with addition of a single oxygen to the thiono-sulphur function, and this, on exposure to water, was shown to give rise to the same breakdown products obtained enzymatically in the microsomal system. In contrast to the phosphorothionates, no breakdown products could be obtained with the corresponding oxons (P=O compounds) either by chemical oxidation or enzymatically, again emphasizing the importance of oxygen addition to the thiono-sulphur as the initial step in the reaction.

Even though desulphuration and monooxygenase-dependent breakdown may proceed through a common oxygenated intermediate, it has been suggested that the two pathways may be catalysed by different mixed function oxidase enzymes, as they appeared to be differentially affected by inducers and inhibitors of the cytochrome P-450 system (Neal, 1967b; Norman *et al.*, 1973). More recently, however, it has been found that an apparently homogeneous preparation of rabbit liver cytochrome P-450 can support the production from parathion of both paraoxon and diethyl phosphorothionate in a reconstituted system (Kamataki *et al.*, 1976), so both pathways can apparently be initiated at a single catalytic site. The question then arises as to what is the controlling factor in partitioning the common intermediate between the two pathways, a point of practical as well as theoretical importance, since desulphuration is an activating pathway whereas breakdown affords protection from acute toxicity. The partition may depend to some extent on the chemical structure of the phosphorothionate and on the relative ease with which sulphur is lost from the intermediate S-oxide, as compared with the 'leaving' aryl group (Ptashne and Neal, 1972). Another important factor may be the extent to (and rapidity with) which the intermediate S-oxide comes in contact with water, something which is presumably facilitated once the metabolite has left the active site of cytochrome P-450. Perhaps desulphuration takes place preferentially within the active centre of the cytochrome and, once the intermediate has diffused away, hydrolytic breakdown is the main fate. Support for the desulphuration taking place within cytochrome P-450 while the reactive S-oxide may diffuse away to react with other targets in the cell is offered by the suicidal inactivation of cytochrome P-450 and toxic cellular damage, which will now be discussed.

1.2.2   Role of reactive sulphur metabolites in the suicidal inhibition of cytochrome P-450 and toxic cellular damage

Nakatsugawa and Dahm (1965) made the important observation that during metabolism of parathion by the microsomal fraction of cockroach fat body the activity of the desulphuration reaction gradually declined, as an unknown sulphur metabolite became bound. Even though these authors suggested a causal connection between these two observations, the orginal findings were not followed up and their toxicological implications were not fully appreciated for a number of years. Williams (1959) had already suggested, however, that desulphuration may be implicated in the toxic effect of phenylthiourea in the rabbit and, in agreement with this, Smith and Williams (1961) had reported a correlation between toxicity to rabbits and degree of desulphuration *in vivo*, when comparing a series of chemically related aryl-thioreas. They suggested a reductive mechanism involving production of the very toxic hydrogen sulphide.

Bond and De Matteis (1969) reported on the liver effects of carbon disulphide (CS_2) in rats. When CS_2 was given orally to starved male rats, a loss of cytochrome P-450 from the liver microsomal fraction was rapidly observed and this was accompanied by reduction in the activity of drug-metabolizing enzymes in the liver (the latter effect was also reported by Freundt and Dreher, 1969). The loss of cytochrome P-450 was too rapid in onset to be due to inhibition of cytochrome P-450 synthesis and too persistent for a direct effect of CS_2 itself, as this compound is volatile and rapidly eliminated. No histological lesions were caused by CS_2 under these conditions. However, when the rats were pretreated with phenobarbitone to increase the activity of the liver cytochrome P-450 system, then CS_2 caused an even greater loss of the cytochrome and this was accompanied, several hours later, by accumulation of water in the liver and by histological lesions in the centrilobular zones (Bond *et al.*, 1969), with pronounced hydropic degenerative changes, and, ultrastructurally, extensive dilation of the RER cisternae in the centrilobular hepatocytes (Butler *et al.*, 1974).

CS_2 was partially metabolized in the intact rat to CO_2 and, after phenobarbitone pretreatment, not only was the liver toxicity of CS_2 enhanced but also the extent of its oxidative conversion to CO_2 was increased. When metabolism was related to the degree of the cytochrome P-450 loss or to the severity of histological changes, a good correlation was found under a variety of experimental conditions. In addition, destruction of cytochrome P-450 could be demonstrated *in vitro* on aerobic incubation of liver microsomes with CS_2 but only when NADPH, the essential cofactor for monooxygenase reactions, was also present (De Matteis and Seawright, 1973). These findings all suggested that CS_2 required metabolism by cytochrome P-450 for both the destruction of the cytochrome (pointing to a suicidal type of inactivation reaction) and also for the more general cellular toxicity of the hepatocytes. Since a marked loss of the spectrally demonstrable cytochrome P-450 could be demonstrated without a corresponding early loss of microsomal haem, the primary target was likely to be the apoprotein moiety (De Matteis and Seawright, 1973): more direct evidence for a suicidal mechanism involving covalent modification of the apoprotein of cytochrome P-450 is discussed below (also see Chapter 7 by Dalvi in Volume 2, Part B of this series).

The conversion of CS_2 to CO_2 can be visualized as a two-stage oxidative desulphuration reaction, analogous to the desulphuration of phosphorothionates already discussed and similarly probably involving intermediary *S*-oxides, with final loss of elemental sulphur. During desulphuration of parathion sulphur had been shown to be liberated in a reactive form (Nagatsugawa and Dahm, 1965; Poore and Neal, 1972) and, in preliminary experiments with phosphorothionates, these were shown to resemble CS_2 in their ability to cause NADPH-dependent destruction of cytochrome P-450 *in vitro* as well as hydropic degeneration in the liver of rats *in vivo* (De Matteis and Seawright, 1973). The possibility that sulphur-related metabolites (such as reactive elemental sulphur) produced during the oxidative desulphuration of several chemicals might cause inactivation of cytochrome P-450 and cellular toxicity was therefore considered (De Matteis and Seawright, 1973; Norman *et al.*, 1974; Hunter and Neal, 1975) and further explored in the experiments described below.

(1) When liver microsomes were incubated in presence of NADPH with either [14]C- or ^{35}S-labelled CS_2 (under conditions, that is, where destruction of cytochrome P-450 was observed), CS_2 was actively desulphurated and a portion of the liberated sulphur became bound covalently to the microsomes (Dalvi *et al.*, 1974; De Matteis, 1974). Although some increase in ^{14}C binding (as compared with values obtained on incubation without NADPH) was also observed, possibly indicating the production of reactive oxides (Fig. 4), the binding of ^{35}S was greatly in excess of that of ^{14}C, so sulphur itself must have become bound. Covalent binding of sulphur to microsomal proteins could also be demonstrated *in vivo* after administration of labelled CS_2 (Järvisalo *et al.*, 1977): as in the *in vitro* experiments, the labelling of ^{35}S was greater than that of ^{14}C and was preferentially stimulated by pretreatment of the animals with phenobarbitone.

(2) A loss of cytochrome P-450 was also shown to occur during the desulphuration of parathion and other chemicals (Norman *et al.*, 1974; De Matteis, 1974; Uchiyama *et al.*, 1975; Hunter and Neal, 1975). In one study (De Matteis, 1974) three pairs of drugs were examined, each pair consisting of a compound containing sulphur (as either P=S or C=S) and the corresponding oxygen

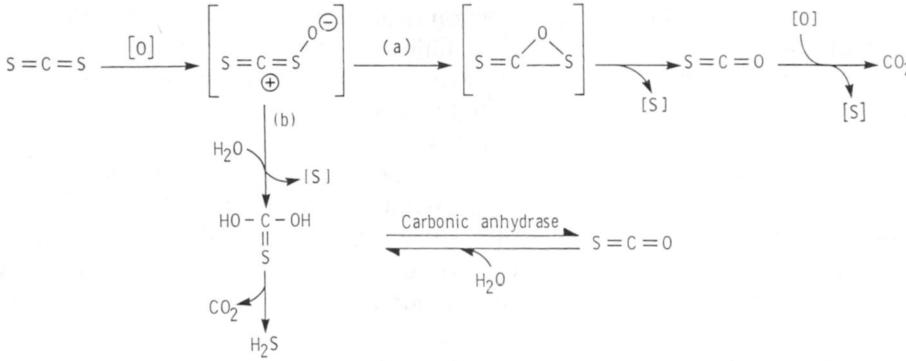

Fig. 4 — Mechanism of oxidative desulphuration of carbon disulphide. Two alternative pathways are illustrated, both involving formation of the S-oxide derivative as the first step. In (a) the S-oxide undergoes cyclization to an oxathiirane intermediate which then loses sulphur and produces COS. The latter can then undergo analogous desulphuration reaction to CO_2. Alternatively (b) the S-oxide may react with water, losing sulphur and producing monothiocarbonic acid which may then decompose to CO_2 and H_2S. Both pathways could produce elemental sulphur in a reactive form. Pathway (a) has been suggested by De Matteis and Seawright (1973) and Catignani and Neal (1975), and pathway (b) more recently by Chengelis and Neal (1987).

analogue. All three sulphur containing drugs (parathion, phenylthiorea and l-naphthyl isothiocyanate) caused a loss of cytochrome P-450) which was dependent on NADPH, whereas the oxygen-containing analogues were all devoid of such activity. In addition, the destruction of cytochrome P-450 caused by parathion could be inhibited by piperonyl butoxide and by replacing air with N_2, and stimulated by either NADH or, more markedly, by NADPH (De Matteis, 1974), all factors previously shown to inhibit and stimulate respectively the oxidative desulphuration of parathion (Nakatsugawa and Dahm, 1967, 1968).

(3) Phosphorothionates were also shown to produce histological evidence of liver damage when given to rats *in vivo*. Parathion has an LD 50 of 5–10 mg/kg (Parke, 1968) because its oxidative derivative is a powerful inhibitor of acetylcholinesterase. Certain analogues of parathion such as O,O-diethyl-O-phenyl phosphorothionate and O,O-diethyl-O-m-trifluoromethylphenyl phosphorothionate are similarly metabolized by the liver, but their general acute toxicity is much lower than that of parathion (Oppenoorth *et al.*, 1971), as their corresponding oxons are less powerful inhibitors of acetylcholinesterase (Aldridge and Davison, 1952). They can therefore be given to rats in much larger doses than parathion. When either of these two phosphorothionates was given to rats pretreated with phenobarbitone they were found to cause — like CS_2 — loss of cytochrome P-450 and centrilobular hydropic degeneration in the liver (Seawright *et al.*, 1976).

The mechanism of suicidal inhibition of cytochrome P-450 by CS_2 and phosphorothionates has to some extent been clarified. In experiments with purified cytochrome P-450 and involving a reconstituted drug-metabolizing system, the cytochrome was found to be the main (if not the only) target of the reactive sulphur liberated during the desulphuration reaction (Kamataki and Neal, 1976). With both parathion and CS_2 nearly half of the total sulphur bound to the microsomes could be released as thiocyanate on treatment with cyanide (Catignani and Neal, 1975; Kamataki and Neal, 1976), suggesting that a considerable proportion of the sulphur produced during desulphuration reacts with sulphydryl groups of cysteine residues in the apoprotein of cytochrome P-450 to form hydrodisulphides. The stoichiometry of binding (i.e. the number of sulphur atoms bound per molecule of inactivated cytochrome P-450) and the exact mechanism by which binding leads to loss of activity have not yet been clarified. Neal (1980) has obtained evidence that, in addition to the sulphur present as hydrodisulphides, sulphur is bound to at least three other unidentified amino acid residues, but since only a portion of the bound sulphur appears to be responsible for inactivation (Morelli and Nakatsugawa, 1978) it is still possible that a single critical amino acid is involved in the inactivation. Loss of microsomal haem has also been reported after treatment with CS_2 or phosphorothionates (Bond and De Matteis, 1969; Halpert et al., 1980) and this may result either from direct destruction of haem by reactive metabolites of these compounds (Halpert et al., 1980) or from reduced affinity of the damaged apoprotein for the prosthetic group, leading to dissociation of haem and accelerated breakdown to bile pigments with accompanying induction of liver haem oxygenase (Järvisalo et al., 1978).

In view of the similarities between CS_2 and phosphorothionates in metabolism and toxicity, it has been suggested that the pathways of desulphuration might also be similar, involving S-oxides and cyclic oxathiirane-type intermediate in both cases and leading, in the case of CS_2, to loss of sulphur in two distinct stages through the monooxygenated intermediates COS (carbonyl sulphide) (De Matteis and Seawright, 1973; Catignani and Neal, 1975; Neal, 1980). In apparent agreement with this proposed pathway (Fig. 4, reaction a) it had been reported that COS was produced from CS_2 by liver microsomes incubated in presence of NADPH and that COS could be further metabolized (again in presence of NADPH) releasing the second sulphur also in a reactive form capable of inhibiting cytochrome P-450 (Dalvi et al., 1975). In more recent studies, however, it has been unexpectedly found that the oxygen present in COS originates not from molecular oxygen (as originally proposed) but from water (Chengelis and Neal, 1987). These authors therefore conclude that the first stage of desulphuration of CS_2, that leading to COS, is not likely to proceed through an oxathiirane intermediate of the kind proposed for parathion; instead, the corresponding S-oxide ($S{=}C{=}S^+{-}O^-$) might suffer attack from water to produce, in addition to reactive sulphur, monothiocarbonate, which will then give rise to COS on exposure to acid (Fig. 4, reaction b). Further work is required to clarify this proposed mechanism, whether the water-catalysed desulphuration now postulated takes place inside or outside the active site of cytochrome P-450 and how it leads to formation of hydrodisulphides in the apoprotein with loss of activity of cytochrome P-450.

Several other sulphur-containing chemicals have been reported to cause inhibition of the drug-metabolizing system and loss of cytochrome P-450 either on incubation with microsomes in presence of NADPH or after their administration to experimental animals in vivo. These include compounds such as disulfiram (Antabuse), diethyldithiocarbamates and bis(ethyl xanthogen) which liberate CS_2 in the body and might therefore be acting through CS_2 itself (Stripp et al., 1969; Hunter and

Neal, 1975; Fiala *et al.*, 1977). Several other phosphorothionates, in addition to the ones already mentioned, are similarly active (Uchiyama *et al.*, 1975), many exhibiting high selectivity for lung cytochrome P-450 (Imamura *et al.*, 1983; Gandy and Imamura, 1985; Verschoyle and Aldridge, 1987). It is interesting to note that, as observed with isolated liver microsomes (De Matteis, 1974), only organophosphorus compounds with a thiono-sulphur function (P=S) were found to be acutely inhibitory to lung cytochrome P-450, the corresponding P=O analogues being relatively inactive, even when they contained one or more thiolate (*S*-alkyl) sulphur in the molecule, although this structural feature made them generally toxic to the lung. The thiolate-containing compounds also apparently required activation by cytochrome P-450 for their lung toxicity, as pretreatment of rats with a thionate analogue (inhibitory of cytochrome P-450 function) afforded protection against subsequent dosing with a thiolate (Verschoyle and Aldridge, 1987). So both the thiono and thiolate sulphur functions can be activated by cytochrome P-450 to reactive derivatives, but the consequences of the two activation processes and the properties of the reactive derivatives may be different in the two cases. With thiono-sulphur compounds cytochrome P-450 is the main target, possibly because the corresponding S-oxide is unstable and breaks down readily, eliminating reactive sulphur in close proximity of the enzyme. In contrast, with thiolate compounds the primary target is not the activating enzyme (cytochrome P-450) but must lie somewhere else in the cell; in this case more stable oxides may be formed which can diffuse away and interact with targets at some distance from cytochrome P-450.

Some phosphorothionates also contain thioether sulphur functions and evidence for oxidative attack of this sulphur has also been obtained. Purified preparations of the flavin-containing monooxygenase (EC 1.14.13.8) rapidly oxidized the thioether sulphur atom of many organophosphorus and carbamate pesticides to the corresponding sulphoxide (Hajjar and Hodgson, 1980). One such organophosphorus substrate, Disulfaton $(C_2H_5O)_2P(S)-S-CH_2-CH_2-S-C_2H_5$, possesses three sulphur atoms, the thione, the thiol and the thioether, but the latter sulphur was the only sulphur to be oxidized and the corresponding sulphoxide the only product formed in these *in vitro* experiments involving the purified flavoprotein.

In conclusion, phosphorothionates and related compounds are subject to oxidative attack by monooxygenation and two different enzyme systems appear to be involved in the oxidation of their sulphur atoms. Cytochrome P-450 is concerned with monooxygenation of the thiono-sulphur function, resulting in oxidative desulphuration to the corresponding oxon and associated production of reactive sulphur. This leads to suicidal inactivation of cytochrome P-450, a response also seen with thiocarbonyl compounds (such as CS_2) which are also metabolized by oxidative desulphuration. Organophosphorus compounds may also contain one or more *S*-alkyl or thiolate sulphur atoms. These are also probably metabolized by cytochrome P-450, resulting in reactive derivatives which do not inactivate the enzyme but are responsible for a more general cellular toxicity. Finally, a thioether sulphur atom, when present in a phosphorothionate or carbamate, is actively metabolized by the flavin-containing monooxygenase, an enzyme which is distinct from cytochrome P-450 and which exhibits selectivity for the thioether sulphur atom (see Chapter 3 by Ziegler in Volume 2, Part A of this series).

1.3 PATHWAYS OF BREAKDOWN OF ORGANOPHOSPHORUS COMPOUNDS NOT INVOLVING MONOOXYGENATION

1.3.1 Interaction of organophosphorus compounds with esterases

Once the phosphorothionates have been desulphurated by cytochrome P-450 to the corresponding phosphates or oxons, these can interact with a number of esterases in several organs and tissues, a general reaction of great importance in the development of neurotoxicity but one which is also involved in the detoxification of organophosphates.

In a paper which can be considered as a classic in this field, Aldridge (1953) distinguished esterases (those which can interact with paraoxon and other organophosphates) in two distinct classes, depending on whether they can hydrolyse the organophosphates (class A) or are inhibited by them (class B). Although purely operational, this classification has stood the test of time and is still very useful in the field (Walker and Mackness, 1983). Aldridge proposed that the basic mechanism underlying the organophosphate–enzyme interaction may be the same in both cases (Fig. 5), involving first the formation of a Michaelis complex, followed by reaction of the organophosphate with the enzyme: in this step the leaving group (p-nitrophenol, in the case of paraoxon) is lost and the phosphoryl moiety becomes attached to the enzyme, phosphorylating it. Finally the phosphorylated enzyme reacts with water so that the phosphoryl moiety is lost, the native esterase is regenerated and a new

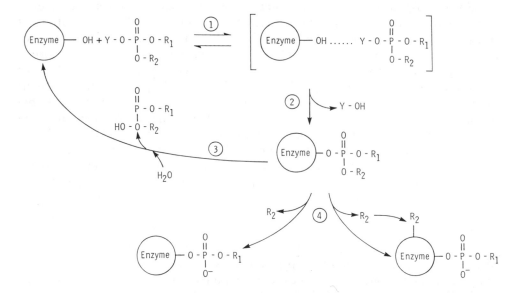

Fig. 5 — Interaction of an organophosphate with a B esterase, showing phosphorylation of the enzyme followed by dephosphorylation (reactivation) or by aging (with two possible fates for the alkyl group released during the aging reaction). The following steps are numbered: (1) formation of a Michaelis complex; (2) phosphorylation of the enzyme; (3) reactivation, a very slow step unless accelerated by a reactivator (oxime); (4) aging reaction, during which one alkyl group (R_2 in the figure) is lost either to the surrounding medium (in most cases) or to other functional groups within the same enzyme (in the case of NTE). Modified from Johnson (1987).

catalytic cycle can be initiated. It is the rate of this last step, involving enzyme dephosphorylation (and reactivation) which varies markedly in the two cases, being very slow with B esterases and extremely rapid with A esterases. Because of this, the latter enzymes hydrolyse the organophosphate very effectively, while the former are blocked half-way in the process and lose activity.

In agreement with this proposal the inhibition of the B esterases by organophosphates has been shown to require catalytic turnover of the enzyme and to exhibit many features (Aldridge and Reiner, 1972) associated with the mechanism-based (or suicidal) enzyme inhibition, including the presence of a covalently modified amino acid residue at the active site. In those cases where this has been examined (see for example, Hartley, 1960), a phosphorylated serine residue has been found at the active site of B esterases, whereas a different amino acid residue (possibly an SH bearing amino acid) appears to be involved at the active site of A esterases (Walker and Mackness, 1983). It is possible that the essential difference between the two classes of esterases (namely the different rates of hydrolysis of the intermediary phosphorylated enzyme) may be largely related to the different amino acid residues which become phosphorylated in the two cases, but this has not yet been established.

Two important consequences flow from this distinction between A and B esterases. The first relates to their importance in inactivating organophosphates. Since A esterases hydrolyse organophosphates to inactive derivatives without losing in the process their enzymatic activity, they are much more important than B esterases in the inactivation of this class of compounds. The second relates to the consequences of the loss of enzymatic activity of the B esterases: if the enzyme which becomes phosphorylated possesses a vital function, or if the stable occupancy of its active site leads indirectly to loss of some important function, then irreversible interaction of an organophosphate with these sites will lead to appearance of symptoms of toxicity. This is not to say that inhibition of B esterases will always be accompanied by signs of toxicity. Blood cholinesterase can be inhibited very significantly without obvious signs of toxicity and in certain insects which have acquired resistance to organophosphorus insecticides the mutation can be ascribed — at least in part — to much greater amounts of 'silent' B esterases (Devonshire and Moores, 1982; Motoyama *et al.*, 1984). These act as a sink for the insecticide, thus scavenging it and preventing its reaction with more vital B esterases, such as acetylcholinesterase. Apart from these extreme cases, however, B esterases are mostly known for their involvement in toxic responses to organophosphorus compounds, while A esterases are generally regarded as protective detoxifying enzymes, as will now be discussed.

1.3.1.1   *Protective role of A esterases*
The role of class A esterases in affording protection against the toxic oxons has recently been reviewed by Walker and Mackness (1987). The serum enzymes which are active on paraoxon have been partially purified from sheep and human serum and shown to be heterogenous, as they consist of at least four different forms, all associated with the high density lipoprotein (HDL) fraction obtained by ultracentrifugation (Mackness *et al.*, 1984, 1985). The various forms could be separated from each other on the basis of the relative hydrolytic activity on paraoxon and other

substrate esters. This suggests the existence in serum of distinct A esterases with overlapping substrate specificities, but more work is necessary before this can be considered as firmly established.

The A esterase activity of human serum, assayed with paraoxon as the substrate, shows a biomodal distribution in human populations (La Du and Eckerson, 1984), although the significance of these variations in activity between different individuals (with respect to susceptibility to organophosphate poisoning) has not yet been explored. Thus, it is clear that as with other drug-metabolizing enzymes, e.g. cytochrome P-450 (Boobis et al., 1985), acetylating enzymes (Evans and White, 1964) and glutathione-S-transferases (Hayes et al., 1987), the A esterases also exhibit multiplicity, overlapping substrate specificity and genetic polymorphism.

In contrast to the considerable A esterase activity of mammalian species, birds possess very much lower activity (Machin et al., 1976; Brealey et al., 1980) and insects even lower activity and possibly none (Walker and Mackness, 1987). There is a broad correlation between activity of these enzymes in the various species and resistance to organophosphates, species with higher esterase activity also exhibiting more resistance to the acute neurotoxic effect of these agents, in agreement with the concept that A esterases afford protection by hydrolysing the toxic inhibitors of acetylcholinesterase to their inactive derivatives. The protective role of carboxylesterases against malathion toxicity will be discussed in section 1.3.1.3.

1.3.1.2   B esterases as targets of organophosphate-induced neurotoxicity: Acetylcholinesterase and neuropathy target esterase

The insecticidal properties of organophosphorus compounds as well as the symptoms of acute toxicity observed in mammals, including the very rapid lethal effect of the related chemicals developed as warfare agents (nerve gases), are generally attributed to inhibition of *acetylcholinesterase* in the central and peripheral nervous system. The clinical picture is dominated by cholinergic symptoms, including profuse lachrymation and salivation, urinary incontinence, muscular weakness and fasciculation and difficulty in respiration (Grob, 1963). Death may follow owing to respiratory paralysis both from depression of the respiratory centres and from peripheral impairment of the respiratory muscles. These symptoms are all mediated by covalent attachment of the organophosphate to a serine residue in the active centre of acetylcholinesterase, leading to inhibition of its activity and build-up of toxic levels of the neurotransmitter. As in all B esterases, the rate of spontaneous dephosphorylation and reactivation is very slow, but reactivation can be markedly accelerated by certain nucleophilic reagents such as the oximes, which have been developed and are employed successfully as antidotes in cases of organophosphate poisoning. With the development of good reactivators, the so-called 'aging' reaction of the inhibited enzyme was discovered, in which an R—O—P bond of the phosphoryl moiety is cleaved, with loss of the alkyl group (R) and generation of a charged monosubstituted phosphoric acid residue still attached to the active site (Fig. 5). Once the enzyme has undergone this 'aging' reaction, its ability to become reactivated by nucleophilic agents, such as the oximes, is lost (Berends, 1987). So the 'aging' phenomenon is of great practical importance as it impairs the effectiveness of reactivating agents, which represent an important part of the therapy of organophosphate poisoning.

The alkyl group lost from the phosphorylated acetylcholinesterase during the 'aging' reaction is released from the enzyme. In contrast during the similar 'aging' reaction of the other B esterase discussed below, the NTE, the leaving alkyl group remains attached to functional groups in the same enzyme, an interesting example of an intramolecular transalkylation reaction, which is thought to be of primary importance (Johnson, 1987) in the pathogenesis of the second type of neurotoxicity to be considered now.

Some organophosphorus compounds also cause a distal neuropathy with a selective damage of large-diameter, long fibres in both the central and peripheral nervous systems, a pathological condition which has been regarded for a long time as the typical example of 'dying-back' axonal neuropathy (Cavanagh, 1973). Unlike the cholinergic symptoms which are experienced early after the exposure, the motor and sensory signs of the neuropathy are delayed, as they take a period of 2–3 weeks to appear. Many species including man have shown this particular toxic reaction, but the hen is the experimental animal of choice as it is very susceptible. A list of organophosphorus pesticides causing delayed neuropathy in hens after a single dose can be found in Johnson (1986) and include several oxons (i.e. P=O compounds of the three classes discussed below), as well as phosphorothionates, the latter of course requiring metabolic activation to the corresponding oxons for their toxicity. Although many phosphorothionates and the corresponding oxons produce both the cholinergic symptoms of the acute intoxication and the delayed neuropathic response, these are clearly distinct effects, can be observed independently of each other (for example the acutely neurotoxic phosphorothionate parathion does not produce delayed neuropathy) and involve interaction with two distinct target enzymes.

Phosphorylation of a B esterase in the nervous system, the so-called NTE has been shown to be the primary lesion in the development of the delayed neuropathy, largely as a result of the work by Martin Johnson and collaborators. The characteristic features of this enzymic site and the way in which the neurotoxic organophosphorus compounds modify it in order to produce the delayed neuropathic response have been recently reviewed (Johnson, 1987).

A carefully devised technique allows the dissection of the relevant enzyme (NTE) from the mixed population of B esterases present in the nervous system by the use of a selective inhibitor, the non-neuropathic organophosphate paraoxon, which will eliminate most esterases, while leaving NTE unaffected. For the delayed neuropathy to occur at least 70% of NTE-dependent activity must be lost after administration of a neurotoxic compound (Johnson and Lotti, 1980). The second essential requirement is that the inhibited NTE must also be able to age, i.e. one of the alkyl group of the phosphorylating moiety attached to its active centre must be cleaved off (Clothier and Johnson, 1979). For chemical reasons this is only possible for organophosphorus compounds where at least one such alkyl group is linked to the phosphorus through an intermediary oxygen or nitrogen atom, as in phosphates, phosphonates and phosphoramidates, all three classes showing examples of compounds with neurotoxic properties. In contrast, when NTE is inhibited by treatment with phosphinates, where both R and R′ are directly linked to the phosphorus atom and cannot be easily cleaved, the phosphinylated NTE cannot age and no delayed neuropathy is observed (Johnson, 1974). Animals so pretreated are also protected from neurotoxic organo-phosphates, persumably because access to the NTE site of the neurotoxic agent is thus prevented as well as the secondary aging reaction which its attachment to the NTE active site would make possible.

NTE appears to be unique among B esterases in that the alkyl group released during the aging process is retained by functional groups within the same enzyme (Williams, 1983). So it is possible (Johnson, 1987) that alkylation of these secondary enzymic sites is the initiating event which sets in motion the neuropathic response. However, the mechanism linking these primary molecular changes to the development of the neuropathy is completely unknown.

1.3.1.3   Metabolism of malathion by carboxylesterases

In the preceding sections (1.3.1.1. and 1.3.1.2) a type of interaction has been discussed between organophosphates and esterases, where the enzymes catalyse cleavage of the bond between their phosphorus atom and a suitable leaving group. Phosphorothionates cannot interact in this way with esterases, probably because their phosphorus atom is not as electrophilic as that of the corresponding phos-phates. Some phosphorothionates, however, among these malathion, also possess carboxylic ester functions and these are subject to enzymatic attack by carboxylester-ases. The products of this reaction (Dauterman and Main, 1966), the corresponding free carboxylic acids, are not toxic (Main and Braid, 1962) as their lipid solubility is markedly reduced and uptake by the cytochrome P-450 system (essential for their conversion into the toxic oxons) is correspondingly diminished. Also, were the corresponding oxons with free carboxylate functions to be formed, these would probably be poor inhibitors of acetyl cholinesterase (Aldridge *et al.*, 1987). Much of the selectivity of malathion in its use as an insecticide, with a wide margin of safety and relatively little toxicity to man and other mammals, is due to these species' possessing very active carboxylesterases, which provide therefore a very effective pathway of detoxification (also see Wester and Cashman, Volume 3 of this series). The effectiveness of this hydrolytic pathway can be judged by the observation (Main and Braid, 1962) that the LD 50 of malathion in the rat (normally approximately 10–12 g/kg of body weight (Umetsu *et al.*, 1977)), can be markedly reduced to values of approximately 7 mg/kg (an increase in toxicity of more than 1000-fold) by pretreatment with tri-2-tolyl phosphate, an inhibitor of carboxylesterases. However, the most compelling evidence for the protective role of carboxylesterases against malathion toxicity in man is that afforded by a toxic epidemic observed in 1976 among workers spraying certain formulation of malathion in Pakistan and by the elucidation of the mechanism of such a widespread toxicity, involving 2800 poisoned workers (Baker *et al.*, 1978). The affected individuals exhibited acute toxicity with cholinergic symptoms of varying severity, leading to death in five cases.

When different toxic samples of formulated malathion were analysed for the presence of impurities, a fairly good correlation was found between toxicity of a given sample and its content in isomalathion (Baker *et al.*, 1978; Aldridge *et al.*, 1979; Miles *et al.*, 1979). Furthermore, laboratory storage at 38°C of formulated malathion increased both its isomalathion content and also its toxicity, suggesting that storage under field conditions in the hot climate of Pakistan may have been responsible for the increased isomalathion content of the unusually toxic samples (Aldridge *et al.*, 1979; Miles *et al.*, 1979), most probably by promoting a chemical isomerization reaction. Finally, mixtures of pure malathion and isomalathion in different proportions were administered to rats and from these studies (Umetsu *et al.*, 1977; Aldridge *et al.*, 1979) it was shown that the toxicity of malathion was clearly potentiated by the isomer isomalation. Another impurity present in the toxic samples of formulated malathion was O,S,S-trimethyl phosphorodithiolate and this too was found to potentiate malathion toxicity in the rat (Pellegrini and Santi, 1972; Umetsu *et al.*, 1977; Aldridge *et al.*, 1979). This second impurity arises during the synthesis of many organophosphorus thionates, including malathion, and may also be produced in excess during storage of malathion under certain conditions (Miles *et al.*, 1980).

Both isomalathion and O,S,S-trimethyl phosphorodithiolate are direct inhibitors of carboxylesterases. In studies where malathion was used as the specific substrate of the esterases Talcott *et al.* (1979a) found both compounds to be potent inhibitors of carboxylesterases from rat liver and plasma, both *in vivo* and *in vitro*. In addition, a partially purified preparation of carboxylesterase from human plasma was inhibited by isomalathion, though the trimethyl phosphorothioate may not have been inhibitory to the liver enzyme (Talcott *et al.*, 1979b). It can therefore be concluded that the potentiation of malathion toxicity responsible for the toxic epidemic in Pakistan was probably due to inhibition of carboxylesterases by impurities (especially isomalathion) produced on storage of certain formulated samples of commercial malathion under the relatively high temperature prevailing in the climate. Instead of being rapidly hydrolysed and inactivated (as it is under normal conditions) malathion would then accumulate in sufficiently high amounts to activate (Fig. 6) a normally minor pathway of metabolism, that to the oxidative derivative malaoxon. This is then the proximal toxin responsible for the acute cholinergic symptoms observed in Pakistan.

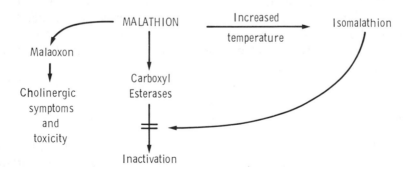

Fig. 6 — Potentiation of malathion toxicity by isomalathion. Role of inhibition of carboxyl esterases.

1.3.2 Glutathione-dependent breakdown

Another important pathway of detoxification of phosphorothionates and corresponding oxons is that represented by glutathione S-transferases. This is a group of related cytosolic enzymes with overlapping substrate specificity which can conjugate to GSH either products of monooxygenation reactions or, directly, foreign chemicals without requirement for a preliminary step of metabolism. The latter is the case with organophosphorus insecticides, including phosphorothionates.

The first clear indication that glutathione S-transferases were involved in the metabolism of organophosphorus pesticides was the demonstration (Shishido and Fukami, 1963; Fukami and Shishido, 1966) that the O-methyl analogue of parathion (methyl parathion) was demethylated in presence of glutathione by cytosolic fractions from rat liver and certain insects. The ability to degrade methyl parathion was lost when the cytosolic fraction was passed through Sephadex G-50 and restored by subsequent addition of glutathione. The corresponding conjugated product, namely S-methyl glutathione, was identified (Hutson et al., 1968; Holingworth, 1969), compatible with the reaction illustrated in Fig. 2D(a), involving cleavage of the O–Me bond.

Although both phosphorothionates and the respective phosphate derivatives can be O-dealkylated by this mechanism, the former are in general better substrates (Shishido et al., 1972; Motoyama and Dauterman, 1977) than the latter. The transferases from both mammalian and insect sources are much more active on O-methyl compounds than on the analogues with larger alkyl groups (O-ethyl and O-isopropyl) (Hollingworth, 1969), and this may account, at least in part, for the differences in toxicity between phosphorothionates with O-alkyl groups of different size (see below).

The cleavage of the aryl phosphate bond of several organophosphates had originally been considered to take place exclusively through a hydrolytic mechanism. However, in more recent studies it has been demonstrated that O-dearylation can also be catalysed by glutathione S-transferases (reaction b in Fig. 2D) and the corresponding products of the conjugation reaction (S-p-nitrophenyl glutathione, in the case of parathion) have also been identified (Shishido et al., 1972; Hollingworth et al., 1973). The transferases involved in O-dearylation appear to be different from those catalysing O-dealkylation (Hollingworth et al. 1973; Motoyama and Dauterman, 1978) and the relative balance between the two glutathione-dependent pathways of breakdown depends on the nature of the O-alkyl group, with O-methyl compounds such as methyl parathion, methyl paraoxon and fenitrothion, O-dealkylation predominating, whereas, with the corresponding O-ethyl compounds, O-dearylation is the major pathway (Shishido and Fukami, 1963; Hollingworth et al., 1973). This is explained, as already mentioned, by the relative ease with which O-methyl groups are transferred to glutathione as compared with O-ethyl groups which are removed only slowly.

The following observations suggest that the glutathione-dependent pathway of O-dealkylation involving the glutathione S-transferases may be important as a detoxification mechanism in mammals. First, O-methyl phosphorothionates (good substrates of the transferases) are less toxic than the corresponding O-ethyl compounds (Sakamato and Nishizawa, 1962). Second, the toxicity of the O-methyl phosphorothionates can be increased more than 10-fold, with the appearance of fatal cholinergic symptoms, by pretreatment of the animals with methyl iodide or diethyl maleate (Hollingworth, 1969), agents which deplete liver glutathione and would therefore be expected to inhibit glutathione S-transferase reactions. Finally, intravenous administration of glutathione has been claimed to be beneficial to patients chronically poisoned with methyl parathion (Matsuda *et al.*, 1972) and possibly also to similarly treated mice (Kawai and Ueda, 1969). This too would be compatible with the importance of the glutathione-dependent O-demethylation pathway for detoxification of phosphorothionates in mammals.

There is also evidence that glutathione-dependent O-demethylation may be the basis for insect resistance to phosphorothionate insecticides, at least in some cases. When strains of house flies resistant or susceptive to azinophosmethyl (an O-methyl phosphorothionate) were compared, the glutathione-dependent degradation of the insecticide by the soluble fraction of fly homogenates was much higher in the resistant strain (Motoyama and Dauterman, 1972). The expected products of the reaction, S-methyl glutathione and the desmethyl moiety of the insecticide, were identified. Further confirmation that in this case the high activity of glutathione S-alkyl transferase was involved in the resistance mechanism was obtained by showing that substitution of O-ethyl groups for the O-methyl groups of azinophosmethyl resulted in a marked decrease in the level of resitance. In this study the resistant strain of house flies possessed very much higher activity of glutathione S-transferase also when the enzyme was assayed with unrelated alkyl and aryl donors (Motoyama and Dauterman, 1974), raising the question of whether resistance resulted from a generalized increase of various transferase enzymes or from the appearance of an enzyme with an unusually wide substrate specificity. A similar basis for the resistance of insects to other insecticides has also been reported (Fukami, 1979), but alternative mechanisms of inactivation of the insecticide, e.g. accelerated breakdown by mixed function oxidases, may be more important in some other cases (Motoyama and Dauterman, 1980).

It can therefore be concluded that glutathione S-transferases are involved in the breakdown of phosphorothionates and related compounds by cleavage of their O-alkyl and O-aryl bonds and transfer of the corresponding leaving groups to glutathione. These reactions result in inactivation of the insecticide and therefore afford protection against their toxicity, contributing significantly to the resistance of mammalian species (and of some insect strains), particularly with O-methyl phosphorothionates which are readily demethylated through this pathway. The molecular basis for the inactivation of these compounds is presumably similar to that already discussed for the carboxylesterase pathway, namely the introduction of a polar group on the molecule, which will discourage oxidative metabolism of the thionate to the toxic oxon and/or interaction of the latter with the target acetylcholinesterase.

ACKNOWLEDGEMENT

I would like to thank Mrs. Rosemary Hill for much skill and patience in typing the manuscript.

REFERENCES

Aldridge, W. N. (1953). Serum esterases. 1. Two types of esterase (A and B) hydrolysing *p*-nitrophenyl acetate, propionate and butyrate, and a method for their determination. *Biochem. J.*, **53**, 110–117.

Aldridge, W. N. and Davison, A. N. (1952). The inhibition of erythrocyte cholinesterase by tri-esters of phosphoric acid. *Biochem. J.*, **51**, 62–70.

Aldridge, W. N. and Reiner, E. (1972). *Enzyme Inhibitors as Substrates*, North Holland, Amsterdam.

Aldridge, W. N., Miles, J. W., Mount, D. L., and Verschoyle, R. D. (1979). The toxicological properties of impurities in malathion. *Arch. Toxicol.*, **42**, 95–106.

Aldridge, W. N., Dinsdale, D., Nemery, B., and Verschoyle, R. D. (1987). Toxicology of impurities in malathion: potentiation of malathion toxicity and lung toxicity caused by trialkyl phosphorothioates. In F. De Matteis and E. A. Lock (eds.), *Selectivity and Molecular Mechanisms of Toxicity*, Macmillan, pp. 265–294.

Baker, E. L., Zack, M., Miles, J. W., Alderman, L., Warren, McW., Dobbin, R. D., Miller, S., and Teeters, W. R. (1978). Epidemic malathion poisoning in Pakistan malaria workers. *Lancet*, **I**, 31–33.

Berends, F.(1987). Mechanisms of ageing of organophosphate-inhibited esterases. In F. De Matteis and E. A. Lock (eds.), *Selectivity and Molecular Mechanisms of Toxicity*, Macmillan, pp. 125–152.

Bond, E. J. and De Matteis, F. (1969). Biochemical changes in rat liver after administration of carbon disulphide with particular reference to microsomal changes. *Biochem. Pharmacol.*, **18**, 2531–2549.

Bond, E. J., Butler, W. H., De Matteis, F., and Barnes, J. M. (1969). Effects of carbon disulphide on the liver of rats. *Brit. J. Ind. Med.*, **26**, 335–337.

Boobis, A. R., Murray, S., Hampden, C. E., Harries, G. C., Huggett, A. C., Thorgiersson, S. S., McManus, M. E., and Davies, D. S. (1985). Enzymic basis for polymorphisms of drug oxidation in man. In A. R. Boobis, J. Caldwell, F. De Matteis and C. R. Elcombe (eds.), *Microsomes and Drug Oxidation*, Taylor and Francis, London, pp. 361–369.

Brealey, C. J., Walker, C. H., and Baldwin, B. C. (1980). "A" esterase activities in relation to the differential toxicity of pirimiphos-methyl to birds and mammals. *Pestic. Sci.*, **11**, 546–554.

Butler, W. H., Hempsall, V., and Magos, L. (1974). Cation movement during carbon disulphide-induced liver injury. *J. Path.*, **113**, 53–59.

Catignani, G. L. and Neal, R. A. (1975). Evidence for the formation of a protein bound hydrodisulfide resulting from the microsomal mixed function oxidase catalysed desulphuration of carbon disulphide. *Biochem. Biophys. Res. Commun.*, **65**, 629–636.

Cavanagh, J. B. (1973). Peripheral neuropathy caused by chemical agents. CRC Crit. Rev. Toxicol., **3**, 365–417.

Chengelis, C. P. and Neal, R. A. (1987). Oxidative metabolism of carbon disulphide by isolated rat hepatocytes and microsomes. *Biochem. Pharmacol.*, **36**, 363–368.

Clothier, B. and Johnson, M. K. (1979). Rapid aging of neurotoxic esterase after inhibition by di-isopropyl phosphorofluoridate. *Biochem. J.*, **177**, 549–558.

Dalvi, R. R., Poore, R. E. and Neal, R. A. (1974). Studies of the metabolism of carbon disulfide by rat liver microsomes. *Life Sci.*, **14**, 1785–1796.

Dalvi, R. R., Hunter, A. L. and Neal, R. A. (1975). Toxicological implication of the mixed-function oxidase catalysed metabolism of carbon disulphide. *Chem. Biol. Interaction*, **10**, 347–361.

Dauterman, W. C. and Main, A. R. (1966). Relationship between acute toxicity and *in vitro* inhibition and hydrolysis of a series of carbalkoxy homologues of malathion. *Toxicol. Appl. Pharmacol.*, **9**, 408–418.

Davison, A. N. (1955). The conversion of Schradan (OMPA) and parathion into inhibitors of cholinesterase by mammalian liver. *Biochem. J.*, **61**, 203–209.

De Matteis, F. (1974). Covalent binding of sulphur to microsomes and loss of cytochrome P-450 during the oxidative desulfuration of several chemicals. *Molec. Pharmacol.*, **10**, 849–854.

De Matteis, F. and Seawright, A. A. S. (1973). Oxidative metabolism of carbon disulphide by the rat. Effect of treatments which modify the liver toxicity of carbon disulphide. *Chem.–Biol. Interactions*, **7**, 375–388.

Devonshire, A. L. and Moores, G. D. (1982). A carboxylesterase with broad substrate specificity causes organophosphorus, carbamate and pyrethroid resistance in peach-potato aphids (*Mayzus persicae*). *Pestic. Biochem. Physiol.*, **18**, 235–246.

Diggle, W. M. and Gage, J. C. (1951). Cholinesterase inhibition by parathion *in vivo*. *Nature (London)*, **168**, 998.

Evans, D. A. P. and White, T. A. (1964). Human acetylation polymorphism. *J. Lab. Clin. Med.*, **63**, 394–403.

Fiala, E. S., Bobotas, G., Kulakis, C., Wattenberg, L. W., and Weisburger, J. H. (1977). Effect of disulfiram and related compounds on the metabolism *in vivo* of the colon carcinogen, 1,2-dimethylhydrazine. *Biochem. Pharmacol.*, **26**, 1763–1768.

Freunt, K. J. and Dreher, W. (1969). Inhibition of drug metabolism by small concentrations of carbon disulphide. *Naunyn-Schmiedebergs Arch. Exp. Path. Pharmak.*, **263**, 208–209.

Fukami, J.-I. (1979). Metabolism of several insecticides by glutathione S-transferase. *Pharmac. Ther.*, **10**, 473–514.

Fukami, J. and Shishido, T. (1966). Nature of a soluble glutathione-dependent enzyme system active in cleavage of methyl parathion to desmethyl parathion. *J. Econ. Entomol.*, **59**, 1338–1346.

Gandy, J. and Imamura, T. (1985). A phosphorothioate isomer protects against pneumotoxicity caused by OOS-trimethyl phosphorothioate (OOS-TMP). *Toxicologist*, **5**, 69 (abstract).

Grob, D. (1963). Anticholinesterase intoxication in man and its treatment. *Handb. Exp. Pharmacol.*, **15**, 989–1028.

Hajjar, N. P. and Hodgson, E. (1980). Flavin adenine dinucleotide-dependent monooxygenase: its role in sulfoxidation of pesticides in mammals. *Science*, **209**, 1134–1136.

Halpert, J., Hammond, D., and Neal, R. A. (1980). Inactivation of purified rat liver cytochrome P-450 during metabolism of parathion (diethyl-p-nitrophenyl phosphorothionate). *J. Biol. Chem.*, **255**, 1080–1089.

Hartley, B. S. (1960). Proteolytic enzymes. *Ann. Rev. Biochem.*, **29**, 45–72.

Hayes, J. D., McLellan, L. I., Stockman, P. K., Howie, A. F., Hussey, A. J., and Beckett, G. J. (1987). Human glutathione-S-transferases: a polymorphic group of detoxification enzymes. in T. J. Mantle, C. B. Pickett and J. D. Hayes (eds.), *Glutathione S-Transferases and Carcinogenesis*, Taylor and Francis, London, pp. 3–18.

Herriott, A. W. (1971). Peroxy acid oxidation of phosphinothioates, a reversal of stereochemistry. *J. Am. Chem. Soc.*, **93**, 3304–3305.

Hollingworth, R. M. (1969). Dealkylation of organophosphorus esters by mouse liver enzymes *in vitro* and *in vivo*. *J. Agr. Food Chem.*, **17**, 987–996.

Hollingworth, R. M., Alstott, R. L., and Litzenberg, R. D. (1973). Glutathione S-aryl transferase in the metabolism of parathion and its analogs. *Life Sci.*, **13**, 191–199.

Hunter, A. L. and Neal, R. A. (1975). Inhibition of hepatic mixed-function oxidase activity *in vitro* and *in vivo* by various thiono-sulfur-containing compounds. *Biochem. Pharmacol.*, **24**, 2199–2205.

Hutson, D. H., Pickering, B. A., and Donninger, C. (1968). Phosphoric acid triester: glutathione alkyl transferase. *Biochem. J.*, **106**, 20P.

Imamura, T., Gandy, J., and Fukuto, T. R. (1983). Selective inhibition of rat pulmonary monooxygenase by O–,O–S,-trimethyl phosphorothioate treatment. *Biochem. Pharmacol.*, **32**, 3191–3195.

Järvisalo, J., Savolainen, H., Elovaara, E., and Vainio, H. (1977). The *in vivo* toxicity of CS_2 to liver microsomes: binding of labelled CS_2 and changes of the microsomal enzyme activities. *Acta Pharmacol. Toxicol.*, **40**, 329–336.

Järvisalo, J., Gibbs, A. H. and De Matteis, F. (1978). Accelerated conversion of heme to bile pigments caused in the liver by carbon disulphide and other sulphur-containing chemicals. *Molec. Pharmacol.*, **14**, 1099–1106.

Johnson, M. K. (1974). The primary biochemical lesion leading to the delayed neurotoxic effects of some organophosphorus esters. *J. Neurochem.*, **23**, 785–789.

Johnson, M. K. (1986). Organo-phosphorus insecticides: a general introduction. *Environmental Health Criteria Document No. 63*, World Health Organisation, Geneva.

Johnson, M. K. (1987). Organophosphate-induced delayed neuropathy: anomalous data lead to advances in understanding. In F. De Matteis and E. A. Lock (eds.), *Selectivity and Molecular Mechanisms of Toxicity*, Macmillan, pp. 27–58.

Johnson, M. K. and Lotti, M. (1980). Delayed neurotoxicity caused by chronic feeding of organophosphates requires a high point of inhibition of neurotoxic esterase. *Toxicol. Lett.*, **5**, 99–102.

Kamataki, T. and Neal, R. A. (1976). Metabolism of diethyl *p*-nitrophenyl phosphorothionate (Parathion) by a reconstituted mixed-function oxidase enzyme system: studies of the covalent binding of the sulphur atom. *Molec. Pharmacol.*, **12**, 933–944.

Kamataki, T., Lee Lin, M. C. M., Belcher, D. H., and Neal, R. A. (1976). Studies of the metabolism of parathion with an apparently homogenous preparation of rabbit liver cytochrome P-450. *Drug Met. Dispos.*, **4**, 180–189.

Kawai, M. and Ueda, K. (1969). Therapy of organophosphate poisoning with oxime antidote and glutathione. In *Abstracts 4th International Congress on Rural Medicine, Usuda*, p. 10 (cited in Fukami, 1979).

Knaak, J. B., Stahmann, M. A., and Casida, J. E. (1962). Peroxidase and ethylene-diaminetetraacetic acid-ferrous iron-catalyzed oxidation and hydrolysis of parathion. *J. Agr. Food Chem.*, **10**, 154–158.

La Du, B. N. and Eckerson, H. W. (1984). The polymorphic paraoxonase/arylesterase isozymes of human serum. *Fed. Proc.*, **43**, 2338–2441.

Machin, A. F., Anderson, P. H., Quick, M. P., Waddell, D. R., Skibniewska, K. A., and Howells, L. C. (1976). The metabolism of diazinon in the liver and blood of species of varying susceptibility to diazinon poisoning. *Xenobiotica*, **7**, 104.

Mackness, M. I., Hallam, S. D., and Walker, C. H. (1984). "A" esterase activity in the lipoprotein fraction of sheep and human serum. *Biochem. Soc. Trans.*, **12**, 135–136.

Mackness, M. I., Hallam, S. D., Peard, T., Warner, S., and Walker, C. H. (1985). The separation of sheep serum "A" esterase activity into the lipoprotein fraction by ultracentrifigation. *Comp. Biochem. Physiol.*, **82B**, 675–677.

Main, A. R. and Braid, D. E. (1962). Hydrolysis of malathion by ali-esterases *in vitro* and *in vivo*. *Biochem. J.*, **84**, 255–263.

Matsuda, S., Mikami, M., Ohtahi, Y., and Kudo, N. (1972). Interrelation between blood glutathione and cholinesterase activity in methyl parathion poisoning. *Seikagaku*, **44**, 209–302 (cited by Fukami, 1979).

McBain, J. B., Yamamoto, I., and Casida, J. E. (1971a). Mechanism of activation and deactivation of Dyfonate (O-ethyl S-phenyl ethylphosphonodithioate) by rat liver microsomes. *Life Sci.*, **10**, 947–954.

McBain, J. B., Yamamoto, I., and Casida, J. E. (1971b). Oxygenated intermediate in peracid and microsomal oxidations of the organophosphonothionate insecticide Dyfonate. *Life Sci.*, **10**, 1311–1319.

Miles, J. W., Mount, D. L., Steiger, M. A. and Teeters, W. R. (1979). The S-methyl isomer content of stored malathion and fenitrothion water-dispersible powders and its relationship to toxicity. *J. Agric. Food Chem.*, **27**, 421–425.

Miles, J. W., Mount, D. L., and Churchill, F. C. (1980). The effect of storage on the formation of minor components in malathion powders. In F. Sanchez-Rasero (ed.), *Proc. of the Collaborative International Pesticides Analytical Council Meeting, Series 2*, Cambridge, pp. 176–192.

Morelli, M. A. and Nakatsugawa, T. (1978). Inactivation *in vitro* of microsomal oxidases during parathion metabolism. *Biochem. Pharmacol.*, **27**, 293–299.

Motoyama, N. and Dauterman, W. C. (1972). *In vitro* metabolism of azinphos-methyl in susceptible and resistant houseflies. *Pestic. Biochem. Physiol.*, **2**, 113–122.

Motoyama, N. and Dauterman, W. C. (1974). The role of non-oxidative metabolism in organophosphorus resistance. *J. Agr. Food Chem.*, **22**, 350–355.

Motoyama, N. and Dauterman, W. C. (1977). Purification and properties of housefly glutathione S-transferase. *Insect Biochem.*, **7**, 361–369.

Motoyama, N. and Dauterman, W. C. (1978). Multiple forms of rat liver glutathione S-transferases: specificity for conjugation of O-alkyl and O-aryl groups of organophosphorus insecticides. *J. Agr. Food Chem.*, **26**, 1296–1301.

Motoyama, N. and Dauterman, W. C. (1980). Glutathione S-transferases: their role in metabolism of organophosphorus insecticides. *Rev. Biochem. Toxicol.*, **2**, 49–69.

Motoyama, N., Kao, L. R., Lin, P. T., and Dauterman, W. C. (1984). Dual role of esterases in insecticide resistance in the green rice leafhopper. *Pestic. Biochem Physiol.*, **21**, 139–147.

Murphy, S. D. and DuBois, K. P. (1957). Enzymatic conversion of dimethoxy ester of benzotriazine dithiophosphoric acid to an anticholinesterase agent. *J. Pharmacol. Ther.*, **119**, 572–583.

Nakatsugawa, T. and Dahm, P. A. (1965). Parathion activation enzymes in the fat body microsomes of the American cockroach. *J. Econ. Entomol.*, **58**, 500–509.

Nakatsugawa, T. and Dahm, P. A. (1967). Microsomal metabolism of parathion. *Biochem. Pharmacol.*, **16**, 25–38.

Nakatsugawa, T. and Dahm, P. A. (1968). Degradation and activation of parathion analogues by microsomal enzymes. *Biochem. Pharmacol.*, **17**, 1517–1528.

Neal, R. A. (1967a). Studies on the metabolism of diethyl 4-nitrophenyl phosphorothionate (parathion) *in vitro. Biochem. J.*, **103**, 183–191.

Neal, R. A. (1967b). Studies of the enzymatic mechanism of metabolism of diethyl 4-nitrophenyl phosphorothionate (parathion) by rat liver microsomes. *Biochem. J.*, **105**, 289–297.

Neal, R. A. (1980). Microsomal metabolism of thiono-sulhur compounds: mechanisms and toxicological significance. *Rev. Biochem. Toxicol.*, **2**, 131–171.

Norman, B. J., Vaughn, W. K., and Neal, R. A. (1973). Studies of the mechanism of metabolism of diethyl *p*-nitrophenyl phosphorothionate (parathion) by rabbit liver microsomes. *Biochem. Pharmacol.*, **22**, 1091–1101.

Norman, B. J., Poore, R. E., and Neal, R. A. (1974). Studies of the binding of sulfur released in the mixed-function oxidase-catalysed metabolism of diethyl *p*-nitrophenyl phosphorothionate (parathion) to diethyl *p*-nitrophenyl phosphate (paraoxon). *Biochem. Pharmac.*, **23**, 1733–1744.

Oppenoorth, F. J., Voerman, S., Welling, W., Houx, N. W. H., and Van der Oudenweyer, J. W. (1971). Synergism of insecticidal action by inhibition of microsomal oxidation with phosphorothionates. *Nature New Biol.*, **233**, 187–188.

Parke, D. V. (1968). *The Biochemistry of Foreign Compounds*, Pergamon, Oxford, p. 203.

Pellegrini, G. and Santi, R. (1972). Potentiation of toxicity of organophosphorus compounds containing carboxylic ester functions toward warm-blooded animals by some organophosphorus impurities. *J. Agric. Food Chem.*, **20**, 944–950.

Poore, R. E. and Neal, R. A. (1972). Evidence for extrahepatic metabolism of parathion. *Toxicol. Appl. Pharmacol.*, **23**, 759–768.

Ptashne, K. A. and Neal, R. A. (1972). Reaction of parathion and malathion with peroxytrifluoroacetic acid, a model system for the mixed function oxidase. *Biochemistry*, **11**, 3224–3228.

Ptashne, K. A., Wolcott, R. M., and Neal, R. A. (1971). Oxygen-18 studies on the chemical mechanisms of the mixed function oxidase catalysed desulfuration and dearylation reactions of parathion. *J. Pharmacol. Exp. Therap.*, **179**, 380–395.

Sakamoto, H. and Nishizawa, Y. (1962). Preparation and biological properties of phenylphosphorothioates. *Agr. Biol. Chem.*, **26**, 252–260.

Seawright, A. A., Hrdlicka, J., and De Matteis, F. (1976). The hepatotoxicity of O,O-diethyl, O-phenyl phosphorothionate (SV_1) for the rat. *Brit. J. Exp. Path.*, **57**, 16–22.

Shishido, T. and Fukami, J. (1963). Studies on the selective toxicities of organic phosphorus insecticides. II. The degradation of ethyl parathion, methyl parathion, methyl paraoxon and sumithion in mammal, insect and plant. *Botyu-kagaku*, **28**, 69–76 (cited in Motoyama and Dauterman, 1980).

Shishido, T., Usui, K., Sato, M., and Fukami, J. (1972). Enzymatic conjugation of diazinon and diazoxon in rat liver and cockroach fat body. *Pestic. Biochem. Physiol.*, **2**, 51–63.

Smith, R. L. and Williams, R. T. (1961). The metabolism of arylthioureas. Parts I, II, III, IV and V. *J. Medic. Pharmaceut. Chem.*, **4**, 97–176.

Snyder, J. P. (1974). Oxathiiranes. Differential orbital correlation effects in the electrocyclic formation of sulfur-containing three-membered rings. *J. Am. Chem. Soc.*, **96**, 5005–5007.

Stripp, B., Green, F. E., and Gillette, J. R. (1969). Disulfiram impairment of drug metabolism by rat liver microsomes. *J. Pharmac. Exp. Ther.*, **170**, 347–354.

Talcott, R. E., Mallipudi, N. M., and Fukuto, T. R. (1979a) Inactivation of esterases by impurities isolated from technical malathion. *Toxicol. Appl. Pharmacol.*, **49**, 107–112.

Talcott, R. E., Denk, H., and Mallipudi, N. M. (1979b). Malathion carboxylesterase activity in human liver and its inactivation by isomalathion. *Toxicol. Appl. Pharmacol.*, **49**, 373–376.

Uchiyama, M., Yoshida, T., Homma, K., and Hongo, T. (1975). Inhibition of hepatic drug-metabolizing enzymes by thiophosphate insecticides and its drug toxicological implications. *Biochem. Pharmacol.*, **24**, 1221–1225.

Umetsu, N., Grose, F. H., Allahyari, R., Abu-El-Haj, S., and Fukuto, T. R. (1977). Effect of impurities on the mammalian toxicity of technical malathion and acephate. *J. Agric. Food Chem.*, **25**, 946–952.

Verschoyle, R. D. and Aldridge, W. N. (1987). The interaction between phosphorothionate insecticides, pneumotoxic trialkyl phosphorothiolates and effects on lung 7-ethyoxycoumarin O-deethylase activity. *Arch. Toxicol.*, **60**, 311–318.

Walker, C. H. and Mackness, M. I. (1983). Esterases: problems of identification and classification. *Biochem. Pharmacol.*, **32**, 3265–3269.

Walker, C. H. and Mackness, M. I. (1987). "A" esterases and their role in regulating the toxicity of organophosphates. *Arch. Toxicol.*, **60**, 30–33.

Williams, R. T. (1959). *Detoxification Mechanism* (2nd edn.), Chapman and Hall, London, p. 181.

Williams, D. G. (1983). Intramolecular group transfer is a characteristic of neurotoxic esterase and is independent of the tissue source of the enzyme. *Biochem. J.*, **209**, 817–829.

2

Thioamides

John R. Cashman
Department of Pharmaceutical Chemistry and Liver Center, University of
California, San Francisco, CA 94143, USA

SUMMARY
1. Biotransformation of the thioamide moiety involves two sequential S-oxygenations, affording S-oxide and S,S-dioxide as major metabolites.
2. Thioamide S-oxygenations may be mediated by both the main mammalian monooxygenases, i.e. cytochrome P-450 and the flavin-containing monooxygenases.
3. Whereas the first oxidation is a detoxication pathway, S,S-dioxides (or sulphenes) are chemically extremely reactive and are probably responsible for the toxicity of thioamides.
4. This review examines the molecular mechanisms of toxicity of thioacetamide, thiobenzamide and related thioamides.

2.1   INTRODUCTION
Non-physiological thioamide-containing compounds are widely distributed in the environment and are known to initiate a number of toxic states in experimental animals and man. The thioamide functionality is also present in drugs, chemicals and agricultural agents and exposure of man and animal to these agents has resulted in inadvertent toxicity. Several clinically useful drugs (i.e. antithyroid, antibiotic, antineoplastic and hypnotic agents) which contain a thioamide group have limited use owing to toxic side effects. In essentially all cases examined, the toxicity of thioamides has been closely associated with the formation of oxidative metabolites. Up until recently, neither the enzymatic basis for bioactivation of thioamides to reactive species nor the detailed biochemical properties of the oxidative metabolites leading to the toxic responses of thioamides were known. Detailed studies of the enzymology, chemistry and toxicity of the oxidative metabolites of thiomides have emerged and form the basis for an unusually clear understanding of the chemical and metabolic factors involved in the toxicity of this class of chemicals.

2.2 THIOAMIDES

Thiocetamide (**I**) is a carcinogenic, teratogenic and hepatotoxic chemical (Neal and Halpert, 1982; McCann et al., 1975). In addition thioacetamide induces necrosis of the proximal tubule and thymic cortical cells. Metabolism studies have implicated cytochrome P-450 (Porter and Neal, 1978; Hunter et al., 1977) and the flavin-containing monooxygenase (Ziegler, 1980) in the bioactivation of thioacetamide to thioacetamide S-oxide and further thioacetamide S,S-dioxide, a highly reactive acylating agent which possesses sufficient chemical reactivity to account for the observed covalent binding to biomacromolecules (Dryoff and Neal, 1981). Cysteine can reduce thioacetamide S-oxide to the parent thioamide (de Ferreyra et al., 1980) and thioacetamide toxicity may result from at least two competing pathways: (a) oxidative bioactivation to a species which covalently binds to macromolecules and impairs cellular function and (b) reduction of S-oxide intermediates.

Ethaniamide (**II**) is an antitubercular agent (Shepard et al., 1985) that parallels thioacetamide in its metabolism. In the rat, mouse and dog, ethionamide is rapidly converted to ethionamide-S-oxide and the major metabolite is 2-ethylisonicotina-mide (Johnston et al., 1967; Bieder and Mazeau, 1962). Like other thioamides, ethionamide-S-oxide has more biological activity than ethionamide (Prema and Gopinathan, 1976), but a high incidence of hepatoxicity is observed for ethionamide (Cartel et al.,1983) and this has lead to its disuse. A fuller account of the metabolism and pharmacokinetics of ethionamide and related drugs is given in Volume 3 of this series (Chapter 6, Part B).

Quazepam (**III**) is a relatively new thioamide-containing benzodiazepine with sedative hypnotic properties (Kales et al., 1980). In mouse, hamster (Hilbert et al., 1984) and man (Zampaglione et al., 1985), quazepam is extensively metabolized, the first metabolic product arising from substitution of oxygen for sulphur in a manner similar to thioacetamide and ethionamide, discussed in greater detail below.

The thioamide Prefix (2,6-dichlorothiobenzamide) (**IV**) finds use as a herbicide (Griffiths et al., 1966) and since the major metabolite of this compound, 2,6-dichlorobenzonitrile, is also a herbicide it would seem that metabolic desulphuration may be involved in its mechanism of action.

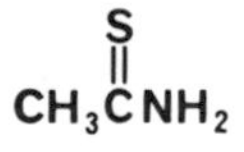

(**I**)

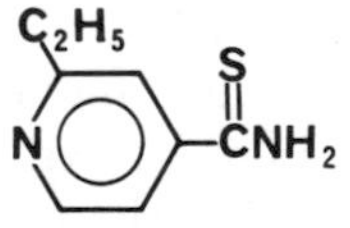

(**II**)

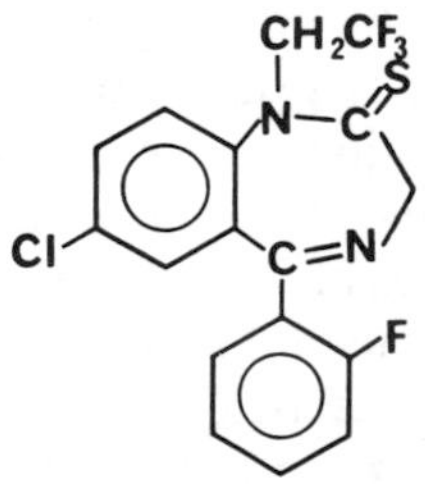

(**III**)

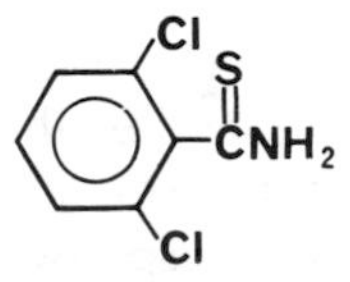

(**IV**)

Thiobenzamide (Fig. 1) is a chemical with antibacterial (Waisser *et al.*, 1982) and immunosuppressant properties (Pasquinelli *et al.*, 1985). Thiobenzamide is an effective antitubercular agent even in strains of mycobacterium tuberculosis which are resistant to isoniazide, ethionamide and thioacetazone (Drasata *et al.*, 1981). Generally, the biological activity and especially the hepatoxicity of a series of substituted thiobenzamides is enhanced by electron-donating *para*-substituents (Pasquinelli *et al.*, 1985; Drasata *et al.*, 1981; Hanzlik *et al.*, 1978, 1980; Hanzlik, 1982; Cashman *et al.*, 1983). It is this feature which allows a relatively simple organic chemical such as thiobenzamide to be an especially good model agent to study the relationship of oxidative bioactivation, reactive metabolites and covalent binding to the biological changes which follow administration to small animals (Hanzlik, 1982).

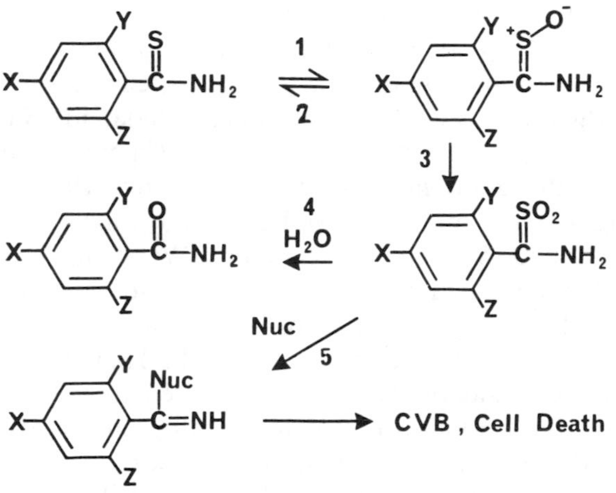

Fig. 1 — Metabolic scheme for thiobenzamide (TB). Step 1: bioactivation of TB to TBSO is facilitated by *para* substituents which are electron donating (i.e. X=CH$_3$O, CH$_3$, etc.). Step 2: enzymatic or non-enzymatic reduction of TBSO is facilitated by *para* substituents which are electron withdrawing (i.e. X=Cl, CF$_3$, etc.). Step 3: enzymatic and non-enzymatic oxygenation of TBSO is facilitated by electron-rich derivatives (i.e. X=CH$_3$O, H, Cl) and abolished by electron-deficient or sterically hindered ones (i.e. CF$_3$ or 2,6-dichloro(Prefix)-TBSO, Y,Z=Cl, respectively). Step 4: acylation of TBSO$_2$ with H$_2$O produces a non-toxic metabolite. Step 5: nucleophilic attack of important biomacromolecules results in covalent binding, and presumably impaired cellular function, resulting in cell death.

2.3   CHEMICAL PROPERTIES OF THIOAMIDES AND THEIR OXIDATION PRODUCTS

Compared with their oxidized derivatives, thioamides are relatively stable toward hydrolysis, rearrangement or other reactions. While few examples of the chemistry of thioamide *S*-oxide or *S,S*-dioxides have been reported, semi-empirical molecular orbital and CNDO calculations suggest that sulphinic acids (sulphenes) and sulphenic acids (sulphines) are reactive species (Zwanenburg, 1982). Thioamide-*S,S*-dioxides are highly reactive species which must be generated and used *in situ*, while the mono *S*-oxides, by contrast, are often stable materials (Walter and Bauer, 1975).

The hydrolysis of thioacetamide in dilute acid solution gives acetamide and H_2S (Butler *et al.*, 1958) but in concentrated acid formation of ammonium ion and thioacetic acid predominates (Edward and Wong, 1979). In base, hydrolysis of thioacetamide gives thioacetate and ammonia (Butler *et al.*, 1958). Metal ions such as Pb^{+2}, Cd^{2+} or Co^{2+} accelerate the hydrolytic release of H_2S from thioacetamide, and the latter is used to precipitate these metal ions for gravimetric analysis. In this context it is interesting to note that a thioamide analog of Z–Gly–Phe is a substrate for carboxypeptidase A (Bartlett *et al.*, 1982).

S-oxidation of thioamides proceeds by way of electrophilic attack by an oxidant (Cashman and Hanzlik, 1982). Oxidation of thioacetamide (Walter and Curts, 1960; Hillhouse *et al.*, 1986) and thiobenzamide (Cashman, 1982; Kitamura, 1938; Walter and Bauer, 1975) with H_2O_2 produces a relatively stable crystalline *S*-oxide. A mechanism involving formation of an oxathiirane intermediate probably does not contribute to desulphuration of thioamide *S*-oxide as it does for CS_2 (Catignani and Neal, 1975; Dalvi and Neal, 1978; Dalvi *et al.*, 1974) or phosphorothionates (Lee *et al.*, 1976) since, for thioamide *S*-oxides, elevated temperatures are required to effect this transformation (Oae, 1981). Secondary thioamide *S*-oxides have been detected (Walter, 1961) and in some cases isolated (Penney *et al.*, 1985). The *N*-substituent prefers the Z-orientation which interferes with the intramolecular hydrogen bonding required for *S*-oxide stabilization (Walter, 1960). In some cases isolable tertiary thioamide *S*-oxides are obtainable (Walter and Bauer, 1976). As the rates of H_2O_2-catalysed thioamide oxidation decrease from primary to tertiary thioamides, the ρ values become progressively larger and more negative (Cashman and Hanzlik, 1982). In contrast to these results, the Hammett plot for the much slower H_2O_2-catalyzed oxidation of *para*-substituted thiobenzamide *S*-oxides is markedly different; the ρ value is essentially zero despite the very low oxidation rates. This implies that a different mechanism of reaction is at work for this reaction: HOO^- can act as a nucleophile on electrophilic *S*-oxide sulphur. Thus, thioamide *S*-oxide sulphur may participate in nucleophilic as well as electrophilic reactions (Cashman and Hanzlik, 1982).

As a model enzyme system for monooxygenase-catalysed *S*-oxidations 4a-hydroperoxy flavins have been observed to *S*-oxidize a number of thioamides and other sulphur-containing compounds efficiently (Doerge and Corbett, 1984; Miller, 1982; Bruice, 1983; Ball and Bruice, 1980). The course of the reaction of thioamide *S*-oxides with nucleophilic species depends on the nature of the reagent as well as on the substituents attached to the sulphine carbon. For example, thioacetamide-*S*-oxide is readily reduced by cysteine, H_2S and other thiols without enzymatic assistance.

The acid-catalysed hydrolysis of thioacetamide-*S*-oxide is rationalized to involve *S*-oxide protonation and concomitant water attack followed by loss of thioperoxide anion to give the carbonyl equivalent. Thiobenzamide *S*-oxide, in contrast, is hydrolysed in dilute acid to the heterocyclic compound 3,5-diphenyl-1,2,4-thiadiazole (Cashman and Hanzlik, 1982). A similar product (Hector's base) is obtained from the exhaustive oxidation of phenyl thiourea (Butler *et al.*, 1978). The basic hydrolysis of thioacetamide *S*-oxide produces acetonitrile and sulphate (Walter, 1960). Likewise, base-catalysed decomposition of thiobenzamide *S*-oxide and 2-chloro-6-methylthiobenzamide *S*-oxide leads to benzonitrile and 2-chloro-6-

methylbenzonitrile respectively (Cashman and Hanzlik, 1982). Presumably, elimination of 'SOH$_2$' efficiently converts thioamide S-oxides to nitriles. Several thiols,
such as N-acetylcysteine and Cleland's reagent, were observed to reduce thiobenzamide S-oxide and 2-chloro-6-methylthiobenzamide S-oxide to their parent thioamides. This reaction is strongly accelerated by electron withdrawing groups on the
ring (Cashman and Hanzlik, 1982; Cashman *et al.*, 1983). Since a number of
thioamide S-oxides have been observed to undergo reduction *in vivo* (de Ferreya
et al., 1980; Prema and Gopinathan, 1976; Becker and Walter, 1965) this may
explain the effect of *meta* or *para* substitution on thiobenzamide hepatotoxicity. The
in vivo thiobenzamide–thiobenzamide S-oxide equilibration would be controlled by
substituents and electron withdrawing groups would decrease the net amount of
material progressing to the reactive sulphene species (Cashman *et al.*, 1983; Hanzlik
et al., 1980). This will be discussed in greater detail in the following section.

2.4   THIOBENZAMIDE TOXICITY

2.4.1   Thiobenzamide hepatotoxicity

Compared with thiobenzamide, p-methoxythiobenzamide was shown to be a much
more potent hepatotoxin and p-chlorothiobenzamide was much less so (Hanzlik
et al., 1978; Chieli *et al.*, 1979). This sequence of relative toxicity might be the result
of electronic substituent effects on (a) the rates of obligatory metabolic activation
steps, (b) the relative reactivities of toxic metabolites, or (c) the relative rates of
detoxication steps. To investigate this aspect more thoroughly, a series of *meta-* or
para substituted thiobenzamides was prepared and the hepatotoxicity was estimated
by plasma glutamic pyruvic transaminase (GTP) or plasma bilirubin responses to
assess their relative toxicities. Hepatotoxicity reponses (i.e. GPT or bilirubin)
followed a strict Hammett-type dependence (Lowry and Richardson, 1976) on the
electronic properties of the *meta* or *para* substituent, toxicity increasing dramatically
with increasing electron donation ($\rho = -3.4$, -4.15 and -1.4 and -2.0 respectively)
(Cashman *et al.*, 1983; Hanzlik *et al.*, 1980). Such a pattern is consistent with
enzymatic S-oxygenation as an obligatory bioactivation step, as has been suggested
for thioacetamide (Porter and Neal, 1978; Hunter *et al.*, 1977; Porter *et al.*, 1982)
carbon disulphide (Dalvi and Neal, 1978; De Matteis and Seawright, 1973) and other
thiourea derivatives (De Matteis, 1974; Boyd and Neal, 1976).

The herbicide 2,6-dichlorothiobenzamide (Prefix) is not hepatotoxic in rats. This
apparent lack of toxicity could result from the electronic substituent effect of the two
chlorines, or it might also arise from their steric effect, since they force the thioamide
group into a plane essentially orthogonal to the plane of the phenyl ring, thereby
eliminating the possibility of conjugation or resonance interaction between the
aromatic system and the thioamide group (Lowry and Richardson, 1976). Since 2,6-
disubstituted thiobenzamides bearing either electron-donating or electron-withdrawing groups were found to be non-hepatotoxic (Hanzlik *et al.*, 1980) the
substituent effects in this case must be caused by steric factors rather than electronic
effects of the kind described above. Mono *ortho*-substituted thiobenzamides are also
non-hepatotoxic (Cashman *et al.*, 1983). Steric hindrance by *ortho* substituents to

formation of a reactive intermediate or interference with nucleophilic attack on a reactive intermediate effectively eliminates the hepatotoxicity of thiobenzamide. Not all thiobenzamides tested conform to the structure–toxicity correlation found above. For example, p-hydroxythiobenzamide, a strongly electron donating substituted thiobenzamide, is devoid of hepatotoxicity (Hanzlik et al., 1980). It is possible that this compound is metabolized in vivo by rapid conjugation processes such as O-sulphation or O-glucuronidation in a similar fashion as observed for the p-hydroxy substituted pneumotoxin phenylthiourea (Scheline et al., 1961), leading to non-toxic metabolites.

As a probe of the possible requirement for further metabolic activation of thiobenzamide S-oxides, the effect of para substituents in the hepatotoxicity (i.e. GPT values) of thiobenzamide S-oxide demonstrated a ρ value of −3.23, being nearly identical to that for the parent thiobenzamide series (Hanzlik et al., 1980; Malvaldi, 1977; Chieli et al., 1980). In contrast, the plasma bilirubin response to the S-oxides is essentially independent of the substituents although this response in the parent thiobenzamide does vary according to the substituent electronic character. The ρ value for thiobenzamide S-oxide hepatotoxicity as assessed by GPT values suggests that the overall toxic response is dependent on a subsequent biotransformation step which is strongly dependent on electronic substituents in a manner consistent with the requirement for a thioamide S,S-dioxide metabolite. Thus, in general, the hepatotoxic response to thiobenzamide S-oxide is qualitatively similar to but more rapid and/or intense than that due to thiobenzamide (Hanzlik et al., 1980; Chieli et al., 1980).

Although thiobenzamide itself does not appear to cause hydrothorax in rats, thiobenzamide S-oxide causes perivascular lung oedema and pleural effusion (Cashman et al., 1982).

2.4.2 Pneumotoxicity of N-substituted thiobenzamides

In contrast to the hepatotoxic effects of primary thiobenzamides, several N-substituted thiobenzamides have been found to be potent lung toxins in rats and mice (Cashman et al., 1982). Like the arylthioureas, N-methylthiobenzamide produces pleural effusion and pulmonary oedema (Penney et al., 1985). In agreement with the results of thiobenzamide which suggest that it is the S,S-dioxide metabolite which is responsible for thiobenzamide hepatotoxicity, biotransformation of N-methylthiobenzamide in the lung may be similarly involved in the expression of its toxicity (Gottschall et al., 1985). N-methylthiobenzamide S-oxide, the major metabolite of N-methylthiobenzamide, produced lung damage which was qualitatively identical to that produced by N-methylthiobenzamide. N-methylthiobenzamide S-oxide pneumotoxicity is, however, a more intense toxic response then N-methylthiobenzamide in terms of lethality and [^{14}C]thymidine incorporation when N-methylthiobenzamide is administered intravenously (Gottschall et al., 1985).

In contrast to the pronounced hepatotoxicity of para-substituted thiobenzamides, only limited hepatic injury occurred in rats treated with N-methylthiobenzamide at doses which were decidedly pneumotoxic (Cashman, 1982).

2.5 THIOAMIDE METABOLISM

Oxidative bioactivation has been implicated in the associated toxicities of thio-amides. For instance, thioacetamide S-oxide (Hunter et al., 1977) and thiobenza-mide S-oxide (Hanzlik et al., 1980) are more hepatotoxic and ethionamide S-oxide has more antitubercular activity than their corresponding thioamides (Prema and Gopinathan, 1976). The thioamide Prefix (2,6-dichlorothiobenzamide) finds use as a herbicide and the major metabolite, 2,6-dichlorobenzonitrile, also a herbicide, suggests the involvement of oxidative desulphuration in its mechanism of action (Griffiths et al., 1966). The antibacterial (Waisser et al., 1982) and antitubercular (Bartos et al., 1985) activity of a series of substituted thiobenzamides was shown to be enhanced by electron donating para substituents.

In vivo ^{3}H-labelled thioacetamide was shown to be converted to sulphate, thioacetamide S-oxide and acetamide (Rees et al., 1966). In addition, covalently labelled protein, nucleic acid and phospholipid were isolated from rats administered radiolabelled thioacetamide. In rabbits, thioacetamide S-oxide was identified as a metabolite of thioacetamide, and the two were found to be interconvertible. Cysteine can reduce the S-oxide to the thioamide in vitro (Ammon et al., 1967) and this may explain the protective effect of cysteine pre-administration towards the hepatotoxicity of thioacetamide (de Ferreyra et al., 1979). Covalent binding of thioacetamide S-oxide is greater than that of thioacetamide and parallels the apparent half-life in various tissues (Porter et al., 1982). In the presence of microsomes, H_2O_2 oxidation of thioacetamide S-oxide leads to a reactive interme-diate (presumably thioacetamide S,S-dioxide) which acylates the ε-amino group of protein lysine to form N-ε-acetyllysine; it can also acylate water, leading to the formation of acetamide (Dryoff and Neal, 1981).

While the metabolism of ethionamide has not been examined in as much detail as thioacetamide, certain parallels between the two compounds suggest a common metabolic pathway (Shepard et al., 1985; Johnston et al., 1967). In the rat, mouse and dog, ethionamide S-oxide is rapidly converted to ethionamide and the major ultimate metabolite is 2-ethylisonicotinamide (Bieder et al., 1963).

A variety of tissues from a number of species S-oxygenate thiobenzamide to a significant extent (Cashman and Hanzlik, 1981). Thiobenzamide S-oxide formation was developed as a photometric assay for liver, lung and kidney microsomes from rabbits, mice and rats. All of these tissues demonstrate efficient S-oxide formation. In addition, both hog (Poulsen, 1981) and human (McManus et al., 1987) liver microsomes S-oxygenate thiobenzamide. The flavin-containing monooxygenase (Ziegler, 1980) mediates the oxidative desulphuration of thiobenzamide which proceeds by two sequential S-oxygenations (Cashman and Hanzlik, 1981; Hanzlik and Cashman, 1983) although the first of these oxygenations may also be affected by certain isozymes of cytochrome P-450 (Tynes and Hodgson, 1983) or even by 'reactive oxygen species' (Younes, 1985; Katori et al., 1981). Benzamide is also formed as a minor product during the initial microsomal oxidation of thiobenzamide. However, benzamide is the sole product arising from microsomal oxidation of thiobenzamide S-oxide (Hanzlik and Cashman, 1983).

The overall microsomal metabolism of thiobenzamide can be summarized as shown in Fig. 1. Initial S-oxygenation of thiobenzamide is mediated by the flavin-containing monooxygenase although a contribution from cytochromes P-450 may be observed depending on the method of microsome preparation (Cashman and Hanzlik, 1981), the tissue selected (Levi *et al.*, 1982) and the pretreatment of the animal employed (Cashman, 1987; McManus *et al.*, 1987). The data of Table 1 show that either electron-donating or electron-withdrawing *para*-substituted thiobenza-mides as well as *ortho*- disubstituted thiobenzamides have similar kinetic parameters for *S*-oxygenation; there is no Hammett substituent or steric dependence on the initial enzymatic *S*-oxygenation.

Table 1 — Substituent effects on the microsomal S-oxidation of 4-substituted and 2,6-disubstituted thiobenzamide derivatives[a]

Thiobenzamide	σ^b	K_m^c (μm)	V_{max}^d
4-CH$_3$O–	−0.27	9.3	5.2
4-CH$_3$–	−0.17	21.6	7.4
4-H–	0.0	6.1	9.5
4-Cl–	0.23	14.6	2.4
4-CF$_3$–	0.54	11.5	13.4
4-NO$_2$–	0.78	10.2	2.0
2,6-(CH$_3$)$_2$–	−0.34	47.2	2.4
2-Cl,6-CH$_3$–	0.06	13.0	1.3
2,6–(Cl)$_2$–	0.46	64.6	4.4

[a]Initial rates were determined photometrically, pH 7.4, 35°C.
[b]values for σ are the sum of values for Hammett *para*-substituent constants (Lowry and Richardson, 1976).
[c]Conditions are as described (Cashman and Hanzlik, 1981).
[d]Units are nmol/min/mg microsomal protein.

In contrast, the microsomal biotransformation of thiobenzamide *S*-oxide is dramatically influenced by substituents. Whereas thiobenzamide *S*-oxide and its *p*-methyl or *p*-chloro derivatives are detectably oxidized (see Table 2), other thiobenzamide *S*-oxides with strongly electron-withdrawing or sterically hindering substituents are inert to active microsomal oxygenation (Cashman, 1982).

The covalent binding of radiolabelled thiobenzamide and 2-chloro-6-methylthio-benzamide was studied (Hanzlik, 1984) in rats and, as expected, thiobenzamide metabolites became bound to liver proteins to a large extent, while smaller but significant amounts became bound to lung and kidney proteins. 2,6-disubstitution of thiobenzamide reduced the amount of covalently bound material to a very low level in all three tissues examined. It would appear that *ortho* substitution on the thiobenzamide nucleus blocks the covalent binding and thus eliminates the asso-ciated hepatoxicity.

N-methylthiobenzamide is a pneumotoxicant that appears to require two succes-sive oxygenations for the expression of its pneumotoxicity (Penney *et al.*, 1985). *N*-methylthiobenzamide is extensively *S*-oxidized in rat liver and lung microsomes (Gottschall *et al.*, 1985). In the lung, the flavin-containing monooxygenase appears

Table 2 — Effect of substituents on the microsomal
oxidation of thiobenzamide S-oxides[a]

Thiobenzamide S-oxide	K_m (µm)	V_{max}[b]
4-CH$_3$–	16.0	1.2
4-H–	12.3	0.5
4-Cl–	28.6	2.2
4-CF$_3$–	No reaction	
4-NO$_2$–	No reaction	
2-CH$_3$,6-CH$_3$–	No reaction	
2-Cl,6-CH$_3$–	No reaction	
2-Cl,6-Cl–	No reaction	

[a]Means of 3–5 determinations with microsomes from one rat. Initial
rates determined photometrically under conditions as given (Cashman and Hanzlik, 1981).
[b]Units are nmol/min mg microsomal protein.

to play a greater role in S-oxidation while, in the liver, oxidations appear to be
mediated by both cytochromes P-450 and the flavin-containing monooxygenase. In
microsomes, N-methylthiobenzamide S-oxide is converted to N-methylbenzamide
and covalently bound metabolites in a 1:1 ratio, probably via enzymatic oxidation to
the highly electrophilic S,S-dioxide. The extensive level of covalent binding seen for
N-methylthiobenzamide is not observed for N-tertbutylthiobenzamide and this
latter compound is much less potent a pneumotoxicant than N-methylthiobenzamide
(Hanzlik, 1984). The reason why N-substitution shifts the toxicity of thiobenzamide
from liver to lungs is unknown, but the decrease in pneumotoxic potency with larger
N-substituents may simply be due to a steric effect on the ability of reactive sulphene
metabolites to acylate proteins.

2.6 SUMMARIZING REMARKS

There is now general agreement that the toxicity of thioamides stems from products
of their oxidative biotransformation rather than from the parent thioamide or by the
hydrolytic generation of H$_2$S. The major pathway for biotransformation of the
thioamide moiety involves two sequential S-oxygenations mediated by the flavin-
containing monooxygenase although the first of these oxygenations may also be
effected by certain isozymes of cytochrome P-450. The first oxidation produces a
thioamide S-oxide which is relatively stable although readily reduced by thiols in a
process which probably constitutes a detoxication pathway. Further oxidation of
thioamide S-oxides leads to an S,S-dioxide or sulphene metabolite which is extremely reactive chemically and which acylates water (to form amides) or cellular
nucleophiles (to form covalently bound materials), the latter of which presumably
impairs cellular function.

ACKNOWLEDGEMENTS

Work in the general area of sulphur metabolism has been financially supported by the March of Dimes Basil O'Connor Starter Scholar Research Award No. 5-558 and the California State Toxic Substances Research and Training Program. The author acknowledges the stimulating discussions and collaborations of Drs. D. Gottschall, R. P. Hanzlik, D. Penney, G. Traiger and D. M. Ziegler.

REFERENCES

Ammon, R., Berninger, H., Haas, H. S., and Laudsberg, I. (1967). Thioacetamid-sulfoxid, ein Stoffwechselprodukt des Thioacetamids. *Arzn. Forsch.*, **17**, 521–525.

Ball, S. and Bruice, T. C. (1980). Oxidations of amines by a 4a-hydroperoxyflavin. *J. Amer. Chem. Soc.*, **102**, 6498–6503.

Bartlett, P. A., Spear, K. L., and Jacobsen, N. E. (1982). A thioamide substrate of carboxypeptidase A. *Biochemistry*, **21**, 1608–1611.

Bartos, F., Findejsova, J., El Zein, K., Mollin, J., and Waisser, K. (1985). Antimitotic activity of thiobenzamides substituted on nitrogen or aromatic group. *Ceskoslov. Farm.*, **34**, 135–138.

Becker, V. and Walter, W. (1965). Die Wirkung von Thioacetylverbindungen auf das Leberparenchym im Tierexperiment. *Acta Hepato. Splenol.*, **12**, 129–133.

Bieder, A. and Mazeau, L. (1962). Etude du metabolisme de l'ethionamide chez l'homme. *Ann. Pharmacol. Fr.*, **20**, 211–219.

Bieder, A., Brunnel, P. and Mazeau, L. (1963). Dosage simultané de l'ethionamide et de son metabolite, le sulfoxyde d'ethionamide. *Ann. Pharmacol. Fr.*, **21**, 375–379.

Boyd, M. R. and Neal, R. A. (1976). Studies of the mechanisms of toxicity and development of tolerance to the pulmonary toxin, α-naphthylthiourea (ANTU). *Drug. Metabol. Disp.*, **4**, 314–322.

Bruice, T. C. (1983). Leaving group tendencies and the rate of monooxygen donation by H_2O_2, ROOH and peroxycarboxylates. *J. Chem. Soc. Chem. Commun.*, 14–15.

Butler, E. A., Peters, D. G., and Swift, E. H. (1958). Hydrolysis reactions of thioacetamide in aqueous solution. *Analyt. Chem.*, **30**, 1379–1383.

Butler, A. R., Glidewell, C., and Liles, D. C. (1978). The constitution of Hector's base; X-ray crystal and molecular structure. *Chem. Commun.*, 652–654.

Cartel, J., Millan, J., Guelpa-Lauras, C., and Grosset, J. H. (1983). Hepatitis in leprosy patients treated by a daily combination of rifampin and a thioamide of clapsone. *Int. J. Leprosy*, **51**, 461–465.

Cashman, J. R. (1982). Toxicity, metabolism and chemistry of thiobenzamide and derivatives. *Ph.D. Thesis*, University of Kansas.

Cashman, J. R. (1987). A convenient radiometric assay for flavin-containing monooxygenase activity. *Analyt. Biochem.*, **160**, 294–300.

Cashman, J. R. and Hanzlik, R. P. (1981). Microsomal oxidation of thiobenzamide. A photometric assay for the flavin-containing monooxygenase. *Biochem. Biophys. Res. Commun.*, **98**, 147–153.

Cashman, J. R., and Hanzlik, R. P. (1982). Oxidation and other reactions of thiobenzamide derivatives of relevance to their hepatotoxicity. *J. Org. Chem.,* **47**, 4645–4650.

Cashman, J. R., Traiger, G. J., and Hanzlik, R. P. (1982). Pneumotoxic effects of thiobenzamide derivatives. *Toxicology,* **23**, 85–93.

Cashman, J. R., Parikh, K. K., Traiger, G. J. and Hanzlik, R. P. (1983). Relative hepatotoxicity of ortho and meta monosubstituted thiobenzamides in the rat. *Chem.-Biol. Interact.,* **45**, 341–347.

Catignani, G. L. and Neal, R. A. (1975). Evidence for the formation of a protein bound hydrodisulfide resulting from microsomal mixed-function oxidase desulfuration of carbon disulfide. *Biochem. Biophys. Res. Commun.,* **65**, 629–636.

Chieli, E., Malvaldi, G., and Tongiani, R. (1979). Early biochemical liver changes following thiobenzamide poisoning. *Toxicology,* **13**, 101–109.

Chieli, E., Malvaldi, G., and Segnini, D. (1980). The hepatotoxicity of thiobenzamide S-oxide. *Toxicol Lett.,* **7**, 175–180.

Dalvi, R. R. and Neal, R. A. (1978). Metabolism *in vivo* of carbon disulfide to carbonyl sulfide and carbon dioxide in the rat. *Biochem. Pharmacol.,* **27**, 1608–1613.

Dalvi, R. R., Poore, R. E., and Neal, R. A. (1974). Studies of the metabolism of carbon disulfide by rat liver microsomes. *Life Sci.,* **14**, 1785–1796.

de Ferreyra, E. C., de Fenos, O. M., Bernacchi, A. S., De Castro, C. R., and Castro, J. A. (1979). Therapeutic effectiveness of cystamine and cysteine to reduce liver cell necrosis induced by several hepatotoxins. *Toxicol. Appl. Pharmacol.,* **48**, 221–229.

de Ferreyra, E. C., De Fenos, O. M., and Castro, J. A. (1980). Effect of different chemicals on thioacetamide-induced liver necrosis. *Toxicology,* **16**, 205–211.

De Matteis, F. (1974). Covalent binding of sulfur to microsomes and loss of cytochrome P-450 during the oxidative desulfuration of several chemicals. *Molec. Pharmacol.,* **10**, 849–854.

De Matteis, F. and Seawright, A. (1973). Oxidative metabolism of carbon disulfide by the rat: effect of treatments which modify the liver toxicity of carbon disulfide. *Chem.-Biol. Interact.,* **7**, 375–388.

Doerge, D. R. and Corbett, M. D. (1984). Hydroperoxyflavin-mediated oxidations of organosulfur compounds. *Molec. Pharmacol.,* **26**, 348–352.

Drasata, J., Loskot, J., and Waisser, K. (1981). Biological side effects of potential antitubercular agents. *Ceskoslov. Farm.,* **30**, 266–269.

Dryoff, M. C. and Neal, R. A. (1981). Identification of the major protein adduct formed in rat liver after thioacetamide administration. *Cancer Res.,* **41**, 3430–3435.

Edward, J. T. and Wong, S. C. (1979). Effect of acid concentration on the partitioning of tetrahedral intermediate in the hydrolysis of thioacetanilide. *J. Amer. Chem. Soc.,* **101**, 1807–1814.

Gottschall, D. W., Penney, D. A., Traiger, G. J., and Hanzlik, R. P. (1985). Oxidation of N-methylthiobenzamide and N-methylthiobenzamide S-oxide by liver and lung microsomes. *Toxicol. Appl. Pharmacol.,* **78**, 332–341.

Griffiths, M. H., Moss, J. A., Rose, J. A., and Hathaway, D. E. (1966). The comparative metabolism of Prefix in the dog and rat. *Biochem. J.,* **98**, 770–781.

Hanzlik, R. P. (1982). Effects of substituents on the reactivity and toxicity of chemically reactive intermediates. *Drug Metab. Rev.,* **13**, 207–234.

Hanzlik, R. P. (1984). Chemistry of covalent binding: studies with bromobenzene and thiobenzamide. In: *Biological Reactive Intermediates,* Vol. III, Plenum, New York, NY, pp. 31–40.

Hanzlik, R. P. and Cashman, J. R. (1983). Microsomal metabolism of thiobenzamide and thiobenzamide S-oxide. *Drug Metabol. Dispos.,* **11**, 201–205.

Hanzlik, R. P., Cashman, J. R., and Traiger, G. J. (1980). Relative hepatoxicity of substituted thiobenzamides and thiobenzamide S-oxides in the rat. *Toxicol. Appl. Pharmacol.,* **55**, 260–272.

Hanzlik, R. P., Vyas, K. P., and Traiger, G. J. (1978). Substituent effects on the hepatotoxicity of thiobenzamide derivatives in the rat. *Toxicol Appl. Pharmacol.,* **46**, 685–694.

Hilbert, J., Pramanik, B., Symchowicz, S., and Zampaglione, N. (1984). The disposition and metabolism of a hypnotic benzodiazepine, quazepam, in the hamster and mouse. *Drug Metabol. Dispos.,* **12**, 452–459.

Hillhouse, J. H., Blair, I. A., and Field, L. (1986). Thiono compounds. 7. Oxidation of thioamides in relation to adverse biological effects. *Phosphorus Sulfur,* **26**, 169–184.

Hunter, A. L., Holscher, M. A., and Neal, R. A. (1977). Thioacetamide-induced hepatic necrosis. I. Involvement of the mixed-function oxidase enzyme system. *J. Pharmacol. Exp. Ther.,* **200**, 439–448.

Johnston, J. P., Kane, P. O., and Kibby, M. R. (1967). The metabolism of ethionamide and its sulfoxide. *J. Pharm. Pharmacol.,* **19**, 1–9.

Kales, A., Scharf, M. B., Soldatos, C. R., Bixler, E. O., Bianchi, S., and Schweitzer, S. K. (1980). Quazepam: a new benzodiazepine hypnotic intermediate-sleep laboratory evaluation. *J. Clin. Pharmacol.,* **20**, 184–192.

Katori, E., Nagano, T., Kunieda, T., and Hirobe, M. (1981). Facile desulfurization of thiocarbonyl groups to carbonyls by superoxide. A model of metabolic reactions. *Chem. Pharm. Bull.,* **29**, 3075–3077.

Kitamura, R. (1938). The synthesis of a new type compound, thioperimidsäure. *J. Pharm. Soc. Japan,* **58**, 246–249.

Lee, P. W., Allahyari, R., and Fukuto, T. R. (1976). Stereospecificity in the metabolism of the chiral isomers of fonofos by mouse liver microsomal mixed function oxidase. *Biochem. Pharmacol.,* **25**, 2671–2674.

Levi, P. E., Tynes, R. E., Sabourin, P. J., and Hodgson, E. (1982). Is thiobenzamide a specific substrate for the microsomal FAD-containing monooxygenase? *Biochem. Biophys. Res. Commun.,* **107**, 1314–1323.

Lowry, R. H. and Richardson, K. S. (1976). Some fundamentals of physical organic chemistry. In: *Mechanism and Theory in Organic Chemistry,* Harper and Row, New York, NY, pp. 60–71.

Malvaldi, G. (1977). Liver changes following thiobenzamide poisoning. *Experientia,* **33**, 1200–1201.

McCann, J. Choi, E., Yamasaki, E., and Ames, B. N. (1975). Detection of carcinogens as mutagens in the Salmonella/microsome test: assay of 300 chemicals. *Proc. Natl. Acad. Sci.,* **72**, 5135–5139.

McManus, M. E., Stupans, I., Burgess, W., Koenig, J. A., Hall, P. de la M. and Birkett, D. J. (1987). Flavin-containing monooxygenase activity in human liver microsomes. *Drug metabol. Dispos.*, **15**, 256–261.

Miller, A. (1982). A model for FAD-containing monooxygenase: the oxidation of thioanisole derivatives. *Tetrahedron Lett.*, **23**, 753–756.

Neal, R. A. and Halpert, J. A. (1982). Toxicology of thiono-sulfur compounds. *Ann. Rev. Pharmacol. Toxicol.*, **22**, 321–329.

Oae, S. (1981). Personal communication.

Pasquinelli, P., Bruschi, F., Saviozzi, M., and Malvadi, G. (1985). Immunosuppressive effect and promoter activity for hepatic carcinogenesis of thiobenzamide. *Boll. Soc. It. Biol. Sper.*, **61**, 61–66.

Penney, D. A., Gottschall, D. W., Hanzlik, R. P., and Traiger, G. J. (1985). The role of metabolism in N-methylthiobenzamide-induced pneumotoxicity. *Toxicol. Appl. Pharmacol.*, **78**, 323–331.

Porter, W. R. and Neal, R. A. (1978). Metabolism of thioacetamide and thioacetamide S-oxide by rat liver microsomes. *Drug. Metabol. Dispos.*, **6**, 379–388.

Porter, W. R., Gudzinowicz, M. J., and Neal, R. A. (1982). Thioacetamide-induced hepatic necrosis. Pharmacokinetics of thioacetamide and thioacetamide S-oxide in the rat. II. *J. Pharmacol. Exp. Ther.*, **208**, 386–391.

Poulsen, L. L. (1981). Organic sulfur substrates for the microsomal flavin-containing monooxygenase. In: E. Hodgson, J. R. Bend, and R. M. Philpot (eds.), *Reviews in Biochemical Toxicology*, Vol. 3, Elsevier, New York, NY, pp. 33–49.

Prema, K. and Gopinathan, K. P. (1976). Distribution, induction and purification of monooxygenase catalyzing sulfoxidation of drugs. *Biochem. J.*, **25**, 1299–1307.

Rees, K. R., Rowland, G. F., and Varcoe, J. S. (1966). The metabolism of tritiated thioacetamide in the rat. *Int. J. Cancer*, **1**, 197–203.

Scheline, R. R., Smith, R. L., and William, R. T. (1961). The metabolism of arythioureas. II. The metabolism of ^{14}C- and ^{35}S-labelled 1-phenyl-2-thiourea and its derivatives. *J. Med. Pharm. Chem.*, **4**, 109–135.

Shepard, C. C., Jenner, P. J., Ellard, G. A., and Lancaster, R. D. (1985). An experimental study of the antileprosy activity of a series of thioamides in the mouse. *Int. J. Leprosy*, **53**, 587–594.

Tynes, R. E. and Hodgson, E. (1983). Oxidation of thiobenzamide by the FAD-containing monooxygenase and cytochrome P-450-dependent monooxygenases of liver and lung microsomes. *Biochem. Pharmacol.*, **32**, 3419–3428.

Waisser, K., Celadnik, M., Palat, K., and Odlerova, Z. (1982). Antituberculotics. The effect of test microbiological media on the evaluation of structure–antituberculotic activity relationships of thiobenzamides. *Ceskoslow. Farm.*, **31**, 303–307.

Walter, W. (1960). Oxydationsprodukte von Thiocarbonsäure-amiden, I. Thioacetamid S-Oxyd. *Lieb. Ann. Chem.*, **633**, 35–42.

Walter, W. (1961). Oxydationsprodukte von Thiocarbonsäure-amiden II. *Lieb. Ann. Chem.*, **633**, 49–56.

Walter, W. and Bauer, O. H. (1975). Über die Oxidationsprodukte von Thiocarbonsäuremiden, XXXII. *Liebigs Ann. Chem.*, 305–310.

Walter, W. and Bauer, O. H. (1976). Eine einfache Darstellungsmethode für tertiäre Thioamide S-oxide. *Liebigs Ann. Chem.*, 1584–1597.

Walter, W. and Curts, J. (1960). Oxydationsprodukte primaren Thioamide. *Chem. Ber.*, **93**, 1511–1523.

Younes, M. (1985). Involvement of reactive oxygen species in the microsomal S-oxidation of thiobenzamide. *Experientia*, **41**, 479–480.

Zampaglione, N., Hilbert, J. M., Ning, J., Chung, M., Gural, R., and Symchowicz, S. (1985). Disposition and metabolic fate of [^{14}C]- quazepam in man. *Drug Metab. Dispos.*, **13**, 25–29.

Ziegler, D. M. (1980). Microsomal flavin-containing monooxygenase. Oxygenation of nucleophilic nitrogen and sulfur compounds. In: W. B. Jakoby (ed.), *Enzymatic Basis of Detoxication*, Vol. 1, Academic Press, New York, NY, pp. 201–227.

Zwanenburg, B. (1982). The chemistry of sulfines. *Reueil.*, **101**, 1–19.

3

Thiocarbamides

G. G. Skellern
Division of Pharmaceutical Chemistry, Department of Pharmacy, University of
Strathclyde, Glasgow G1 1XW, UK

SUMMARY

1. Those physical and chemical properties of thioureas, 2-thioimidazoles and
   2-thiouracils which are pertinent to the understanding of their biochemistry,
   pharmacology and toxicology are reviewed. This includes discussion of their
   ionizability, complexation with metal ions, oxidation, reduction, alkylation and
   acylation.
2. Metabolic thiocarbamide desulphuration *via* S-oxygenated intermediates to
   either oxygen analogues or formamidines is examined. Conjugation reactions, in
   particular thiocarbamide glucuronidation, ribosylation and methylation, are also
   described.
3. The various enzyme systems with which the thiocarbamides interact are
   reviewed, with a special emphasis on the peroxidases, monooxygenases and
   copper-containing enzymes.
4. While noting that thiocarbamides are goitrogenic, the section on the toxicity of
   thiocarbamides deals mainly with the relationship between toxic responses and
   the metabolic activation of thioureas, including ethylenethiourea.

3.1 INTRODUCTION

Thiourea (**1**), the simplest of the thiocarbamides, is the progenitor of this class of
compounds. The thiocarbamide moiety is present in a variety of alicyclic and
heterocyclic compounds including 2-thioimidazoles, 2-thiohydantoins and 2-thiopyr-
imidines (Fig. 1).

Ergothioneine (**2**), the betaine of 2-thio-L-histidine, is unique in that it is the only
known naturally occurring thiocarbamide. It is present in ergot and in appreciable
amounts in mammalian blood cells and in the seminal fluid of some species (Stowell,

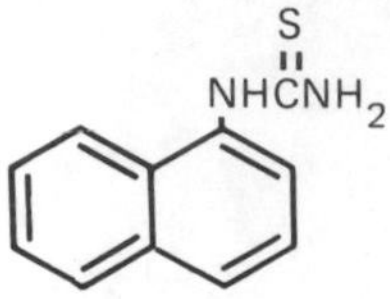

(3)

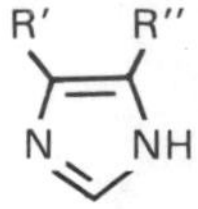

(5)

a $R' = H$

$R'' = CH_2CH_2CH_2CH_2NHC(=S)NHMe$

b $R' = Me$

$R'' = CH_2SCH_2CH_2NHC(=S)NHCH_3$

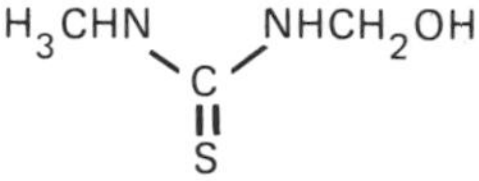

(6)

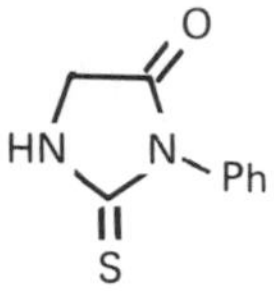

(7)

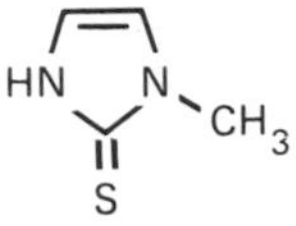

(8)

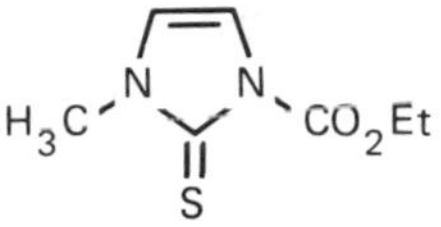

(9)

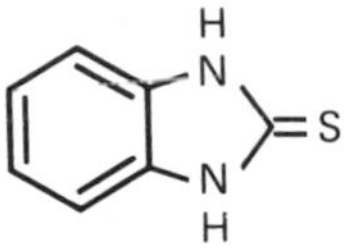

(10)

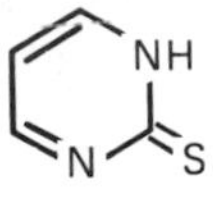

(11)

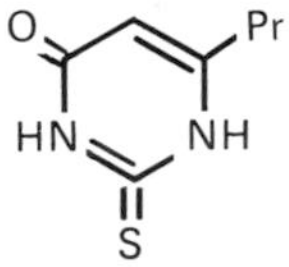

(12)

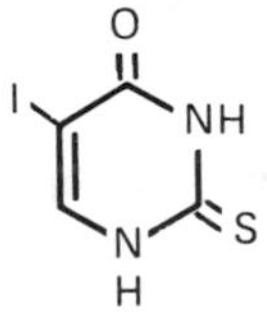

(13)

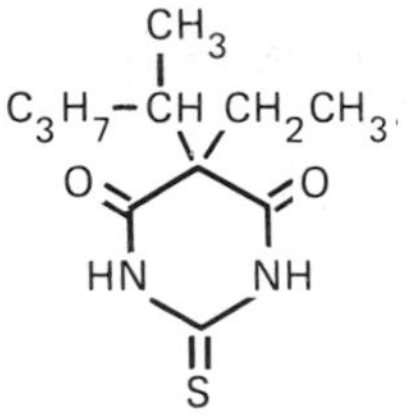

(14)

Fig. 1 — Some typical thiocarbamides.

1961). There has been some debate as to whether its presence in the body is due to dietary intake or to its biosynthesis *in situ*. Current evidence shows that it is not synthesized in the body, but derived from food (Kawano *et al.*, 1982).

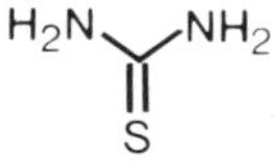

(1)

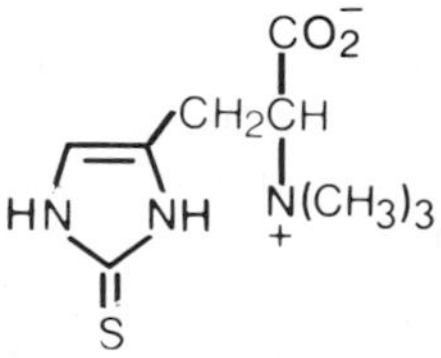

(2)

The synthetic thiocarbamides have a wide variety of uses, ranging from intermediates in synthetic reactions to therapeutic agents (Table 1). The cyclic thiocarbamides, 2-thioimidazoles, 2-thiouracils and 2-thiobarbiturates, are usually prepared
by the condensation of a thiourea with an appropriate compound. The antibacterial
properties of thiourea have been known for some time (Reid, 1963a). Currently the
N-hydroxymethyl derivative (6) of N-methylthiourea, which is a 'masked' formaldehyde compound, is used as a lavaging agent after abdominal surgery (Pickard,
1972).

Table 1 — Thiocarbamides and some of their uses

Compound	Use	Reference
Thioureas		
Thiourea (**1**)	Synthesis of other thiocar-bamides	Reid (1963a, 1965)
1-Naphthylthiourea (**3**)	Rodenticide	Richter (1945)
Ethylenethiourea (**4**)	Compounding of rubber	
Burimamide (**5a**) and metiamide (**5b**)	Lead compounds in the development of the H_2-antagonist cimetidine	Ganellin and Durant (1981)
N-Methyl-N'-(hydroxy methyl)thiourea (**6**) (noxythiolin)	Topical bactericide	Pickard (1972)
2-Thiohydantoins (**7**)	Amino acid sequencing of proteins	Edman (1950)
2-Thioimidazoles		
1-Methyl-2-thioimida-zole (methimazole) (**8**)	Treatment of hyperthy-roidism (antithyroid agents)	Solomon (1986)
1-Carbethoxy-3-methyl-2-thioimidazole (**9**) (carbimazole)		
2-Thiobenzimidazole (**10**)	Rubber processing	El Dareer *et al.* (1984)
2-Thiouracils		
6-*n*-Propyl-2-thiouracil (**12**)	Treatment of hyperthyr-oidism (antithyroid agent)	Solomon (1986)
5-Iodo-2-thiouracil (**13**)	Diagnosis of melanoma	Franken *et al.* (1986)
2-Thiobarbiturates		
Thiopentone (**14**)	Anaesthetic	Morgan *et al.* (1981)

Because of its toxicity, considerable interest has been shown in ethylenethiourea (**4**) which, apart from its use in the rubber industry, is a contaminant (eqn. (1)) (Lopatecki and Newton, 1952; Bontoyan and Looker, 1973), an environmental degradation product (Vonk, 1971; Sijpesteijn and Vonk, 1975) and a metabolite (Jordan and Neal, 1979) of the ethylene dithiocarbamates, Maneb (**15**), Zineb and Nabam (also see Chapter 4 on thiocarbamates, Volume 1, Part B in this series):

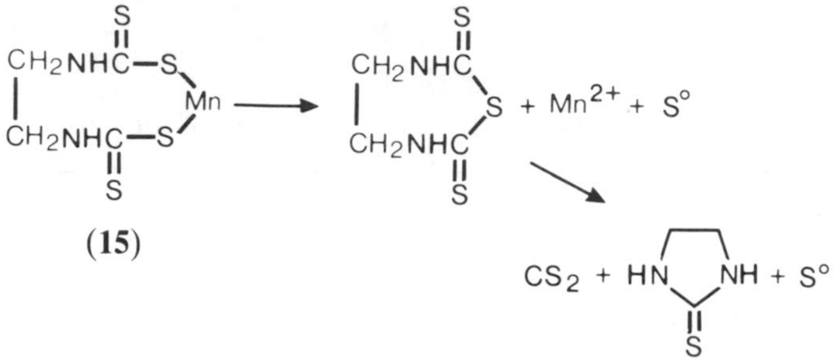

$$(1)$$

(**15**)   (**4**)

Nearly all the thiocarbamides are goitrogenic and affect thyroidal iodine metabolism to a lesser or greater extent (Astwood, 1943; Astwood *et al.*, 1945; Stanley and Astwood, 1947; Seifter and Ehrich, 1948; Searle *et al.*, 1950, 1951). This observation led to the development of the thiocarbamides currently used in the treatment of hyperthyroidism. Methimazole (**8**) and its 3-carbethoxy derivative (**9**) are 2-thioimidazoles, and propyl-2-thiouracil (**12**) is a 2-thiouracil (see Volume 3 of this series).

Apart from their antithyroid activity, exogenous 2-thiouracil and its derivatives are incorporated into the pigment melanin during its biosynthesis (Whittaker, 1971; Farishian and Whittaker, 1979; Dencker *et al.*, 1981; Fairchild et al., 1982; Wätjen *et al.*, 1982). Consequently [^{125}I]5-iodo-2-thiouracil (**13**) is at present being examined as a potential melanoma diagnostic agent (Broxterman *et al.*, 1983; Franken *et al.*, 1986).

This chapter will deal initially with those chemical properties of the thiocarbamides which are pertinent to understanding the subsequent discussion of their biochemistry, pharmacology and toxicology.

3.2 PHYSICAL AND CHEMICAL PROPERTIES

Early reviews on the thioureas (Reid, 1963a), 2-thiohydantoins (Ware, 1950; Reid, 1963b; Edward, 1966), 2-thioimidazoles (Hofmann, 1953a) and 2-thiopyrimidines and derivatives (Brown, 1962, 1970; Reid, 1963c, 1965) are comprehensive and deal with their synthesis, chemistry and uses.

Thiocarbamides are a heterogeneous class of compounds, differing in their hydrophilicity, lipophilicity, ionizability and reactivity. Although some generalizations can be made when comparing one group with another, this may not always be possible. Dissimilarities between the groups are not unexpected since the compounds within this class range from water-soluble thioureas to the highly lipophilic heterocyclic aromatic 2-thiopyrimidine (**11**) and its derivatives.

3.2.1 Acidity and basicity

In aqueous solution the tautomeric thiocarbamides exist predominantly in the thione form, in equilibrium with a very small amount of the thioenol:

(16)   (2)

The preference for the thione tautomer (Katritzky and Lagowski, 1963; Bojarska-Olejnik *et al.*, 1985), which is even greater for 5-membered than 6-membered thiocarbamide heterocycles, is attributed in part to the π electron stabilization of the thione form (Kjellin and Sandstrom, 1969). Compared with thioamides, thiocarbamides are more nucleophilic (Janssen, 1962). This increased nucleophilicity is attributable to the donation of electrons by the second nitrogen atom adjacent to the thiocarbonyl group, thereby increasing the π electron density.

Although the basicity of the double-bonded sulphur atom is either very weak or completely suppressed (Table 2), ease of proton loss from the thiocarbamide moiety can vary by several orders of magnitude for the heterocyclic thiocarbamides: for example, 2-thioimidazole (16) has $pK_a=11.6$ whereas 2-thiopyrimidine (11) has $pK_a=7.04$.

Table 2 — Dissociation constants of some simple thiocarbamides

Compound	pK_{aH^+} (S-protonated)	pK_{a1}^a	pK_{a2}
Thiourea (1)	-1.5^b, -1.19^i	15.0^b	
Ethylenethiourea (4)	-1.9^i		
2-Thiohydantoin		8.9^c	
2-Thioimidazole (16)	-1.6^d	11.6^d	
1-Methyl-2-thioimidazole (8; methimazole)	-2.0^d	11.9^d, 11.22^e, 11.38^f	
Ergothioneine (2)	-2.3^d	10.8^d	
2-Thiobenzimidazole (10)	2.6^e	9.8^e	
2-Thiopyrimidine (11)	1.40^g	7.04^g	
2-Thiouracil	$>2.3^e$, -4.16^g	7.65^g, 7.73^e, 7.75^h	
6-n-Propyl-2-thiouracil (12)	-4.22^g	8.25^g	
2-Thiobarbituric acid	-4.39^g	3.70^g	7.89^g
Thiopentone (14)	-7.15^g	7.58^g	

[a]pK for deprotonation of thiocarbamide.
[b]Ganellin and Durant (1981).
[c]Edward and Nielson (1957).
[d]Stanovnik and Tisler (1964a).
[e]Foye and Lo (1972).
[f]Hanlon and Shuman (1975).
[g]Stanovnik and Tisler (1964b).
[h]Mautner and Clayton (1959).
[i]Janssen (1962).

All the thiocarbamides are amphoteric and in strong acids are *S*-protonated (Stanovik and Tisler, 1964a and b). While thiourea (**1**), the class progenitor, is highly water soluble, weakly amphoteric and over the pH range 2–12 is completely undissociated, the dioxothiopyrimidine, thiobarbituric acid, is sparingly soluble in water and ionized at pH values greater than 7. It also forms water-soluble salts.

2-Thiopyrimidine (**11**) is more acidic than its 5-membered ring analogue, 2-thioimidazole (**16**). The increased resonance stabilization of the anion would explain the greater acidity of 2-thiopyrimidine. Compare (**17**) with (**18**).

(**17**) 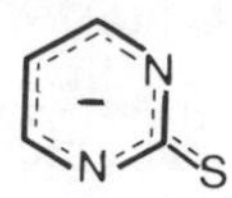   (**18**)

Greater charge delocalization would also account for the difference in the acidities of the mono-oxo-derivatives of these compounds, namely 2-thiohydantoin (**19**) and 2-thiouracil (**20**).

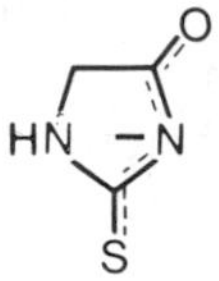

(**19**) 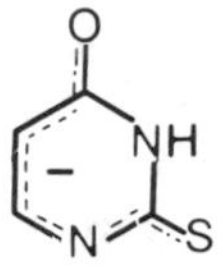   (**20**)

Assignment of the dissociable group is not easy since the oxo-derivatives possess a carbonyl group which although it is more highly polarized than the thiocarbonyl group will contribute to the stabilization of the anion. Substitution at either the nitrogen atoms or carbon atoms adjacent to the thiocarbamide moiety may affect the pK_a value depending on the size of substituent and its electron donating or withdrawing properties (Kjellin and Sandstrom, 1969).

S-Alkylation of, for example, 4,5-dimethyl-2-thioimidazole (**21**) maintains the sulphur in the thiol form, thereby either restoring (Table 3) or increasing the basicity of the compound.

Table 3 — Effect of *S*-substitution on the basicity of
4,5-dimethyl-2-thioimidazole (**21**)

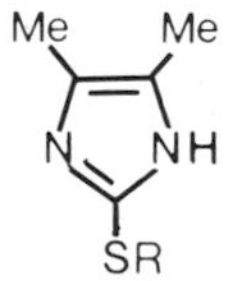

	R=H(**21**)	R=CH$_3$
Basicity pK_a (proton gained)	−1.2	6.85

From Kjellin and Sandstrom (1969).

3.2.2 Complexation with metal ions

In common with simple thiols, thiourea and the 5- and 6-membered ring thiocarbamide heterocycles form stable complexes with ions of the first row transition elements and with other ions (Table 4).

Table 4 — Thiocarbamide complexation and salt formation

Compound	Complexin *a* ion
Thioureas	76 ions cited (Reid, 1963a (review)); Cu(II), Mn(II), Cr(III) (Askalini and Bailey, 1969); Cu(II) (Eaton and Zaw, 1971); Zn(II) (Gattegno *et al.*, 1973); Cu(II) (Griffith *et al.* 1978)
2-Thiohydantoins	Heavy metal ion complexes (Reid, 1963b (review))
2-Thioimidazoles	Cu(II), Al(III), Fe(III) (Foye and Lo, 1972); Cu(II) (Hanlon and Shuman, 1975); Re(V) (Kotegov *et al.*, 1975); Cu(II), Zn(II), Pd(II), Ag(I), Cd(II) (Lenarcik and Wiśniewski, 1977); Hg(II) (Buncel *et al.*, 1982; Norris *et al.*, 1983); Pd(II), Pt(II) (Dehand and Jordanov, 1976); Ga, Re, Mo (Cooper *et al.*, 1986); Fe(I) (Basosi *et al.*, 1978).
2-Thiouracils	Heavy metal ion complexes (Reid, 1965 (review covers 2-thiopyrimidine complexes)); Cu(II), Pb(II), Cd(II), Ni(II), Zn(II) (Garrett and Weber, 1970, 1971, 1972); Cu(II), Al(III), Fe(III) (Foye and Lo, 1972); Fe(I) (Basosi *et al.*, 1978)
2-Thiobarbiturates	Heavy metal ion complexes (Reid, 1963c)

Complex formation has been known for some time and the nature of some of the complexes has been thoroughly studied. There has, however, been a renewed interest in thiocarbamide complexes, especially those with Cu^{2+} ions, since thiocarbamides inhibit particular copper-containing enzymes, e.g. β-dopamine hydrolase (Von Voigtlander and Moore, 1970; Stolk and Hanlon, 1973) and mushroom tyrosinase (Andrawis and Kahn, 1986).

Thyroid peroxidase, which participates in the biosynthesis of the thyroid hormones, and other haemoprotein peroxidases are iron-containing enzymes which are inhibited by thiocarbamides (section 3.4.1). Complexation has been proposed as a possible mechanism of inhibition (Garrett and Weber, 1970, 1971, 1972).

The nature of the complex, whether it is 1:1 or greater, depends on the reaction conditions, in particular on the type and concentration of thiocarbamide and metal ion and pH of the solution. In general sulphur and nitrogen are the donor atoms. The order of complexing ability of 2-thiouracils with metal ions has been shown to be $Cu^{2+}>Pb^{2+}>Cd^{2+}\gg Ni^{2+}=Zn^{2+}$ (Garrett and Weber, 1972). In another study with Cu^{2+}, Al^{3+} and Fe^{3+} (Foye and Lo, 1972) some correlation was observed between the stability of the thiono compound metal complex and antimicrobial activity.

Complexation is the probable mechanism whereby administration of thiocarbamide to animals offers protection against Cd^{2+}-induced teratogenesis (Mayumi et al., 1982) and organo-mercurial poisoning (Szabo et al., 1974).

3.2.3 Oxidation

More detailed reviews of the chemistry of the thiol group (Jocelyn, 1972) and the oxidation of cyclic thiones (Loosmore and McKinnon, 1976) are available. Although thiocarbamides are oxidized by a variety of reagents, this review will focus on oxidation reactions of biological relevance. The nature of the product of thiocarbamide oxidation depends on whether the sulphur atom is constrained in the thione form or is able to exist as the thioenol. It is important to note that thiocarbamides are not always as readily oxidized as simple thiols.

3.2.3.1 Disulphide formation

5-Methylthiobarbituric acid (Nishikawa, 1940), ergothioneine (2) (Heath and Toennies, 1957) and thiourea (1) (Griffith et al., 1978) are oxidized to their corresponding disulphides in aqueous solutions containing Cu^{2+}:

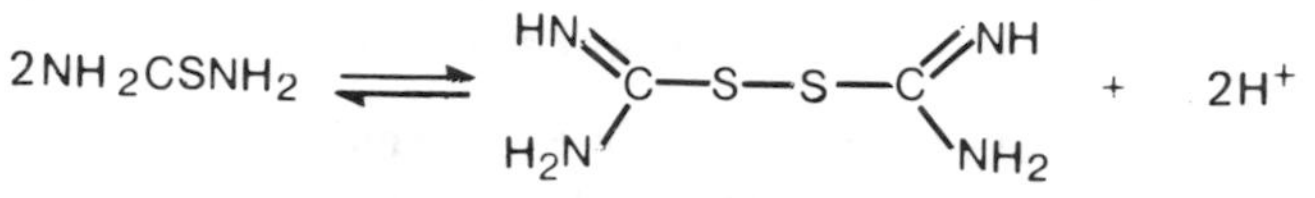

$$2NH_2CSNH_2 \rightleftharpoons \underset{H_2N}{\overset{HN}{>}}C-S-S-C\underset{NH_2}{\overset{NH}{<}} + 2H^+ \qquad (3)$$

$$\textbf{(1)} \qquad\qquad\qquad \textbf{(22)}$$

These reactions are pH sensitive and probably involve the initial formation of the thiocarbamide–metal ion complex prior to oxidation. Aqueous solutions containing metal ions such as Cu^{2+} and Fe^{2+} set up redox systems. 2-Thiohistidine and its betaine, ergothioneine (2) are oxidatively desulphurated to histidine and herzynine (23) respectively (histidine betaine) by Fe^{3+} (Hofmann, 1953b).

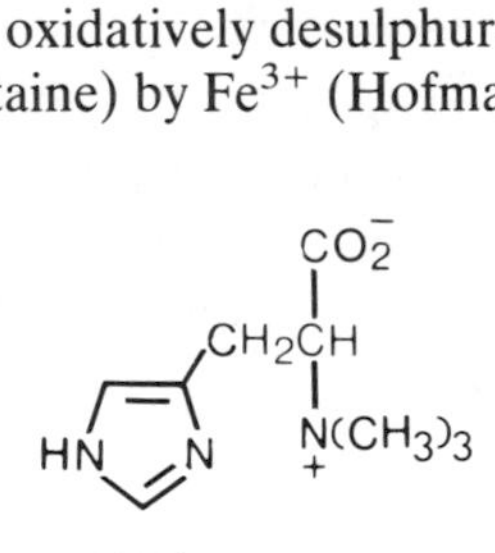

$$\underset{HN\quad N}{\boxed{}}\overset{\overset{\displaystyle CO_2^-}{|}}{\underset{\underset{\displaystyle N(CH_3)_3}{|}}{CH_2CH}}$$

$$\textbf{(23)}$$

Garrett and Weber (1971), however, concluded from their studies that any 2-thiouracil disulphide formed in alkaline solution containing Cu^{2+} and 2-thiouracil was due to auto-oxidation without complexation.

Thiocarbamide oxidation to a disulphide in the presence of metal ions could involve a free radical mechanism, as has been proposed for the oxidation of thiols (Jocelyn, 1972). The fact that bis[1-methylimidazolyl-(2)]disulphide (24) is formed following the reaction of 1-methyl-2-thioimidazole (8) and diphenylpicrylhydrazyl, a stable free radical, is consistent with the production of a thiyl free radical (Berg,

1971b). These observations have formed the basis of spectrophotometric assays for the thiocarbamide antithyroid drugs methimazole (8) and propylthiouracil (12) (Berg, 1971a, c).

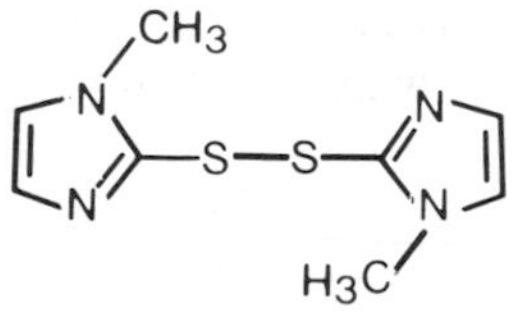

(24)

A free radical mechanism with methimazole (8) acting as a scavenger has been suggested to explain the formation of the bis[methylimidazolyl-(2)-]disulphide (24) in reaction mixtures containing thyroxine, H_2O_2, horseradish peroxidase and methimazole (Björkstén, 1966). Recently Taylor *et al.* (1984) observed that methimazole (8) interacts directly with hydroxyl (OH·) and iodine (I_2^-) radicals and proposed that free radical scavenging may in part explain the inhibitory activity of this compound on the thyroid peroxidase system.

Thiocarbamide disulphides have been regarded as the initial products in the oxidation sequence which may ultimately result in desulphuration of the thiocarbamide. Apart from oxidation to the disulphide in the presence of metal ions, solutions of either iodine or dilute hydrogen peroxide also readily oxidize the thiocarbamides to disulphides (Table 5). The nature of the reaction and the species generated prior to disulphide formation depend primarily on the oxidizing agent.

Table 5 — Some typical thiocarbamide disulphides

Compound	Oxidizing agent	Reference
Thiourea (1)	H_2O_2	Barnet (1910)
	H_2O_2–H^+	Toennies (1937)
	H_2O_2–H^+	Preisler and Berger (1947)
Ethylenethiourea (4)	HOCl–NaOH	Marshall and Singh (1977)
	H_2O_2–CH_3OH	Freedman and Corwin (1949)
4(5)-Carbethoxy-4(5)methyl-2-thioimidazole	Iodine[a]	Balaban and King (1927)
Ergothioneine (2)	H_2O_2–H^+	Heath and Toennies (1957)
1-Methyl-2-thioimidazole (8)	Iodine[a]	Berg (1971b)
2-Thiouracil	Iodine–NaOH	Miller *et al.* (1945)
6-*n*-Propyl-2-thiouracil (12)	Iodine–NaOH	Desbarats-Schönbaum *et al.* (1972)
	Iodine[a]	Lindsay *et al.* (1974a)

[a] I_2–KI solution.

Oxidation of thiols by iodine has formed the basis of assays for the analysis of thiols. Sulphenyl iodide (eqn. (4)) formation is the likely initial step (Danehy *et al.*, 1971) in the oxidation of the thiol. Thereafter, depending on the structure of thiol and the concentration of reactants the sulphenyl iodide either oxidizes another thiol molecule (eqn. (5)) or reacts with water to form the sulphenic acid (eqn. (6)):

$$RSH \ + \ I_2 \ \longrightarrow \ RSI \ + \ HI \tag{4}$$

$$RSI \nearrow^{+RSH} \quad RSSR \ + \ HI \tag{5}$$

$$\searrow^{+H_2O} \quad RSOH \ + \ HI \tag{6}$$

It would be reasonable to assume that thiocarbamides are oxidized by iodine in a similar manner.

Iodide cleavage of thyroid protein disulphide bonds to produce a protein sulphenyl iodide species which iodinates tri-iodothyronine and thyroxine precursors had been postulated (Maloof and Soodak, 1963; Jirousek and Cunningham, 1968). Thus it was suggested that the thiocarbamides which inhibit thyroid hormone biosynthesis compete with iodide for disulphide bonds in thyroid peroxidase. Later studies, however, indicated that the sulphenyl iodide group was not the 'active iodinating species' (Jirousek and Pritchard, 1971). Various mechanisms have been proposed whereby the thiocarbamides inhibit thyroid hormone biosynthesis. These are described in section 3.4.1. Desulphuration of thiocarbamide in the presence of halide and a peroxidase system may proceed via a thiocarbamide sulphenyl iodide intermediate (Morris *et al.*, 1962).

Since hydrogen peroxide is also a substrate for thyroid peroxidase, an appreciation of the oxidation of thiocarbamides by hydrogen peroxide should provide insight into their peroxidase inhibitory activity. With hydrogen peroxide as oxidizing agent it is proposed that the thiocarbamide moiety in thiourea (Hoffmann and Edwards, 1977) and 2-thioimidazoles (Suszka, 1980) acts as a nucleophile towards the electrophilic oxidizing agent, and that initially a sulphenic acid is produced (eqn. (7)) which reacts with another thiocarbamide molecule to produce the disulphide (eqn. (8)). Thus the formation of an unstable sulphenic acid is postulated for both the iodine and peroxide oxidation of thiocarbamides. Its subsequent oxidation will be discussed later.

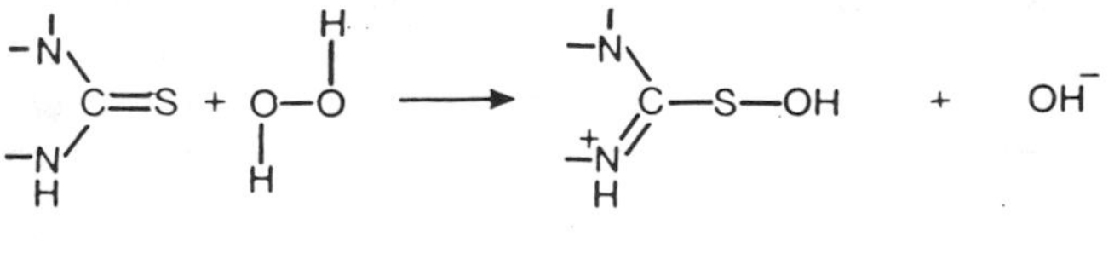

$$\begin{array}{c} -\overset{|}{N} \\ \diagdown \\ \diagup \ C{=}S \ + \ \overset{H}{\underset{H}{\overset{|}{O{-}O}}} \ \longrightarrow \ \overset{-\overset{|}{N}}{\underset{-\overset{+}{\underset{H}{N}}}{\diagdown}} C{-}S{-}OH \ + \ OH^- \end{array} \tag{7}$$

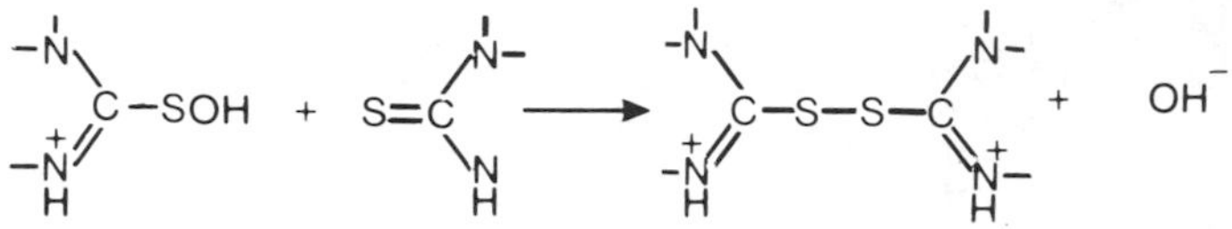

$$\text{(structure)} \quad C\text{-SOH} \;+\; S{=}C \quad \longrightarrow \quad C\text{-S-S-}C \;+\; OH^- \qquad (8)$$

Disulphides are relatively stable in the solid state but vary in stability in aqueous solution. Although relatively stable in acid solution, dithioformamidine (**22**) is unstable in alkaline solution and may either be hydrolysed to the free thiol (Heath and Toennies, 1957) or decompose with loss of sulphur forming cyanamide (Toennies, 1937):

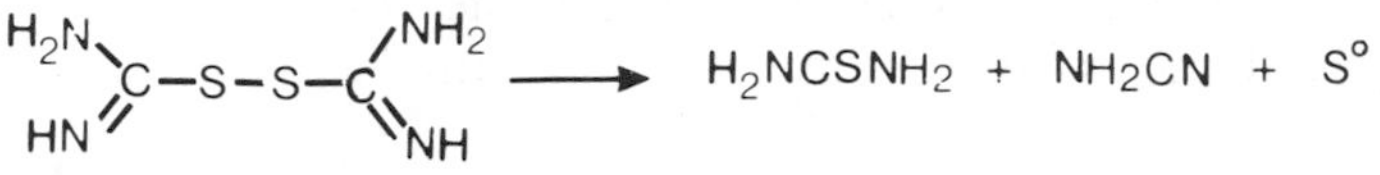

$$\begin{array}{c} H_2N \\ \quad\;\; C\text{-S-S-}C \\ HN \end{array} \begin{array}{c} NH_2 \\ \\ NH \end{array} \quad \longrightarrow \quad H_2NCSNH_2 \;+\; NH_2CN \;+\; S^\circ \qquad (9)$$

(**22**)   (**1**)

Dithioformamidine (**22**) and cystine (**25**), in aqueous solution, undergo an exchange reaction yielding a mixed disulphide, S-guanylthiocysteine (**26**) (Toennies, 1937):

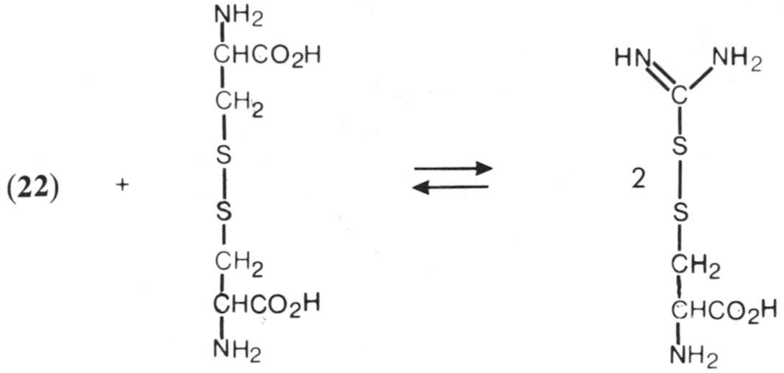

$$\text{(22)} \;+\; \begin{array}{c} NH_2 \\ CHCO_2H \\ CH_2 \\ S \\ S \\ CH_2 \\ CHCO_2H \\ NH_2 \end{array} \quad \rightleftharpoons \quad 2 \begin{array}{c} HN \quad NH_2 \\ C \\ S \\ S \\ CH_2 \\ CHCO_2H \\ NH_2 \end{array} \qquad (10)$$

(**25**   (**26**)

Similarly, ergothioneine (**2**) and its disulphide also form mixed disulphides with cysteine and glutathione (Heath and Toennies, 1957). Although isolable, S-guanyl-thiocysteine (**26**), ergothioneine–cysteine and ergothioneine–glutathione are more unstable in aqueous solution than the parent thiocarbamide disulphides.

Ergothioneine (**2**), 2-thioimidazole (**16**) 1-methyl-2-thioimidazole (**8**) and thiourea (**1**) also undergo thiol–disulphide exchange reactions with the 2,2′- and 4,4′-dipyridyl disulphides, producing mixed disulphides (Carlsson *et al.*, 1974).

3.2.3.2 S-oxygenated thiocarbamides
Sulphenic acid formation has been implicated in the oxidation of thiocarbamides to disulphides in both hydrogen peroxide and iodine (section 3.2.3.1). These S-monoxides Scheme 1, which are reactive, are readily oxidized to their correspond-

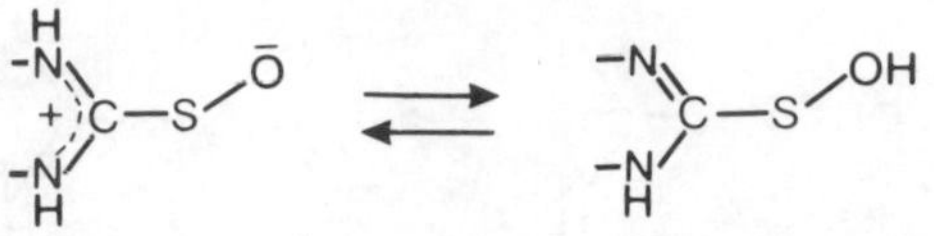

Scheme 1

ing sulphinic (S-dioxide) and sulphonic (S-trioxide) acids (Fig. 2). Sulphenic acids of thiocarbamides are extremely unstable. There is evidence, however, for the formation of the sulphenic acids of substituted thioureas (Walter and Randau, 1969a; Poulson *et al.*, 1979), including ethylenethiourea (**4**), under controlled experimental conditions with hydrogen peroxide in neutral solution. Although relatively stable for 24 h at 0–2°C in neutral solution, they rapidly decompose in acid or alkaline media. Savolainen and Pyysalo (1979), however, report the isolation of the sulphenic acid (**27**) of ethylenethiourea (**4**), after its irradiation with ^{60}Co γ-rays.

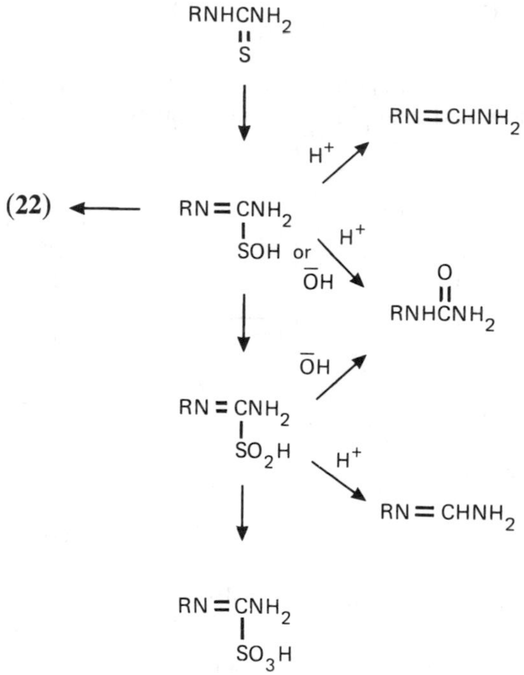

Fig. 2 — Sequential oxidation of thioureas (data from Maryanoff *et al.*, 1986).

Dioxygenated thiocarbamides (sulphinic acids), which vary in stability, are isolable in particular instances (Table 6). Since they are intermediates in the sequential S-oxidation of thiocarbamides to their corresponding sulphonic acids, they are readily oxidized and thus are moderately strong reducing agents (Stirling, 1971). In basic conditions thiourea sulphinic acids decompose to urea, while in acid solution they eliminate sulphur dioxide, yielding a formamidine (Walter and Randau, 1969b). Thiourea sulphinic acid in alkaline media reduces Cd^{2+} whilst being converted to urea (McGill and Lindström, 1977).

Table 6 — Thiocarbamide-S-dioxide (sulphinic acid)

Compound	Oxidizing agent	Reference
Thiourea (**1**)[a]	H_2O_2	Barnet (1910)
	H_2O_2	Walter and Randau (1969b)
1-Methyl- and 1-phenyl-thiourea (**33**)	H_2O_2	Poulson et al. (1979)
Ethylenethiourea (**4**)	H_2O_2 (neutral)	Poulson et al. (1979) Marshall and Singh (1977)
4(5)-Methyl-2-thioimidazole	H_2O_2 (neutral)	Balaban and King (1927)
6-n-Propyl-2-thiouracil (**12**)	H_2O_2 (1% NH_4OH)	Lindsay et al. (1978)
	H_2O_2 (1 M KOH)	Kariya et al. (1986)

[a]See also Stirling (1971) for review.

Other studies (Marshall and Singh, 1977; Marshall, 1979) with ethylenethiourea (**4**) have demonstrated that 2- imidazolinylsulphinate (**28**) is a key intermediate in the oxidation of ethylenethiourea (**4**) and that the reaction conditions dictate the relative amounts of 2-imidazoline hydrosulphate (**29**) and ethyleneurea (**30**) formed, since the pathway of 2-imidazolinylsulphinate (**28**) decomposition depends on the concentration of base (Fig. 3). It is implied that 2-imidazolinylsulphonate (**31**), which is formed in small amounts, is not a key intermediate in the production of ethyleneurea (**30**).

The sulphinic acid of propylthiouracil is also oxidized further to its corresponding sulphonic acid, sulphate, and propyluracil, the oxygen analogue of propythiouracil (Lindsay et al., 1978).

Although 4(5)-methyl-2-thioimidazole is also oxidized to an unstable isolable sulphinic acid, it preferentially decomposes to 4(5)-methylimidazole plus sulphite (Balaban and King, 1927). Similarly, substituted 2-thiopyrimidines are oxidized via sulphinates (Evans et al., 1956) to the corresponding substituted pyrimidines (Hunt et al., 1959). Presumably these reactions are favoured because the aromaticity of the

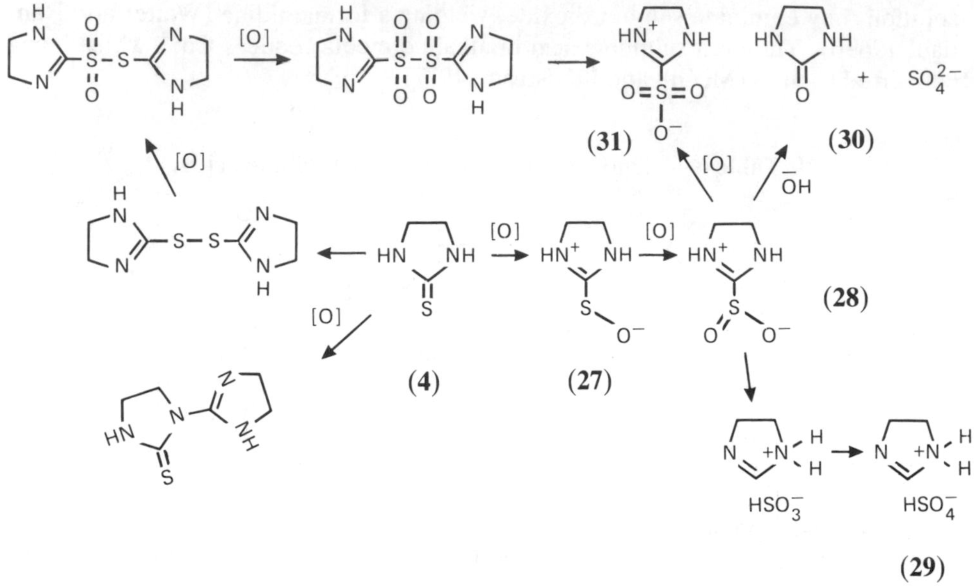

Fig. 3 — Oxidation of ethylenethiourea (**4**) by hydrogen peroxide and hypochlorite (data from Marshall and Singh, 1977; Marshall, 1979).

imidazole and pyrimidine rings is restored. 2-Thiouridines are oxidatively desulphurated to their corresponding pyrimidinones (Ogihara and Mitsunobu, 1982).

Facile formation of mono-, di- and tri-oxide intermediates is postulated for the S-oxidation of 1,3-dimethyl-2-thioimidazole by hydrogen peroxide, where the sulphur atom is constrained in the thione form (Karkhanis and Field, 1985). 1,3-Dimethyl-2-imidazoline hydrosulphite (**32**) is the isolable product of this reaction.

(**32**)

The sulphonic acid derivatives of the thiocarbamides are isolable and relatively stable (Table 7). It would thus appear that the reactive sulphinic acids are converted more readily into desulphurated products than are their corresponding sulphonic acids.

Superoxide anion ($O_2^{\cdot-}$) is a biological oxidizing agent which has been implicated in a variety of deleterious biochemical transformations. To date there are only a few reports on the superoxide oxidation of thiocarbamides. Interestingly, in a study of thiourea and mono-substituted thioureas, Crank and Makin (1984) showed that these compounds were desulphurated to their corresponding cyanamides, while in another study with $O_2^{\cdot-}$, thioureas, 2-thiouracils and 2-thiobarbital (Katori et al., 1981) were converted into their oxygen analogues.

Table 7 — Thiocarbamide-*S*-trioxides

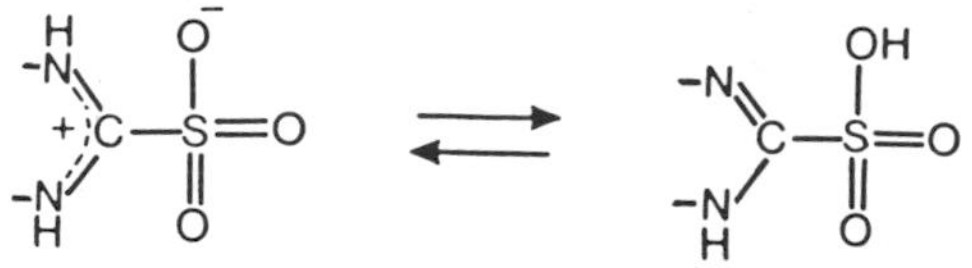

Compound	Oxidizing conditions	Reference
Thioureas	H_2O_2–CH_3COOH	Walter and Randau (1969c)
1-Methyl- and 1-phenylthiourea (**33**)	H_2O_2–CH_3COOH	Poulson *et al.* (1979)
Ethylenethiourea (**4**)	H_2O_2 in CCl_4 at 0°C	Marshall and Singh (1977)
	H_2O_2–CH_3COOH	Poulson *et al.* (1979)
4(5)-Methyl-2-thioimidazole	H_2O_2	Balaban and King (1927)
1-Methyl-2-thioimidazole (**8**)	H_2O_2	Comrie and Skellern (unpublished results)
6-*n*-Propyl-2-thiouracil (**13**)	H_2O_2 (1% NH_4OH) $KMnO_4$ (50 mM)	Lindsay *et al.* (1978) Kariya *et al.* (1986)

3.2.4   Reduction

Thiocarbamides can also be desulphurated by reducing agents. Reduction of substituted thioureas yields formamidines (Kashima *et al.*, 1986). Similarly, 2-thiopyrimidines (Brown, 1962; Reid, 1965) and 2-thioimidazoles (Hofmann, 1953a) are reduced to their corresponding pyrimidines and imidazoles. 2-Thiohydantoins are also reductively desulphurated (Reid, 1963a).

3.2.5   Alkylation and acylation

Thioureas (Reid, 1963a), thiohydantoins (Reid, 1963b), thioimidazoles (Hofmann, 1953a), thiopyrimidines and derivatives (Reid, 1965; Brown, 1962, 1970) all readily form thioether derivatives in the presence of base and alkyl halide. Whether substitution is on the sulphur atom or a nitrogen atom depends on the reaction conditions (Neelakantan and Thyagarajan, 1969).

Similarly, acylation of thiocarbamides can result in the production of either the *S*- or the *N*-acyl derivative (Lawson and Morley, 1956; Baker, 1958). For example, in the presence of base, 1-methyl-2-thioimidazole (**8**) is *N*-acylated by ethylchloroformate (eqn. (11)) to yield carbimazole (**9**) while with non-basic conditions the *S*-acyl derivative is produced (eqn. (12)):

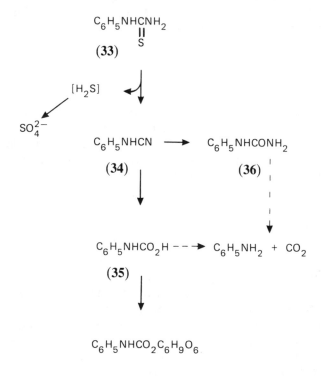

3.3 THIOCARBAMIDE METABOLISM

3.3.1 Oxidation reactions

Early studies with phenylthiourea (**33**) (Scheline *et al.*, 1961) and other *N*-substituted thioureas (Williams, 1961; Smith and Williams, 1959, 1961a, b, c) in rat and rabbit indicated that there was a correlation between the extent of desulphuration and toxicity. Although after ^{35}S-phenylthiourea administration $^{35}SO_4^{2-}$ was the

Fig. 4 — Biotransformation of phenylthiourea (**33**) (data from Scheline *et al.*, 1961).

major sulphur-containing metabolite, it was postulated that, because phenylcar-
bamic acid (**35**) and phenylurea (**36**) were produced, desulphuration was reductive
and that hydrogen sulphide is the toxic metabolite (Fig. 4).

Subsequent studies (Poulson *et al.*, 1979), however, implicated *S*-oxides as
putative reactive metabolites, since labile sulphenic and sulphinic acids were characterized as products of the oxidation of phenylthiourea (**33**) and ethylenethiourea (**4**)
catalysed by hepatic flavin-containing monooxygenase. In another study 2-imidazoli-
nyl sulphenate (**27**) was isolated and characterized as a major urinary metabolite in
mice which were administered ethylenethiourea (**4**) (Savolainen and Pyysalo, 1979).

Sulphenic acids of thiocarbamides, some of which are reported to have been
synthesized and isolated, are susceptible either to further oxidation to the more
stable sulphinic acid which can be hydrolysed to the corresponding urea or to
conversion into a formamidine or urea, depending on the reaction conditions
(section 3.2.3.2) Interestingly, however, phenylcyanamide (**34**) was the product of
the superoxide desulphuration of phenylthiourea (**33**) (Crank and Makin, 1984);
S-dioxide formation is proposed. Phenylcyanamide (**34**) was reported to be an *in vivo*
metabolite of phenylthiourea (**33**) (Scheline *et al.*, 1961). Thiourea disulphides have
also been implicated as intermediates in cyanamide formation (Davidson *et al.*,
1979).

The two thioimidazoles ergothioneine (**2**) and methimazole (**8**) are desulphurated to herzynine (**23**), the betaine of histidine (Wolf *et al.*, 1961), and to
1-methylimidazole (**37**) (Poulson *et al.*, 1974; Lee and Neal, 1978; Skellern and
Steer, 1981).

Desulphuration of methimazole (Poulson *et al.*, 1974) is *via* putative sulphenic
and sulphinic acid intermediates which are labile and non-isolable, the latter being
subsequently hydrolysed to the imidazole. Neal (1980) proposed that the *S*-monox-
ide of methimazole decomposes to the imidazole carbene plus sulphur monoxide,
and that disproportionation of sulphur monoxide yields sulphur dioxide and reactive
atomic sulphur. If the thiocarbamide *S*-dioxides are the reactive metabolites then
their relative stabilities may offer an explanation for the differing toxicities of these
compounds (Poulson *et al.*, 1979).

Not only are the thiocarbamides accumulated in the thyroid gland (Marchant
et al., 1972) where they inhibit thyroid peroxidase and consequently the biosynthesis
of the hormones thyroxine (T_4) and tri-iodothyronine (T_3), but also they are
desulphurated in the gland to a lesser or greater extent since sulphate is an
intrathyroidal metabolite (Schulman, 1950; Maloof and Soodak, 1957; Marchant *et
al.*, 1972). The sulphinic acid and sulphonic acid metabolites of propylthiouracil are
the only intrathyroidal intermediate *S*-oxidation products of thiocarbamide metabolism to be reported to date (Lindsay *et al.*, 1979), suggesting that the *S*-dioxide of
propylthiouracil is more stable than the *S*-dioxides of the 2-thioimidazoles.

Thiocarbamide desulphuration to yield the oxygen analogue may be *via* nucleophilic attack of the *S*-dioxides, as the sulphur dioxide group is a good leaving group
(section 3.2.3.2). Alternatively, it has been postulated for the rodenticide
1-naphthylthiourea (**3**) that, following *S*-monoxide formation, nucleophilic attack by
the oxygen atom occurs at the thiocarbonyl carbon atom resulting in oxathiiran (**38**)
formation which rearranges to produce 1-naphthylurea and atomic sulphur (Neal,
1980).

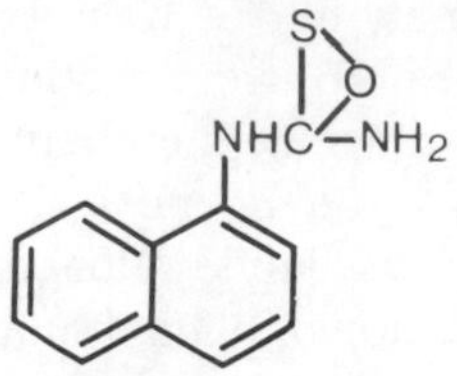

(38)

It is noteworthy that in general the thioureas and 2-thiopyrimidine derivatives are capable of being desulphurated to their corresponding oxygen analogues and to formamidine derivatives (Table 8), while 2-thioimidazoles are apparently only desulphurated to their corresponding imidazole. To date no oxygen analogues of 2-thioimidazoles have been reported as metabolites. These metabolic products are what would be predicted by chemical studies (section 3.2.3.2).

Table 8 — *In vitro* and *in vivo* thiocarbamide desulphuration

Thiocarbamide	Desulphurated products
Thioureas	
Thiourea (**1**)	Urea (Maloof and Spector, 1959; Maloof and Soodak, 1961; Hollinger *et al.*, 1976)
1-Phenylthiourea (**33**)	1-phenylurea (Scheline *et al.*, 1961)
1-Naphthylthiourea (**3**)	1-naphthylurea (Neal, 1980; Lee *et al.*, 1980)
Ethylenethiourea (**4**)	Ethyleneurea (Lyman, 1971; Kato *et al.*, 1977, Jordan and Neal, 1979; Savolainen and Pyysalo, 1979); ethyleneurea, imidazoline, imidazolone (Ruddick *et al.*, 1977; Iverson *et al.*, 1980)
2-Thioimidazoles	
Ergothioneine (**2**)	Herzynine (**23**), betaine of histidine (Wolf *et al.*, 1961).
1-Methyl-2-thioimidazole (**8**, methimazole)	1-methylimidazole (**37**) (Poulson *et al.*, 1974; Lee and Neal, 1978; Skellern *et al.*, 1977; Paterson *et al.*, 1983; Conn *et al.*, 1987); 1-methylimidazole (**37**), methylhydantoin (**40**) (Skellern and Steer, 1981)
2-Thiobenzimidazole (**10**)	Benzimidazole (El Dareer *et al.*, 1984)
2-Thiopyrimidine derivatives	
2-Thiouracil	Uracil (Spector and Shideman, 1959; Morris *et al.*, 1962).
Methyl-2-thiouracil	Methyluracil, 4-oxo-6-methylpyrimidine (Kano *et al.*, 1974)
Propyl-2-thiouracil (**12**)	Propyluracil (Lindsay *et al.*, 1974b, 1979)
5-Ethyl-5-(1-methylbutyl)-2-Thiobarbituric acid (**14**, thiopentone)	Pentobarbitone (Spector and Shideman, 1959; Sharma and Stowe, 1970)

Although 2-thioimidazoles are not metabolized directly to oxygen analogues, a methylhydantoin (**40**) is an *in vivo* metabolite of methimazole (**8**) (Fig. 5). Either 3-methyl-2-thiohydantoin (**39**), a known metabolite of methimazole, or 1-methylimidazole (**37**) (Skellern and Steer, 1981) could be the proximate metabolite since imidazole is known to be metabolized to hydantoin (Mariaggi *et al.*, 1973).

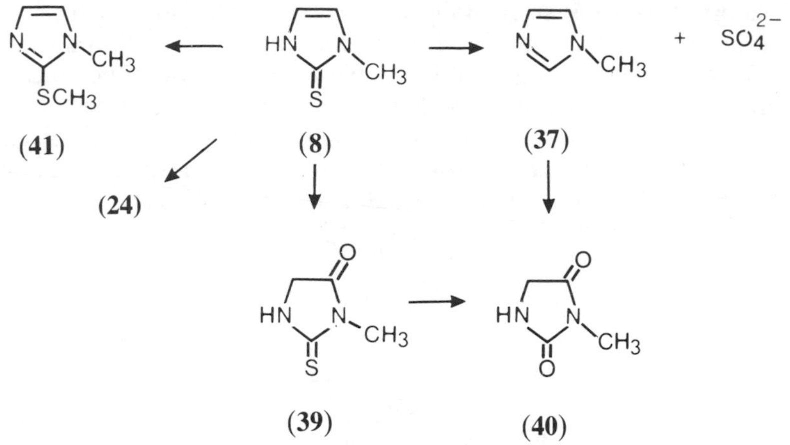

Fig. 5 — Metabolic fate of the thiocarbamide moiety in the 2-thioimidazole, methimazole (**8**) (data from Sitar and Thornhill, 1973; Poulson *et al.*, 1974; Skellern and Steer, 1981; Paterson *et al.*, 1983).

3.3.2   Conjugation reactions

3.3.2.1   *Glucuronidation and ribosylation*

Formation of glucuronides of endogenous and exogenous compounds involves the nucleophilic attack of oxygen-, nitrogen- or sulphur-containing functional groups at the C-1 position of the glucuronic acid moiety of uridine diphosphate glucuronic acid (UDPGA). This reaction is catalysed by membrane-dependent UDP-glucuronyl transferases. Glucuronidation is reputed to be an important metabolic pathway for the 2-thiouracils in man (McGinty *et al.*, 1948; Kampmann *et al.*, 1974; Marchant *et al.*, 1978) and rat (Sitar and Thornhill, 1972, Papapetrou *et al.*, 1972; Lindsay *et al.*, 1974b; Marchant *et al.*,1978). Methimazole (**8**) and thiourea (**1**) are not glucuronidated by a guinea pig hepatic UDP-glucuronyl transferase *in vitro* (Lindsay *et al.*, 1977a) whereas the 2-thiopyrimidine derivatives 2-thiouracil and propyl-thiouracil (**12**) are. This dissimilarity in reactivity is possibly due to differences in ionizability of the thiocarbamides (Table 2), as the 2-thiouracils are more acidic than thiourea (**1**) and methimazole (**8**). Sitar and Thornhill (1973) have reported that a glucuronide of methimazole was excreted by rat after administration of methimazole.

   Although it has not been fully ascertained whether the 2-thiouracils are converted into *O*-, *N*- or *S*-glucuronides, there is strong circumstantial evidence based on chemical and pharmacological observations (Lindsay *et al.*, 1977b) that PTU glucuronide (**42**) is an *S*-glucuronide.

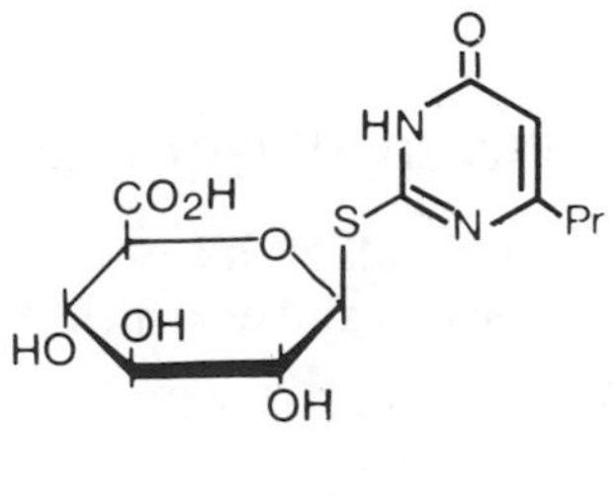

2-Thiouracil, which is structurally similar to the naturally occurring pyrimidine uracil, is incorporated into bacterial, viral (Lindsay *et al.*, 1972) and rat (Yu *et al.*, 1973a) RNA. Initially 2-thiouracil is *N*-ribosylated to form either the nucleotide 2-thiouridine (Strominger and Friedkin, 1954) or the nucleotide 2-thio-5′-UMP (Lindsay *et al.*, 1972; Yu *et al.*, 1973a). Methylthiouracil, propylthiouracil (**12**) and methimazole (**8**) were not substrates for UMP-pyrophosphorylase, which can mediate the formation of 2-thiouridine (**43**) from 2-thiouracil (Lindsay *et al.*, 1972).

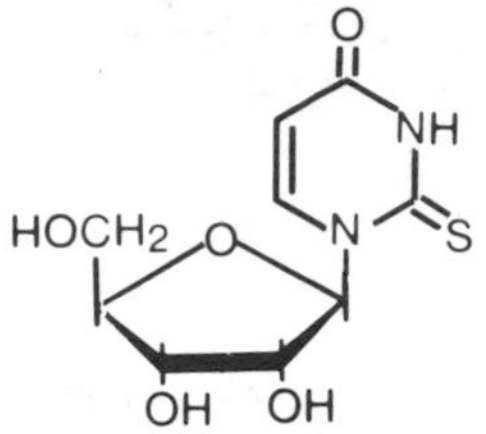

(**43**)

Neither 2-thiouridine (**43**) nor 2-thio-5′-UMP is as effective an antithyroid agent in rat as the parent compound 2-thiouracil (Yu *et al.*, 1973b).

3.3.2.2   *Methylation*
The enzymology and mechanism of *S*-methylation are discussed in detail in Volume 2 of this series. Metabolic methylation of the thiocarbamide moiety is at the electron-rich sulphur atom. 2-thiouracil (Sarcione and Sokal, 1958), 6-methyl-2-thiouracil (Kano *et al.*, 1974) and propylthiouracil (**12**) (Lindsay *et al.*, 1974b) are all metabolized in the rat to their corresponding *S*-methyl metabolites. Although the rat does not *S*-methylate ethylenethiourea (**4**) *in vivo*, *S*-methyl ethylenethiourea is the major urinary metabolite excreted by the cat (Iverson *et al.*, 1980). While 2-thiouracil, propylthiouracil (**12**) and 6-methyl-2-thiouracil were *S*-methylated by a mouse kidney *S*-methyltransferase preparation (Lindsay *et al.*, 1975), the 2-thio-imidazole methimazole (**8**) was not. The 2-thioimidazoles methimazole (**8**) and 2-thiobenzimidazole (**10**) and propylthiouracil (**12**) are, however, substrates for a rat liver *S*-methyltransferase (Weisiger and Jakoby, 1979) Moreover *S*-methylmethimazole (**41**) (Fig. 5) is a minor rat urinary metabolite of methimazole (Skellern and Steer, 1981).

3.4   ENZYME SYSTEMS INHIBITED BY THIOCARBAMIDES

Thiocarbamides are substrates and inhibitors of a variety of enzyme systems including peroxidases, monooxygenases and enzymes involved in conjugation reactions (Table 9).

Table 9 — Thiocarbamides as enzyme substrates and inhibitors

Enzyme	Thiocarbamide
Peroxidases	
(a) Thyroid peroxidase (TPO)	Thiourea (Maloof and Spector, 1959); thiourea, 2-thiouracil, methimazole (Coval and Taurog, 1967); methimazole, propylthiouracil (Nagasaka and Hidaka, 1976); methimazole, propylthiouracil (Taurog, 1976); methimazole, propylthiouracil. thiouracil (Davidson *et al.*, 1978a); methimazole (Ohtaki *et al.*, 1982); methimazole, propylthiouracil (Engler *et al.*, 1983); methimazole (Magnusson *et al.*, 1984a)
(b) Lactoperoxidase (LPO)	Methyl-2-thiouracil, methimazole (Michot *et al.*, 1979); methimazole, thiobarbituric acid, 2-thiopyrimidine, 2-thiohydantoin, thiourea (Edelhoch *et al.*, 1979); methimazole (Ohtaki *et al.*, 1982); methimazole (Doerge, 1986)
(c) Myeloperoxidase (MPO)	Thiourea, 2-thiouracil (Sörbo and Ljunggren, 1958); ergothioneine, methimazole, 2-thiouracil (Yip and Klebanoff, 1963)
(d) Chloroperoxidase (CPO)	2-thiouracil (Morris *et al.*, 1962); methimazole, 2-thiouracil, thiourea (Taurog and Howells, 1966); thiourea, 2-thiouracil (Morris and Hager, 1966)
(e) Salivary gland peroxidase	Thiourea, 2-thiouracil (Mahajani *et al.*, 1973)
(f) Prostaglandin synthetase	Methimazole, propylthiouracil (Zenser *et al.*, 1983); propylthiouracil, methimazole (Zelman *et al.*, 1884); methimazole (Conn *et al.*, 1988)
Monooxygenases (hepatic)	
(a) Cytochrome P-450	Thiourea, ethylenethiourea, methimazole, 2-thiouracil, 6-methyl-and 6-propyl-thiouracil (Hunter and Neal, 1975); methimazole (Lee and Neal, 1978); methimazole (Kedderis and Rickert, 1985)
(b) Flavin-containing monooxygenase	Methimazole, 1-naphthyl-thiourea, phenylthiourea (Poulsen *et al.*, 1974); methimazole, phenylthiourea, ethylenethiourea (Poulsen *et al.*, 1979); methimazole (Poulsen and Ziegler, 1979); phenylthiourea (Beaty and Ballou, 1980); methimazole phenylthiourea (Hanzlik and Cashman, 1983); thiourea, phenylthiourea, methimazole, 2-thiobenzimidazole, 2-methylthiobenzimidazole (Sabourin and Hodgson, 1984); methimazole, phenylthiourea, 2-thiobenzimidazole, thiourea, 1-naphthylthiourea (Tynes and Hodgson, 1985a); 1-naphthylthiourea, thiourea, ethylenethiourea (Tynes and Hodgson, 1985b)
Copper-containing enzymes	
(a) Dopamine β-hydrolase	Thioureas (Von Voigtlander and Moore, 1970); methimazole (Stolk and Hanlon, 1973)
(b) Tyrosinase	Methimazole (Hanlon and Shuman, 1975); methimazole (Andrawis and Kahn, 1986)
(c) Ceruloplasmin oxidase	Methimazole (Hanlon and Shuman, 1975)
Iodothyronine-5'-deiodinase	2-thiouracils, 2-thiomidazoles (Visser *et al.*, 1979); propylthiouracil (references in Cavalieri and Pitt-Rivers, 1981 (review)); propylthiouracil (Goswami and Rosenberg, 1986); propylthiouracil analogues (Nogimori *et al.*, 1986)
UDP-glucuronyltransferase	2-thiouracil, propylthiouracil (Lindsay *et al.*, 1977a, b)
S-Methyltransferases	2-thiouracil, propylthiouracil, 6-methyl-thiouracil (Lindsay *et al.*, 1975); methimazole, 2-thiobenzimidazole, propylthiouracil (Weisiger and Jakoby, 1979)
Glutathione S-transferases	Propylthiouracil (Yamada and Kaplowitz, 1980; Habig *et al.*, 1984); propylthiouracil, sulphinic and sulphonic acids of propylthiouracil, 2-thiouracil (Kariya *et al.*, 1986)

3.4.1 Peroxidases

Since nearly all the thiocarbamides are goitrogenic (MacKenzie and MacKenzie, 1943a; Astwood, 1943; Astwood *et al.*, 1945; Searle *et al.*, 1951) and capable of inhibiting thyroid hormone biosynthesis to various extents, the interaction of these compounds with thyroid peroxidase and other known peroxidases has evoked a considerable amount of interest. Peroxidases are ubiquitous haemoproteins which, apart from being localized in specialized organs such as the thyroid and salivary glands, are widely distributed in the body in the various blood cell sub-populations (Tenovuo, 1985). The peroxidase systems in leukocytes and salivary glands have an antimicrobial action.

Peroxidases are multi-substrate enzymes, which in the presence of hydrogen peroxide catalyse the oxidation of a variety of compounds that are suitable electron donors. The potential substrates comprise phenols, amines and halide ions. Thiocyanate is included with the halide ions since it acts as a pseudohalide (Thomas, 1985). The peroxidases exhibit different halide ion specificities in that chloroperoxidase and myeloperoxidase catalyse the oxidation of Cl^-, Br^-, SCN^- and I^-, while thyroid peroxidase and lactoperoxidase catalyse the oxidation of Br^-, SCN^- and I^-.

Initially hydrogen peroxide reacts with the native peroxidase to form compound I (oxidized enzyme), and it is compound I which is the oxidizing agent. A sequential transfer of electrons from the donor substrate occurs, with the initial single electron transfer reducing compound I to compound II. Transfer of the second electron to compound II regenerates the native enzyme (Thomas, 1985).

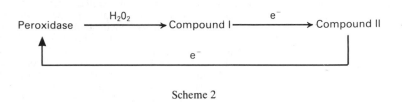

Scheme 2

Based on studies with purified thyroid peroxidase and lactoperoxidase, various mechanisms have been proposed for the thiocarbamide inhibition of these enzymes. Although there are mechanistic differences between these enzyme systems (Ohtaki *et al.*, 1982), they are similar with respect to their inactivation by thiocarbamides.

Firstly, thiocarbamides in the absence of iodide or at low iodide concentrations act as possible suicide substrates of oxidized lactoperoxidase (Doerge, 1986; Edelhoch *et al.*, 1979; Michot *et al.*, 1979) and oxidized thyroid peroxidase (Nagasaka and Hidaka, 1976; Davidson *et al.*, 1978a; Engler *et al.*, 1982) since inactivation is irreversible. Reversible inactivation, however, occurs in the presence of halide (Engler *et al.*, 1983), when the thiocarbamide either competes with halide ion for compound I or reacts with the putative reactive iodinating species (enzyme–I_{ox}). Production of free radicals ($I^{\cdot}$) (Nunez and Pommier, 1982), enzyme-bound iodinium ion (I^+) (Morris and Hager, 1966; Ohtaki *et al.*, 1981) and hypoiodite intermediates [EOI^-] (Magnusson *et al.*, 1984a, b) have all been proposed as the possible reactive iodinating species important to thyroid hormone biosynthesis. The mechanism of iodination and the nature of the iodinating species are dealt with in comprehensive reviews (Taurog, 1976, 1986; DeGroot and Niepomniszcze, 1977; Davidson *et al.*, 1978b; Nunez and Pommier, 1982).

Irrespective of whether the inactivation of the enzyme is reversible or irreversible, the thiocarbamide is oxidized, the extent of oxidation being dependent on the concentration of compound I and halide ion (Fig. 6). Inorganic sulphate is one of the end-products of peroxidase-catalysed oxidation of thiocarbamides (Maloof and Spector, 1959; Taurog, 1976). In only a few studies with peroxidases has an attempt been made to elucidate the nature of the initial product of thiocarbamide oxidation. Moreover, it is not possible to generalize, since the thiocarbamide oxidation products differ in their stability (section 3.2.3).

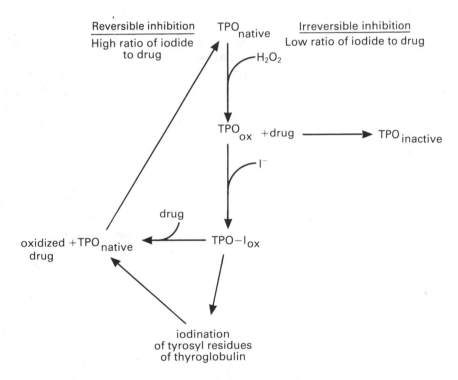

Fig. 6 — Proposed mechanism of thiocarbamide inhibition of thyroid peroxidase: TPO, thyroid peroxidase; TPO$_{ox}$, oxidized thyroid peroxidase; TPO-I$_{ox}$, oxidized iodide enzyme complex (modified from Taurog, 1986).

Formamidine disulphide (**22**) and sulphinic acid are the products of the catalysed oxidation of thiourea by chloroperoxidase (CPO) (Morris and Hager, 1966) and myeloperoxidase (Sörbo and Ljunggren, 1958). Similarly, 2-thiouracil is oxidized by a CPO–H$_2$O$_2$–halide ion system to a disulphide (Morris and Hager, 1966) and propylthiouracil (**12**) by rat thyroid peroxidase to a sulphinic acid (Lindsay *et al.*, 1979). Moreover, oxidative desulphuration yielding the corresponding uracils also occurs. Evidence has been presented which indicates that methimazole (**8**) is oxidized to its disulphide (**24**) by a sheep thyroid peroxidase system (Paterson *et al.*,

1983). Both sulphenyl iodides (Morris and Hager, 1966; Davidson *et al.*, 1978a, 1979; Engler *et al.*, 1983) and sulphenic acids (Engler *et al.*, 1983) have been implicated as the initial thiocarbamide oxidation products of peroxidase-mediated reactions.

3.4.2 Monoxygenases

Microsomal cytochrome P-450 dependent monooxygenases (Lee and Neal, 1978) and flavin-containing monooxygenase (Poulsen *et al.*, 1974, 1979) catalyse the *S*-oxidation of thiocarbamides (Table 9). There is evidence, however, which would suggest that the flavin-containing monooxygenase is mainly responsible for catalysing the *S*-oxidation of sulphur-containing compounds, such as the thiocarbamides (Ziegler, 1984). These enzymes catalyse the initial oxidation of the thiocarbamide sulphur atom to form the putative sulphenic acid (Neal, 1980; Poulson *et al.*, 1979). The labile sulphenic acid of the thiocarbamide can either disproportionate or be catalytically oxidized further to sulphinic and sulphonic acids (Poulson *et al.*, 1974, 1979). The stabilities of the various *S*-oxygenated products depend on the character of the thiocarbamide being studied. For example, the *S*-dioxides of thiourea (**1**) (Walter and Randau, 1969b) and propylthiouracil (**12**) (Lindsay *et al.*, 1978) are relatively stable and can be isolated, while the sulphinic acid of methimazole (**8**) has not been isolated (section 3.2.3.2).

In vivo in rats, pre-treated with either phenobarbitone or 3-methylcholanthrene, particular thiocarbamides deplete hepatic cytochrome P-450-dependent monooxygenases and inhibit benzphetamine metabolism (Hunter and Neal, 1975). Similar effects were observed *in vitro* with hepatic microsomal preparations from pre-treated animals. It is noteworthy that the oxygen analogues of some of the thiocarbamides had no significant effect on the monooxygenase. It was tentatively proposed that these effects were due to initial metabolic activation of the thiocarbamide (see section 3.5).

Since sulphur is susceptible to oxidation, it may be difficult to delineate between chemical- and enzyme-catalysed oxidation *in vivo* and *in vitro* as hydrogen peroxide and other oxidizing agents are generated in cells and subcellular preparations of enzymes.

More detailed reviews on monooxygenase-catalysed *S*-oxidation of thiocarbamides are available (Neal *et al.*, 1977; Neal, 1980; Ziegler, 1980, 1984; also see Chapter 3, Volume 2 of this series).

3.4.3 Copper-containing enzymes

Thiocarbamide complexation with metal ions has been known for some considerable time (section 3.2.2). Thus it is not surprising that thiocarbamides inhibit particular copper-containing enzymes (Table 9). Chelation with copper at the active site of the enzyme, however, may not be the exclusive mechanism of inhibition, since methimazole (**8**) also reacts with a product of 1-hydroxyphenol oxidation catalysed by mushroom tyrosinase forming a putative dihydroxyphenol adduct (Andrawis and Kahn, 1986).

3.4.4 Other enzymes

The enzymic S-glucuronidation and S-methylation of thiocarbamides has been discussed in section 3.3.2 and is dealt with more fully elsewhere in this series (see Volume 2).

The fact that thiocarbamides, in particular those used to treat hyperthyroidism, can inhibit the deiodination of the thyroid hormone thyroxine (T_4) to tri-iodothyronine is of some interest, since a major proportion of this iodothyronine, which is more potent than T_4, is produced by the enzymic deiodination of T_4 (Chopra, 1986).

While Yamada and Kaplowitz (1980) reported that propylthiouracil (12) inhibited glutathione S-transferase and also replaced glutathione as a substrate for this enzyme, others (Habig et al., 1984) found that it only acted as an inhibitor. It is noteworthy that the S-dioxide and S-trioxide of propylthiouracil (12) are more effective inhibitors of glutathione S-transferase than the parent compound (Kariya et al., 1986).

3.5 TOXICITY OF THIOCARBAMIDES

Apart from their goitrogenicity and their inability to inhibit thyroid hormone biosynthesis (Astwood, 1943; Astwood et al., 1945; Stanley and Astwood, 1947; Seifter and Ehrich, 1948; Searle et al., 1950, 1951), an action which has been utilized therapeutically in the development of antithyroid drugs for the treatment of hyperthyroidism (see Chapter 5, Volume 3, Part B of this series), thiocarbamides exhibit different toxicities depending on the species of animal being studied (see Neal et al., 1977; Neal and Halpert, 1982).

Hypersensitivity reactions are the most common toxic reactions observed in patients receiving the 2-thioimidazole and 2-thiouracil drugs for hyperthyroidism (Bouillon, 1980). Drug-induced agranulocytosis occurs in less than 1% of patients receiving those drugs.

The toxicities of the thioureas, including ethylenethiourea (4), will be dealt with in more detail in this section since more metabolic and biochemical information is available for these compounds.

The adverse reactions are attributed to the thiocarbonyl moiety since in some cases the corresponding carbonyl compounds did not cause the same toxic effect (Boyd and Neal, 1976; Hunter and Neal, 1975).

3.5.1 Thioureas

Thiourea (1) (MacKenzie and MacKenzie, 1943b; Dieke and Richter, 1945) and its N-substituted derivatives 1-phenylthiourea (33) (Richter and Clisby, 1942) and 1-naphthylthiourea, (3) (Latta, 1947; Richter, 1952; Meyrich et al., 1972), in addition to being goitrogenic (MacKenzie and MacKenzie, 1943a), produce pulmonary oedema and pleural effusion of varying intensities in susceptible animals. It would appear that this toxic response is not elicited by the thiourea per se but is a consequence of metabolic activation (Cunningham and Hurley, 1972; Hollinger et al., 1976; Boyd and Neal, 1976; Lee et al., 1980). In other earlier studies on the metabolism of substituted thioureas it was concluded that the relative toxicity of these compounds was related to the extent of their desulphuration (Williams, 1961;

Smith and Williams, 1959, 1961c). Symmetrical N,N'-disubstituted thioureas, however, have a much lower toxicity and were not desulphurated by rabbits (Williams, 1961; Smith and Williams, 1961a).

In a systematic study with various mono- and di-N-substituted thioureas and their oxygen analogues it was shown that the type of teratogenic effect induced in rat and mice depended on the nature of substitution (Teramoto *et al.*, 1981). Chronic administration of thiourea (**1**) also induced liver tumours in rat (Fitzhugh and Nelson, 1948). It is noteworthy that ethylenethiourea (**4**), which can be regarded as a symmetrical N,N-disubstituted thiourea, has not been reported to induce pulmonary oedema significantly, although it is goitrogenic, carcinogenic, teratogenic and induces liver tumours (Innes *et al.*, 1969; Ulland *et al.*, 1972; Graham *et al.*, 1973, 1975; Khera and Iverson, 1981) in rat and mice.

Both ^{35}S and ^{14}C radioactivity localize in highly perfused organs, in particular lung, liver and kidney, with the rate of elimination of ^{35}S radiolabel from the organ being slower than for the ^{14}C radiolabel after administration of ^{35}S- or ^{14}C-thiourea (**1**) to rat (Hollinger *et al.*, 1974, 1976; Hirate *et al.*, 1982) and mice (Slanina *et al.*, 1973; Hirate *et al.*, 1983) and similarly radiolabelled 1-phenylthiourea (**33**) (Scheline *et al.*, 1961) and 1-naphthylthiourea (**3**) (Boyd and Neal, 1976; Patil and Radhakrishnamurty, 1977) to rabbit and rat respectively. Not only are the thioureas or their metabolites localized in these organs, but they are also retained in the thyroid gland (Maloof and Soodak, 1957) and blood cells (Williams and Kay, 1945; Hollinger *et al.*, 1974) which contain peroxidases. The radioactivity is covalently bound to tissue macromolecules. Moreover, the extent of binding of either ^{14}C- or ^{35}S-1-naphthylthiourea (**3**) to macromolecules in rat lung and liver *in vitro* is NADPH dependent (Boyd and Neal, 1976), with more ^{35}S label bound than ^{14}C label. There was also a concomitant decrease in cytochrome P-450-dependent monooxygenase activity. Based on these and other observations it has thus been proposed that metabolic activation is a prerequisite for covalent binding. Although S-oxidation is implied the nature of the reactive metabolite which initiates the various toxic responses has not been established.

Thyroid peroxidase mediated trans-sulphuration of the sulphur of thiourea (Maloof and Spector, 1959; Maloof and Soodak, 1960) and other thiocarbamides (Marchant *et al.*, 1978) to thyroid protein is well known since, apart from sulphate, protein-bound ^{35}S is formed with ^{35}S-radiolabelled thiocarbamides *in vitro* and *in vivo*. Thus it is possible that sequential oxidation of the thiourea to its corresponding S-oxides is the initial activation step of these compounds (Ziegler, 1978). It is noteworthy that the S-dioxide of thiourea is more toxic than the parent compound in rats (Zhislin and Ovetskaya, 1972). The S-oxide could react with protein sulphydryl groups to form mixed disulphides. Alternatively, it has been proposed that, following oxidation of the thiourea (**1**), reactive atomic sulphur is released (Lee *et al.*, 1980; Neal, 1980) which reacts with protein sulphydryl groups to form hydrodisulphide residues. Metabolically released atomic sulphur has also been implicated as being responsible for the hepatotoxicity of the thiocarbamide antithyroid drug methimazole (**8**) (Neal, 1980).

Since the covalent binding of the thioureas is NADPH dependent, both a flavin-containing monooxygenase (Ziegler, 1978) and the cytochrome P-450-dependent monooxygenases (Lee *et al.*, 1980; Neal, 1980) have been implicated. However, the possibility that peroxidases and other biological oxidizing agents may also contribute *in vivo* cannot be ignored, since thiocarbamides are oxidatively desulphurated in peroxidase-containing tissues (section 3.4.1).

Not only are there marked species differences with respect to the toxicity of thioureas in that 1-phenyl- (**33**) and 1-naphthylthiourea (**3**) (Richter, 1945) are very toxic to rats, but on exposure to thiourea (**1**) and *N*-arylthioureas mature rats are more susceptible than immature animals (Dieke and Richter, 1945, 1946; Van den Brenk *et al.*, 1976; Hollinger *et al.*, 1976). Pre-treatment of rats with sublethal doses of thioureas induces resistance (Van den Brenk *et al.*, 1976; Hollinger *et al.*, 1976) to subsequent exposure to normally lethal doses.

3.5.2 Ethylenethiourea

Ethylenethiourea (**4**) exhibits a range of toxicities in animals including teratogenicity (section 3.5.1). Since ethylenethiourea (**4**) is a contaminant and degradation product of the ethylene dithiocarbamate fungicides, its toxicity and metabolism have been studied in a variety of species. Ethylenethiourea (**4**) induces teratogenesis in rat (Khera, 1973) but not in rabbit (Khera, 1973), cat (Khera and Iverson, 1978) or mouse (Teramoto *et al.*, 1981). However, recently ethylenethiourea-induced teratogenesis has been reported in mouse (Khera, 1984). The insusceptibility of cat (Iverson *et al.*, 1980) and mouse (Ruddick *et al.*, 1977) may be due to its ability to metabolize ethylenethiourea (**4**) to a greater extent than rat (Table 10). Moreover, the principal biotransformation pathway is *S*-methylation in a cat, while oxidative pathways dominate in rat. While thiourea-induced lung toxicity is due to metabolic activation, it is possible that it is the thiocarbonyl moiety *per se* in ethylenethiourea (**4**) which induces teratogenesis (Ruddick *et al.*, 1976; Khera and Iverson, 1978) and that metabolism inactivates the compound in certain species.

Table 10 — Urinary excretion of ethylenethiourea and its metabolites in various species (percentage of administered dose excreted in urine)

Compound	Species			
	Mouse[a]	Rat[b]	Cat[c]	Cow[d]
Ethylenethiourea (**4**)	17.4	51.8	22.6	7
S-Methylethylenethiourea		—	51.8	
Imidazoline		1.6		
Imidazolone		4.1	2.8	
Ethyleneurea (**30**)	4.0	15.1		18
Polar compounds	13.0	10.2	3.4	43

[a]Dose, 0.05 mmol/kg p.o.; data from Jordan and Neal (1979).
[b]Dose, 0.04 mmol/kg p.o.; data from Iverson *et al.* (1980).
[c]Dose, 0.04 mmol/kg intravenous; data from Iverson *et al.* (1980).
[d]Dose, 20 μg p.o. daily for 14 days data from Lyman (1971).

ACKNOWLEDGEMENTS

I am grateful to Dr G. A. Smail for his advice and Mrs E. Crossan for her assistance in the preparation of this manuscript.

REFERENCES

Andrawis, A. and Kahn, V. (1986). Effect of methimazole on the activity of mushroom tyrosinase. *Biochem. J.*, **235**, 91–96.

Askalani, P. and Bailey, R. A. (1969). Some substituted urea and thiourea complexes of Co(II), Mn(II) and Cr(III). *Can. J. Chem.*, **47**, 2275–2282.

Astwood, E. B. (1943). The chemical nature of compounds which inhibit the function of the thyroid gland. *J. Pharmacol.*, **78**, 79–89.

Astwood, E. B., Bissell, A., and Hughes, A. M. (1945). Further studies on the chemical nature of compounds which inhibit the function of the thyroid gland. *Endocrinology*, **37**, 456–481.

Baker, J. A. (1958). The mechanism of *N*-acylation of 2-mercaptoglyoxalines. *J. Chem. Soc.*, 2387–2390.

Balaban, I. E. and King, H. (1927). Gold and mercury derivatives of 2-thiolglyoxalines. Mechanism of the oxidation of 2-thioglyoxalines to glyoxalines. *J. Chem. Soc.*, 1858–1874.

Barnet, E. de Barry (1910). The action of hydrogen dioxide on thiocarbamides. *J. Chem. Soc.*, 63–65.

Basosi, R., Niccolai, N., and Rossi, C. (1978). Coordination behaviour of antithyroid drugs against the $Fe(I)(NO)_2$ group in solution: ESR and FT-NMR study. *Biophys. Chem.*, **8**, 61–69.

Beaty, N. B. and Ballou, D. P. (1980). Transient kinetic study of liver microsomal FAD-containing monooxygenase. *J. Biol. Chem.*, **255**, 3817–3819.

Berg, B. H. (1971a) The reactions of mercaptopurine, 2-thiouracil, 6-methyl-2-thiouracil and 6-propyl-2-thiouracil with diphenylpicrylhydrazyl. *Acta Pharm. Suecica*, **8**, 443–452.

Berg, B. H. (1971b). The reaction of thiamazole with diphenylpicrylhydrazyl. *Acta Pharm. Suecica*, **8**, 431–442.

Berg, B. H. (1971c) The spectrophotometric determination of some pharmaceutical thiols with diphenylpicrylhydrazyl. *Acta Pharm. Suecica*, **8**, 453–460.

Björkstén, F. (1966). Do thionamide antithyroid compounds act as free radical scavengers? *Biochim. Biophys. Acta*, **127**, 265–268.

Bojarska-Olejnik, E., Stefaniak, L., Witanowski, M., Hamdi, B. T., and Webb, G. A. (1985). Applications ^{15}N-NMR to a study of tautomerism in some monocyclic azoles. *Mag. Res. Chem.*, **23**, 166–169.

Bontoyan, W. R. and Looker, J. B. (1973). Degradation of commercial ethylene bisdithiocarbamate formulations to ethylenethiourea under elevated temperature and humidity. *J. Agr. Food Chem.*, **21**, 338–341.

Bouillon, R. (1980). Thyroid and antithyroid drugs. In M. N. G. Dukes (ed.), *Meyler's Side Effects of Drugs*, 10th edn., Excerpta Medica, Amsterdam, pp. 782–794.

Boyd, M. R. and Neal, R. A. (1976) Studies on the mechanism of toxicity and of development of tolerance to the pulmonary toxin, α-naphthylthiourea (ANTU). *Drug Met. Disp.*, **4**, 314–322.

Brown, D. J. (1962). Sulphur-containing pyrimidines. In *The Pyrimidines*, Interscience, New York, 272–305.

Brown, D. J. (1970). Sulphur-containing pyrimidines. In *The Pyrimidines*, Supplement 1, Interscience, New York, pp. 202–226.

Broxterman, H. J., van Langevelde, A., Bakker, C. N. M., Boer, H., Journée-de Korver, J. G., Kaspersen, F. M., Kakebeeke-Kemune, H. M., and Pauwels, E. K. J. (1983). Incorporation of [^{125}I]-5-iodo-2-thiouracil in cultured hamster, rabbit and human melanoma cells. *Cancer Res.*, **43**, 1316–1320.

Buncel, E., Norris, A. R., Taylor, S. E., and Racz, W. J. (1982). Metal ion–biomolecule interactions IV. Methylmercury(II) complexes of 1-methylimidazoline-2-thione (methimazole), a potentially useful protective agent in organomercurial intoxication. *Can. J. Chem.*, **60**, 3033–3038.

Carlsson, J., Kierstan, M. P. J. and Brocklehurst, K. (1974). Reactions of L-ergothioneine and some other aminothiones with 2,2^1- and 4,4^1-dipyridyl disulphides and of L-ergothioneine with iodoacetamide. *Biochem. J.*, **139**, 221–235.

Cavalieri, R. R. and Pitt-Rivers, R. (1981). The effects of drugs on the distribution and metabolism of thyroid hormones. *Pharmacol. Rev.*, **33**, 55–80.

Chopra, I. J. (1986). Nature, sources, and relative biologic significance of thyroid circulating hormones. In S. H. Ingbar and L. E. Braverman (eds.), *Werner's The Thyroid: a Fundamental and Clinical Text*, 5th edn., Lippincott, Philadelphia, pp. 136–152.

Conn, I. G., Ferguson, M. M., Skellern, G. G., Sweeney, D., and Steer, S. T. (1988) Co-oxidation of arachidonic acid and 1-methyl-2-thioimidazole (methimazole) by prostaglandin endoperoxide synthetase. *Pharmacology*, **36**, 145–150.

Cooper, D. A., Rettig, S. J., Storr, A., and Trotter, J. (1986). The 2-mercapto-1-methylimidazolyl moiety as a bridging ligand in complexes of gallium, rhenium and molybdenum. *Can. J. Chem.*, **64**, 1643–1651.

Coval, M. L. and Taurog, A. (1967). Purification and iodinating activity of hog thyroid peroxidase. *J. Biol. Chem.*, **242**, 5510–5523.

Crank, G. and Makin, M. I. H. (1984). A new method for converting thiourea and monosubstituted thioureas into cyanamides: desulphurisation by superoxide ion. *J. Chem. Soc., Chem. Commun.*, 53–54.

Cunningham, A. L. and Hurley, J. V. (1972). α-Naphthyl-thiourea-induced pulmonary oedema in the rat: a topographical and electron microscopic study. *J. Path.*, **106**, 25–35.

Danehy, J. P., Egan, C. P., and Switalski, J. (1971). The oxidation of organic divalent sulfur by iodine. III. Further evidence for sulfenyl iodides as intermediates and for the influence of structure on the occurrence of cyclic intermediates in the oxidation of thiols. *J. Org. Chem.*, **36**, 2530–2534.

Davidson, B., Soodak, M., Neary, J. T., Strout, H. V., Kieffer, J. D., Mover, H., and Maloof, F. (1978a). The irreversible inactivation of thyroid peroxidase by methylmercaptoimidazole, thiouracil, and propylthiouracil *in vitro* and its relationship to *in vivo* findings. *Endocrinology*, **103**, 871–882.

Davidson, B., Neary, J. T., Strout, H. V., Maloof, F., and Soodak, M. (1978b). Evidence for a thyroid peroxidase associated "active iodine" species. *Biochim. Biophys. Acta,* **522**, 318–326.

Davidson, B., Soodak, M., Strout, H. V., Neary, J. T., Nakamura, C., and Maloof, F. (1979). Thiourea and cyanamide as inhibitors of peroxidase: the role of iodide. *Endocrinology,* **104**, 919–924.

DeGroot, L. J. and Niepomniszcze, H. (1977). Biosynthesis of thyroid hormone: basic and clinical aspects. *Metabolism,* **26**, 665–718.

Dehand, J. and Jordanov, J. (1976). Complexes of Pt(II), Pd(II), Rh(I) and Rh(III) with nitrogen and sulphur-containing heterocyclic ligands of biological interest. Synthesis, characterisation and influence of pH. *Inorg. Chim. Acta,* **17**, 37–44.

Dencker, L., Larsson, B., Olander, K., Ullberg, S. and Yokota, M. (1981) Incorporation of thiouracil and some related compounds into growing melanin. *Acta Pharmacol. Toxicol.,* **49**, 141–149.

Desbarats–Schönbaum, M. L., Endrenyi, L., Koves, E., Schönbaum, E., and Sellers, E. A. (1972). On the action and kinetics of propylthiouracil. *Eur. J. Pharmacol.,* **19**, 104–111.

Dieke, S. H. and Richter, C. P. (1945) Acute toxicity of thiourea to rats in relation to age, diet, strain and species variation. *J. Pharm. Exp. Ther.,* **83**, 195–202.

Dieke, S. H. and Richter, C. P. (1946). Age and species variation in the acute toxicity of alpha-naphthylthiourea. *Proc. Soc. Exp. Biol. Med.,* **62**, 22–25.

Doerge, D. R. (1986). Mechanism-based inhibition of lactoperoxidase by thiocarbamide goitrogens, *Biochemistry,* **25**, 4724–4728.

Eaton, D. R. and Zaw, K. (1971) Nuclear magnetic resonance studies of Co(II) complexes of thiourea and related ligands. *Can. J. Chem.,* **49**, 3315–3326.

Edelhoch, H., Irace, G., Johnson, M. L., Michot, J. L., and Nunez, J. (1979). The effects of thioureylene compounds (goitrogens) on lactoperoxidase activity. *J. Biol. Chem.,* **254**, 11822–11830.

Edman, P. (1950). Method for determination of the amino acid sequence in peptides. *Acta Chem. Scand.,* **4**, 283–293.

Edward, J. T. (1966). Thiohydantoins. In N. Kharasch and C. Y. Meyers (eds.), *The Chemistry of Organic Sulfur Compounds*, Vol. 2, Pergamon, New York, pp. 287–309.

Edward, J. T. and Nielsen, S. (1957) Thiohydantoins. 1. Ionisation and ultraviolet absorption. *J. Chem. Soc.,* 5075–5079.

El Dareer, S. M., Kalin, J. R., Tillery, K. F. and Hill, D. L. (1984). Disposition of 2-mercaptobenzimidazole in rats dosed orally or intravenously. *J. Toxicol. Environm. Health,* **14**, 595–604.

Engler, H., Taurog, A., and Nakashima, T. (1982). Mechanism of inactivation of thyroid peroxidase by thioureylene drugs. *Biochem. Pharmacol.,* **31**, 3801–3806.

Engler, H., Taurog, A., Luthy, C., and Dorris, M. L. (1983). Reversible and irreversible inhibition of thyroid peroxidase-catalyzed iodination by thioureylene drugs. *Endocrinology,* **112**, 86–95.

Evans, R. M., Jones, P. G., Palmer, P. J. and Stephens, F. F. (1956). The preparation of 4-amino- and other pteridines. *J. Chem. Soc.,* 4106–4113.

Fairchild, R. G., Packer, S., Greenberg, D., Som, P., Brill, A. B., Fand, I., and McNally, W. P. (1982). Thiouracil distribution in mice carrying transplantable melanoma. *Cancer Res.*, **42**, 5126–5132.

Farishian, R. A. and Whittaker, J. R. (1979). Tyrosine utilization by cultured melanoma cells: analysis of melanin biosynthesis using [^{14}C]tyrosine and [^{14}C]thiouracil. *Arch. Biochem. Biophys.*, **198**, 449–461.

Fitzhugh, O. G. and Nelson, A. A. (1948). Liver tumours in rats fed thiourea or thioacetamide. *Science,* **108**, 626–628.

Foye, W. O. and Lo, J. (1972). Metal-binding abilities of antibacterial heterocyclic thiones. *J. Pharm. Sci.*, **61**, 1209–1212.

Franken, N. A. P., van Langevelde, A., Pauwels, E. K. J., Journée-de Korver, J. G., Bakker, C. N. M., and Oosterhuis, J. A. (1986). Radio-iodine-labelled-5-iodo-2-thiouracil: a potential radiopharmaceutical for establishing the viability of ocular melanoma after radiation therapy. *Nuc. Med. Comm.*, **7**, 797–809.

Freedman, L. D. and Corwin, A. H. (1949). Oxidation–reduction potentials of thiol-disulphide systems. *J. Biol. Chem.*, **181**, 601–621.

Ganellin, C. R. and Durant, G. J. (1981). Histamine H$_2$-receptor agonists and antagonists. In M. E. Wolf (ed.), *Burger's Medicinal Chemistry*, Part III, 4th edn. Wiley, New York, pp. 487–551.

Garrett, E. R. and Weber, D. J. (1970). Metal complexes of thiouracils. I. Stability constants by potentiometric titration studies and structures of complexes. *J. Pharm. Sci.*, **59**, 1383–1398.

Garrett, E. R. and Weber, D. J. (1971). Metal complexes of thiouracils. II. Solubility analyses and spectrophotometric investigation. *J. Pharm. Sci.*, **60**, 845–853.

Garrett, E. R. and Weber, D. J. (1972). Metal complexes of thiouracils. III. Polarographic studies and correlations among complex stabilities, thiouracil structures and biological activities. *J. Pharm. Sci.*, **61**, 1241–1252.

Gattegno, D., Giuliani, A. M., Bossa, M., and Ramunni, G. (1973). Substituted thioureas. Part I. Study of trimethylthiourea and its complexes with Zinc(II). *J. Chem. Soc. Dalton*, 1399–1403.

Goswami, A. and Rosenberg, I. N. (1986). Iodothyronine 5′-deiodinase in brown adipose tissue: thiol activation and propylthiouracil inhibition. *Endrocrinology*, **119**, 916–923.

Graham, S. L., Hansen, W. H., Davis, K. J., and Perry, C. H. (1973). Effects of one-year administration of ethylenethiourea upon the thyroid of the rat. *J. Agric. Food. Chem.*, **21**, 324–329.

Graham, S. L., Davis, K. J., Hansen, W. H., and Graham, C. H. (1975). Effects of prolonged ethylenethiourea ingestion on the thyroid of the rat. *Fd. Cosmet. Toxicol.*, **13**, 493–499.

Griffith, E. A. H., Spofford III, W. A., and Amma, E. L. (1978). Crystal and molecular structure of a colored intermediate from the reaction of tetramethyl-thiourea and copper (2+) (dichlorobis(tetramethylthiourea) copper(II)). *Inorg. Chem.*, **17**, 1913–1917.

Habig, W. H., Jakoby, W. B., Guthenberg, C., Mannervik, B., and Vander Jagt, D. L. (1984). 2-Propylthiouracil does not replace glutathione for the glutathione transferase. *J. Biol. Chem.*, **259**, 7409–7410.

Hanlon, D. P. and Shuman, S. (1975). Copper ion binding and enzyme inhibitory properties of the antithyroid drug methimazole. *Experientia,* **31**, 1005–1006.

Hanzlik, R. P. and Cashman, J. R. (1983). Microsomal metabolism of thiobenzamide and thiobenzamide S-oxide. *Drug Met. Disp.,* **11**, 201–205.

Heath, H. and Toennies, G. (1957). The preparation and properties of ergothioneine disulphide. *Biochem. J.,* **68**, 204–210.

Hirate, J., Watanabe, J., Iwamoto, K., and Ozeki, S. (1982). Distribution of thiourea following intravenous and oral administration to rats. *Chem. Pharm. Bull.,* **30**, 3319–3327.

Hirate, J. Watanabe, J., and Ozeki, S. (1983). The change of disposition kinetics for thiourea accompanied by growth in mice. *J. Pharm. Dyn.,* **6**, 323–330.

Hoffmann, M. and Edwards, J. O. (1977). Kinetics and mechanism of the oxidation of thiourea and N,N^1-dialkylthioureas by hydrogen peroxide. *Inorg. Chem.,* **16**, 3333–3338.

Hofmann, K. (1953a). The oxo- and hydroxyimidazoles and their sulfur analogues. In *Chemistry of Heterocyclic Compounds,* Vol. 6, *Imidazole and its Derivatives,* Part I, Interscience, New York, pp. 55–110.

Hofmann, K. (1953b) The imidazole carboxylic and sulphonic acids. *ibid,* pp. 175–211.

Hollinger, M. A., Giri, S. N., Alley, M., Budd, E. R., and Hwang, F. (1974). Tissue distribution and binding of radioactivity from ^{14}C-thiourea in the rat. *Drug Met. Disp.,* **2**, 521–525.

Hollinger, M. A., Giri, S. N., and Budd, E. (1976). A pharmacodynamic study of [^{14}C]thiourea toxicity in mature, immature, tolerant and non-tolerant rats. *Toxicol. App. Pharmacol.,* **37**, 545–556.

Hunt, R. R., McOmie, J. F. W., and Sayer, E. R. (1959). Pyrimidines X. Pyrimidine 4:6 dimethylpyrimidine and their 1-oxides. *J. Chem. Soc.,* 525–530.

Hunter, A. L. and Neal, R. A. (1975). Inhibition of hepatic mixed-function oxidase activity *in vitro* and *in vivo* by various thiono-sulphur-containing compounds. *Biochem. Pharmacol.,* **24**, 2199–2205.

Innes, J. R. M., Ulland, B. M., Valerio, M. G., Petrucelli, L., Fishbein, L., Hart, E. R., and Pallotta, A. G. (1969). Bioassay of pesticides and industrial chemicals for tumorigenicity in mice: a preliminary note. *J. Nat. Cancer Inst.,* **42**, 1101–1114.

Iverson, F., Khera, K. S., and Hierlihy, S. L. (1980). *In vivo* and *in vitro* metabolism of ethylenethiourea in the rat and the cat. *Toxicol. Appl. Pharmacol.,* **52**, 16–21.

Janssen, M. J. (1962). Physical properties of organic thiones. Part IV. The basicity of the thiocarbonyl group in various thiones. *Rec. Trav. Chim. Pays-Bas,* **81**, 650–660.

Jirousek, L. and Cunningham, L. W. (1968). Stimulation of thiouracil binding and the iodination system in beef thyroid microsomes. *Biochim. Biophys. Acta,* **170**, 160–171.

Jirousek, L. and Pritchard, E. T. (1971). On the chemical iodination of tyrosine with protein sulfenyl iodide and sulfenyl periodide derivatives: the behaviour of thiol protein–iodine systems. *Biochim. Biophys. Acta,* **243**, 230–238.

Jocelyn, P. C. (1972). *Biochemistry of the SH Group,* Academic Press, London.

Jordan, L. W. and Neal, R. A. (1979). Examination of the *in vivo* metabolism of Maneb and Zineb to ethylenethiourea (ETU) in mice. *Bull. Environm. Contam.*

Toxicol., **22**, 271–277.

Kampmann, J. P., Skovsted, L. and Lund, B. (1974). The kinetics of propylthiouracil in euthyroid and hyperthyroid subjects. *Endocrinol. Experiment.*, **8**, Abstract **63**, 215.

Kano, K., Uetake, A., Shimizu, S., Nitta, K., and Yamamoto, Y. (1974). Studies on the metabolism of 2-thio-6-methyluracil in rats. *Yakugaku Zasshi*, **94**, 332–337.

Kariya, K., Sawahata, T., Okuno, S. and Lee, E. (1986). Inhibition of hepatic glutathione transferases by propylthiouracil and its metabolites. *Biochem. Pharmacol.*, **35**, 1475–1479.

Karkhanis, D. W. and Field, L. (1985). Thiono compounds. 5. Preparation and oxidation of some thiono derivatives of imidazoles. *Phosphorus Sulphur*, **22**, 49–57.

Kashima, C., Shimizu, M., Eto, T., and Omote, Y. (1986). A convenient synthesis of *N*-substituted formamidines by desulfurization of the corresponding thioureas. *Bull. Chem. Soc. Jpn.*, **59**, 3317–3319.

Kato, Y., Odanaka, Y., Teramoto, S., and Matano, O. (1977). Metabolic fate of ethylenethiourea in pregnant rats. *Bull. Environm. Contam. Toxicol.*, **16**, 546–555.

Katori, E., Nagano, T., Kunieda, T., and Hirobe, M. (1981). Facile desulfurization of thiocarbonyl groups to carbonyls by superoxide. A model of metabolic reactions. *Chem. Pharm. Bull.*, **29**, 3075–3077.

Katritzky, A. R. and Lagowski, J. M. (1963). Protropic tautomerism of heteroaromatic compounds: II. Six-membered rings. In A. R. Katritzky (ed.), *Advances in Heterocyclic Chemistry*, Vol. 1, Academic Press, London, pp. 339–437.

Kawano, H., Otani, M., Takeyama, K., Kawai, Y., Mayumi, T., and Hama, T. (1982). Studies on ergothioneine. VI. Distribution and fluctuations of ergothioneine in rats. *Chem. Pharm. Bull.*, **30**, 1760–1775.

Kedderis, G. L. and Rickert, D. E. (1985). Loss of rat liver microsomal cytochrome P.450 during methimazole metabolism: role of flavin-containing monooxygenase. *Drug Met. Disp.*, **13**, 58–61.

Khera, K. S. (1973). Ethylenethiourea: teratogenicity study in rats and rabbits. *Teratology*, **7**, 243–252.

Khera, K. S. (1984) Ethylenethiourea-induced hindpaw deformities in mice and effects of metabolic modifiers on their occurrence. *J. Toxicol. Environ. Health*, **13**, 747–756.

Khera, K. S. and Iverson, F. (1978). Toxicity of ethylenethiourea in pregnant cats. *Teratology*, **18**, 311–314.

Khera, K. S. and Iverson, F. (1981). Effects of pretreatment with SKF-525A, *N*-methyl-2-thioimidazole, sodium phenobarbital, or methyl cholanthrene on ethylenethiourea-induced teratogenicity in rats. *Teratology*, **24**, 131–137.

Kjellin, G. and Sandstrom, J. (1969). Tautomeric cyclic thiones IV. The thione-thiol equilibrium in some azoline-2-thiones. *Acta Chem. Scan.*, **23**, 2888–2899.

Kotegov, K. V., Zegzhda, G. D., Amindzhanov, A. A., and Kukushkin, Yu. N. (1975) Potentiometric study of complex formation by rhenium (V) with 1-methylimidazole-2-thiol. *Russ. J. Inorg. Chem.*, **20**, 63–65.

Latta, H. (1947). Pulmonary edema and pleural effusion produced by acute α-naphthylthiourea poisoning in rats and dogs. *Bull. John Hopkins Hosp.*, **80**, 181–197.

Lawson, A. and Morley, H. V. (1956). 2-Mercaptoglyoxalines. X. The acylation of 2-mercaptoglyoxalines. *J. Chem. Soc.*, 1103–1108.

Lee, P. W. and Neal, R. A. (1978) Metabolism of methimazole by rat liver cytochrome P.450-containing monooxygenase. *Drug Met. Disp.*, **6**, 591–600.

Lee, P. W., Arnau, T., and Neal, R. A. (1980) Metabolism of α-naphthylthiourea by rat liver and rat lung microsomes. *Toxicol. Appl. Pharmacol.*, **53**, 164–173.

Lenarcik, B. and Wiśniewski, M. (1977). Stability and structure of transition metal complexes with azoles in aqueous solutions. XII. 1-methyl-2-mercaptoimidazole complexes of Cu(II), Zn(II), Pd(II), Ag(I) and Cd(II). *Roczniki Chem.*, **51**, 1625–1631.

Lindsay, R. H., Tillery, C. R., and Yu, M. W. (1972). Conversion of the antithyroid drug 2-thiouracil to 2-thio-5'-UMP by UMP pyrophosphorylase. *Arch. Biochem. Biophys.*, **148**, 466–474.

Lindsay, R. H., Aboul-Enein, H. Y., Morel, D., and Bowen, S. (1974a). Synthesis and antiperoxidase activity of propylthiouracil derivatives and metabolites. *J. Pharm. Sci.*, **63**, 1383–1386.

Lindsay, R. H., Hill, J. B., Kelly, K., and Vaughn, A. (1974b). Excretion of propylthiouracil and its metabolites in rat bile and urine. *Endocrinology*, **94**, 1689–1698.

Lindsay, R. H., Hulsey, B. S., and Aboul-Enein, H. Y. (1975). Enzymatic S-methylation of 6-n-propyl-2-thiouracil and other antithyroid drugs. *Biochem. Pharmacol.*, **24**, 463–468.

Lindsay, R. H., Cash, A. G., Vaughn, A. W., and Hill, J. B. (1977a). Glucuronide conjugation of 6-n-propyl-2-thiouracil and other antithyroid drugs by guinea pig liver microsomes *in vitro*. *Biochem. Pharmacol.*, **26**, 617–623.

Lindsay, R. H., Vaughn, A., Kelly, K., and Aboul-Enein, H. Y. (1977b). Site of glucuronide conjugation to the antithyroid drug 6-n-propyl-2-thiouracil. *Biochem. Pharmacol.*, **26**, 833–840.

Lindsay, R. H., Kelly, K., and Hill, J. B. (1978). Oxidation products of the antithyroid drug 6-n-propyl-2- thiouracil. *Ala. J. Med. Sci.*, **15**, 29–41.

Lindsay, R. H., Kelly, K. and Hill, J. B. (1979). Oxidative metabolites of [2-^{14}C]propylthiouracil in rat thyroid. *Endocrinology*, **104**, 1686–1697.

Loosmore, S. M. and McKinnon, D. M. (1976). Oxidation products of cyclic thiones. *Phosphorus Sulfur*, **1**, 185–209.

Lopatecki, L. E. and Newton, W. (1952). The decomposition of dithiocarbamate fungicides with special reference to volatile products. *Can. J. Bot.*, **30**, 131–138.

Lyman, W. R. (1971). The metabolic fate of dithane M-45 (Coordination product of zinc ion and manganous ethylenebisdithiocarbamate). In A. S. Tahori (ed.), *International Symposium on Pesticide Terminal Residues*, Butterworths, London, pp. 243–256.

MacKenzie, C. G. and MacKenzie, J. B. (1943a). Effect of sulphonamides and thioureas on the thyroid gland and basal metabolism. *Endocrinology*, **32**, 185–209.

MacKenzie, J. B. and MacKenzie, C. G. (1943b). Production of pulmonary edema by thiourea in the rat and its relation to age. *Proc. Soc. Exp. Biol. Med.*, **54**, 34–37.

Magnusson, R. P. Taurog, A., and Dorris, M. L. (1984a). Mechanism of iodide-dependent catalatic activity of thyroid peroxidase and lactoperoxidase. *J. Biol. Chem.,* **259**, 197–205.

Magnusson, R. P., Taurog, A., and Dorris, M. L. (1984b). Mechanisms of thyroid peroxidase- and lactoperoxidase-catalyzed reactions involving iodide. *J. Biol. Chem.,* **259**, 13783–13790.

Mahajani, U., Haldar, I., and Datta, A. G. (1973). Purification and properties of an iodide peroxidase from submaxillary gland of goat. *Eur. J. Biochem.,* **37**, 541–552.

Maloof, F. and Soodak, M. (1957). The uptake and metabolism of S^{35}-thiourea and thiouracil by the thyroid and other tissues. *Endocrinology,* **61**, 555–569.

Maloof, F. and Soodak, M. (1960). The transsulfuration of the sulfur of thiourea to protein in thyroid tissue. In *Advances in Thyroid Research, 4th International Goitre Conference, London,* pp. 52–55.

Maloof, F. and Soodak, M. (1961). Cleavage of disulphide bonds in thyroid tissue by thiourea. *J. Biol. Chem.,* **236**, 1689–1692.

Maloof, F. and Soodak, M. (1963). Intermediary metabolism of thyroid tissue and the action of drugs. *Pharmacol. Rev.,* **15**, 43–95.

Maloof, F. and Spector, L. (1959). The desulfuration of thiourea by thyroid cytoplasmic particulate fractions. *J. Biol. Chem.,* **234**, 949–954.

Marchant, B., Alexander, W. D., Lazarus, J. H., Lees, J., and Clark, D. H. (1972). The accumulation of ^{35}S-antithyroid drugs by the thyroid gland. *J. Clin. Endocr. Metab.,* **34**, 847–851.

Marchant, B., Lees, J. F. H., and Alexander, W. D. (1978). Antithyroid drugs. *Pharmac. Ther. B,* **3**, 305–348.

Mariaggi, N., Cornu, A., and Rinaldi, R. (1973). Investigation and identification of imidazole metabolites in the rat, using thin-layer chromatography, infrared spectroscopy, and mass spectrometry, *Analusis,* **2**, 485–488. (*CA* **79** 101123d).

Marshall, W. D. (1979). Oxidative degradation of ethylenethiourea (ETU) and ETU progenitors by hydrogen peroxide and hypochlorite. *J. Agric. Food. Chem.,* **27**, 295–299.

Marshall, W. and Singh, J. (1977). Oxidative inactivation of ethylenethiourea by hypochlorite in alkaline medium. *J. Agric. Food. Chem.,* **25**, 1316–1320.

Maryanoff, C. A., Stanzione, R. C., and Plampin, J. N. (1986). Reactions of oxidized thioureas with amine nucleophiles. *Phosphorus Sulfur,* **27**, 221–232.

Mautner, H. G. and Clayton, E. M. (1959). 2-Seleno barbiturates. Studies of some analogous oxygen, sulphur and selenium compounds. *J. Am. Chem. Soc.,* **81**, 6270–6273.

Mayumi, T., Okamoto, K., Yoshida, K., Kawai, Y., Kawano, H., Hama, T., and Tanaka, K. (1982). Studies on ergothioneine VIII. Preventive effects of ergothioneine on cadmium-induced teratogenesis. *Chem. Pharm. Bull.,* **30**, 2141–2154.

McGill, J. E. and Lindström, F. (1977). Mechanism of reduction of cadmium by aminoiminomethane sulfinic acid in alkaline media. *Anal. Chem.,* **49**, 26–29.

McGinty, D. A., Sharp, E. A., Dill, W. A., Rawson, B. S., and Rawson, R. W. (1948). Excretion studies on thiouracil and its 6-benzyl, 6-*n*-propyl and 6-methyl derivatives in man. *J. Clin. Endocr.,* **8**, 1043–1050.

Meyrich, B., Miller, J., and Reid, L. (1972). Pulmonary edema induced by ANTU or by high or low oxygen-concentrations in rat — an electron microsopic study. *Br. J. Exp. Path.*, **53**, 347–358.

Michot, J. L., Nunez, J., Johnson, M. L., Irace, G., and Edelhoch, H. (1979). Iodide binding and regulation of lactoperoxidase activity toward thyroid goitrogens. *J. Biol. Chem.*, **254**, 2205–2209.

Miller, W. H., Roblin Jr., R. O., and Astwood, E. B. (1945). Studies in chemotherapy XI. Oxidation of 2-thiouracil and related compounds by iodine. *J. Am. Chem. Soc.*, **67**, 2201–2204.

Morgan, D. J., Blackman, G. L., Paull, J. D. and Wolf, L. J. (1981). Pharmacokinetics and plasma binding of thiopental II. Studies at cesarean section. Anesthesiology, **54**, 474–480.

Morris, D. R. and Hager, L. P. (1966). Mechanism of the inhibition of enzymatic halogenation by antithyroid agents. *J. Biol. Chem.*, **241**, 3582–3589.

Morris, D. R., Eberwein, H., and Hager, L. P. (1962). On the mechanism of inhibition of enzymatic halogenation by antithyroid agents. *Life Sci.*, **1**, 321–325.

Nagasaka, A. and Hidaka, H. (1976). Effect of antithyroid agents 6-propyl-2-thiouracil and 1-methyl-2-mercaptoimidazole on human thyroid iodide peroxidase. *J. Clin. Endocrinol. Metab.*, **43**, 152–158.

Neal, R. A. (1980). Microsomal metabolism of thiono-sulfur compounds: mechanisms and toxicological significance. In E. Hodgson, J. R. Bend and R. M. Philpot (eds.), *Reviews in Biochemical Toxicology*, Vol. 2, Elsevier, New York, pp. 131–171.

Neal, R. A. and Halpert, J. (1982). Toxicology of thiono-sulphur compounds. *Ann. Rev. Pharmacol. Toxicol.*, **22**, 321–339.

Neal, R. A., Kamataki, T., Hunter, A. L., and Catignani (1977). Monooxygenase catalysed activation of thiono-sulphur containing compounds to reactive intermediates. In V. Ullrich, I. Roots, A. Hildebrandt, R. W. Estabrook and A. H. Conney *Microsomes and Drug Oxidations*, Pergamon, New York, pp. 467–475.

Neelakantan, P. and Thyagarajan, G. (1969). Methylation of *N*-benzyl-2-mercaptoimidazole. *Indian J. Chem.*, **7**, 189–190.

Nishikawa, T. (1940) Barbituric acid derivatives VI. Bis[5-methyl-4,6-dioxo-2-pyrimidyl]disulphide and its cuprous salt. *J. Chem. Soc. Jpn.*, **61**, 81–88 (*CA* **36** 2864).

Nogimori, T., Braverman, L. E., Taurog, A., Fang, S., Wright, G., and Emerson, C. H. (1986). A new class of propylthiouracil analogs: comparison of 5′-deiodinase inhibition and antithyroid activity. *Endocrinology*, **118**, 1598–1605.

Norris, A. R., Taylor, S. E., Buncel, E., Bélanger-Gariépy, F., and Beauchamp, A. L. (1983). Crystal structures of two methylmercury complexes with 1-methylimidazoline-2-thione. *Can. J. Chem.*, **61**, 1536–1541.

Nunez, J. and Pommier, J. (1982). Formation of thyroid hormones. *Vitam. Horm.*, **39**, 175–229.

Ogihara, T. and Mitsunobu, O. (1982). Desulfurisation of 2-thiouridines by dipotassium diazenedicarboxylate. *Chem. Lett.*, 1621–1624.

Ohtaki, S., Nakagawa, H., Kimura, S. and Yamazaki, I. (1981). Analyses of catalytic intermediate of hog thyroid peroxidase during its iodinating reaction. *J. Biol. Chem.*, **256**, 805–810.

Ohtaki, S., Nakagawa, H., Nakamura, M., and Yamazaki, I. (1982). Reactions of purified hog thyroid peroxidase with H_2O_2, tyrosine, and methylmercaptoimidazole (goitrogen) in comparison with bovine lactoperoxidase. *J. Biol. Chem.*, **257**, 761–766.

Papapetrou, P. D., Marchant, B., Gavras, H. and Alexander, W. D. (1972). Biliary excretion of ^{35}S-labelled propylthiouracil, methimazole and carbimazole in untreated and pentobarbitone pretreated rats. *Biochem. Pharmacol.*, **21**, 363–377.

Paterson, J. R., Hood, H. T. and Skellern, G. G. (1983). The role of porcine thyroid peroxidase and FAD-containing monooxygenase in the metabolism of 1-methyl-2-thioimidazole (methimazole). *Biochem. Biophys. Res. Comm.*, **116**, 449–455.

Patil, T. N. and Radhakrishnamurty, R. (1977). Distribution and excretion of α-naphthylthio-^{14}C]urea in albino rats. *Indian J. Biochem. Biophys.*, **14**, 275–299.

Pickard, R. G. (1972) Treatment of peritonitis with pre- and postoperative irrigation of the peritoneal cavity with noxythiolin solution. *Br. J. Surg.*, **59**, 642–648.

Poulsen, L. L. and Ziegler, D. M. (1979). The liver microsomal FAD-containing monooxygenase. *J. Biol. Chem.*, **254**, 6449–6455.

Poulsen, L. L., Hyslop, R. M., and Ziegler, D. M. (1974). S-oxidation of thioureyelenes catalysed by a microsomal flavoprotein mixed-function oxidase. *Biochem. Pharmacol.*, **23**, 3431–3440.

Poulsen, L. L., Hyslop, R. M., and Ziegler, D. M. (1979). S-Oxygenation of N-substituted thioureas catalyzed by the pig liver microsomal FAD-containing monooxygenase. *Arch. Biochem. Biophys.*, **198**, 78–88.

Preisler, P. W. and Berger, L. (1947). Oxidation–reduction potentials of thiol–dithio systems: thiourea–formamidine disulphide. *J. Am. Chem. Soc.*, **69**, 322–325.

Reid, E. E. (1963a). Thioureas. In *Organic Chemistry of Bivalent Sulphur*, Vol. 5, Chemical Publishing Co., New York, pp. 11–193.

Reid, E. E. (1963b). Thiohydantoin and pseudothiohydantoin. In *Organic Chemistry of Bivalent Sulphur*, Vol. 5, Chemical Publishing Co., New York, pp. 341–394.

Reid, E. E. (1963c). Thiobarbituric acid. In *Organic Chemistry of Bivalent Sulphur.*, Vol. 5, Chemical Publishing Co., New York, pp. 315–321.

Reid, E. E. (1965). Thiopyrimidine and derivatives. In *Organic Chemistry of Bivalent Sulphur*, Vol. 6, Chemical Publishing Co., New York, pp. 227–330.

Richter, C. P. (1945). The development and use of α-naphthylthiourea (ANTU) as a rat poison. *J. Amer. Med. Assoc.*, **129**, 927–931.

Richter, C. P. (1952). Physiology and cytology of pulmonary edema and pleural effusion produced in rats by α-naphthylthiourea (ANTU). *J. Thorac. Surg.*, **23**, 66–91.

Richter, C. P. and Clisby, K. H. (1942). Toxic effects of the bitter-tasting phenylthiocarbamide. *Arch. Pathol.*, **33**, 46–57.

Ruddick, J. A., Williams, D. T. Hierlihy, L., and Khera, K. S. (1976). [14]C]Ethylenethiourea: distribution, excretion, and metabolism in pregnant rats. *Teratology*, **13**, 35–39.

Ruddick, J. A., Newsome, W. H., and Iverson, F. (1977). A comparison of the distribution, metabolism and excretion of ethylenethiourea in the pregnant mouse and rat. *Teratology,* **16**, 159–162.

Sabourin, P. J. and Hodgson, E. (1984). Characterisation of the purified microsomal FAD-containing monooxygenase from mouse and pig liver. *Chem.-Biol. Interact.,* **51**, 125–139.

Sarcione, E. J. and Sokal, J. E. (1958). Detoxication of thiouracil by *S*-methylation. *J. Biol. Chem.,* **231**, 605–608.

Savolainen, K. and Pyysalo, H. (1979). Identification of the main metabolite of ethylenethiourea in mice. *J. Agric. Food. Chem.,* **27**, 1177–1181.

Scheline, R. R., Smith, R. L., and Williams, R. T. (1961). The metabolism of arylthioureas II. The metabolism of ^{14}C and ^{35}S-labelled 1-phenyl-2-thiourea and its derivatives. *J. Med. Pharm. Chem.,* **4**, 109–135.

Schulman Jr. J. (1950). Studies on the metabolism of thiourea II. The metabolic fate of thiourea in the thyroid gland. *J. Biol. Chem.,* **186**, 717–723.

Searle, C. E., Lawson, A., and Hemmings, A. W. (1950). Antithyroid substances. 1. The mercaptoglyoxalines. *Biochem. J.,* **47**, 77–81.

Searle, C. E., Lawson, A., and Morley, H. V. (1951). Antithyroid substances. 2. Some mercaptoglyoxalines, mercaptothiazoles and thiohydantoins. *Biochem. J.,* **49**, 125–128.

Seifter, J. and Ehrich, W. E. (1948). Goitrogenic compounds: Pharmacological and pathological effects. *J. Pharmacol. Exp. Ther.,* **92**, 303–314.

Sharma, R. P. and Stowe, C. M. (1970). Desulfuration as a pathway of thiopental metabolism in large domestic animals. *Pharmacol. Res. Comm.,* **2**, 205–212.

Sijpesteijn, A. and Vonk, J. W. (1975). Decomposition of bis(dithiocarbamates) and metabolism by plants and microorganisms. *Environ. Qual. Saf. Suppl.,* **3**, 57–61 (*CA* **85** 105159s).

Sitar, D. S. and Thornhill, D. P. (1972). Propylthiouracil: absorption, metabolism and excretion in the albino rat. *J. Pharmacol. Exp. Ther.,* **183**, 440–448.

Sitar, D. S. and Thornhill, D. P. (1973). Methimazole: absorption, metabolism and excretion in the albino rat. *J. Pharmacol. Exp. Ther.,* **184**, 432–439.

Skellern, G. G. and Steer, S. T. (1981). The metabolism of [2-^{14}C]methimazole in the rat. *Xenobiotica,* **11**, 627–634.

Skellern, G. G., Knight, B. I., Luman, F. M., Stenlake, J. B., McLarty, D. G., and Hooper, M. J. (1977). Identification of 3-methyl-2-thiohydration, a metabolite of carbimazole, in man. *Xenobiotica,* **7**, 247–253.

Slanina, P., Ullberg, S., and Hammarström, L. (1973). Distribution and placental transfer of ^{14}C-thiourea and ^{14}C-thiouracil in mice studied by whole-body autoradiography. *Acta Pharmacol. Toxicol.,* **32**, 358–368.

Smith, R. L. and Williams, R. T. (1959). The metabolism of substituted thioureas. *Biochem. J.,* **71**, 2P.

Smith, R. L. and Williams, R. T. (1961a). The metabolism of arylthioureas 1. The metabolism of 1,3-diphenyl-2-thiourea (thiocarbanilide) and its derivatives. *J. Med. Pharm. Chem.,* **4**, 97–107.

Smith, R. L. and Williams, R. T. (1961b). The metabolism of arylthioureas III. a) The toxicity of hydrogen sulphide in relation to that of phenylthiourea. b) The protection of rats against the toxic effects of phenylthiourea with 1-methyl-2-phenylthiourea. *J. Med. Pharm. Chem.,* **4**, 137–146.

Smith, R. L. and Williams, R. T. (1961c). The metabolism of arylthioureas IV. *p*-chlorophenyl- and *p*-tolythiourea. *J. Med. Pharm. Chem.*, **4**., 147–162.

Solomon, D. H. (1986). Treatment of Graves' hyperthyroidism. In S. H. Ingbar and L. E. Braverman (eds.), *Werner's The Thyroid: a Fundamental and Clinical Text*, 5th edn., Lippincott, Philadelphia, pp. 987–1014.

Sörbo, B. and Ljunggren, J. G. (1958). The catalytic effect of peroxidase on the reaction between hydrogen peroxide and certain sulfur compounds. *Acta Chem. Scand.*, **12**, 470–476.

Spector, E. and Shideman, F. E. (1959). Metabolism of thiopyrimidine derivatives: thiamylal, thiopental and thiouracil. *Biochem. Pharmacol.*, **2**, 182–196.

Stanley, M. M. and Astwood, E. B. (1947). Determination of the relative activities of antithyroid compounds in man using radioactive iodine. *Endocrinology*, **41**, 66–84.

Stanovnik, B. and Tisler, M. (1964a). Dissociation constants and structure of ergothioneine. *Anal. Biochem.*, **9**, 68–74.

Stanovnik, B. and Tisler, M. (1964b). Contribution to the structure of heterocyclic compounds with thioamide groups. *Arzneimittel-Forsch.*, **14**, 1004–1012.

Stirling, C. J. M. (1971). The sulfinic acids and their derivatives. *Int. J. Sulfur Chem.*, **6**, 277–320.

Stolk, J. M. and Hanlon, D. P. (1973). Inhibition of brain dopamine-β-hydroxylase activity by methimazole. *Life Sci.*, **12**, 417–423.

Stowell, E. C. (1961). Ergothioneine. In N. Kharasch (ed.), *Organic Sulphur Compounds*, Vol. 1, Pergamon, New York, pp. 462–490.

Strominger, D. M. and Friedkin, M. (1954). Enzymatic synthesis of thiouracil riboside and thiouracil desoxyriboside. *J. Biol. Chem.*, **208**, 663–668.

Suszka, A. (1980). Kinetics and mechanism of oxidation of imidazole-2-thiones with hydrogen peroxide. *Polish J. Chem.*, **54**, 2289–2295.

Szabo, S., Kourounakis, P., Kovacs, K., Tuchweber, B., and Garg, B. D. (1974). Prevention of organomercurial intoxication by thyroid deficiency in the rat. *Toxicol. Appl. Pharmacol.*, **30**., 175–184.

Taurog, A. (1976). The mechanism of action of the thioureylene antithyroid drugs. *Endocrinology*, **98**, 1031–1046.

Taurog, A. (1986) Hormone synthesis: thyroid iodine metabolism. In S. H. Ingbar and L. E. Braverman (eds.), *Werner's The Thyroid: a Fundamental and Clinical Text*, Lippincott, Philadelphia, pp. 53–97.

Taurog, A. and Howells, E. M. (1966). Enzymatic iodination of tyrosine and thyroglobulin with chloroperoxidase. *J. Biol. Chem.*, **241**, 1329–1339.

Taylor, J. J., Willson, R. L., and Kendall-Taylor, P. (1984). Evidence for direct interactions between methimazole and free radicals. *FEBS Letts.*, **176**, 337–340.

Tenovuo, J. O. (1985). The peroxidase system in human secretions. In K. M. Pruitt and J. O. Tenovuo (eds.), *The Lactoperoxidase System: Chemistry and Biological Significance*, Dekker, New York, pp. 101–122.

Teramoto, S., Kaneda, M., Aoyama, H. and Shirasu, Y. (1981). Correlation between the molecular structure of *N*-alkylthioureas and their teratogenic properties. *Teratology*, **23**, 335–342.

Thomas, E. L. (1985). Products of lactoperoxidase-catalyzed oxidation of thiocyanate and halides. In K. M. Pruitt and J. O. Tenovuo (eds.), *The Lactoperoxidase System: Chemistry and Biological Significance*, Dekker, New York, pp. 31–53.

Toennies, G. (1937). Relations of thiourea, cysteine, and the corresponding disulphides. *J. Biol. Chem.*, **120**, 297–313.

Tynes, R. E. and Hodgson, E. (1985a). Catalytic activity and substrate specificity of the flavin-containing monooxygenase in microsomal systems: characterisation of the hepatic pulmonary and renal enzymes of the mouse, rabbit, and rat. *Arch. Biochem. Biophys.*, **240**, 77–93.

Tynes, R. E. and Hodgson, E. (1985b). Magnitude of involvement of the mammalian flavin-containing monooxygenase in the microsomal oxidation of pesticides, *J. Agr. Food Chem.*, **33**, 471–479.

Ulland, B. M., Weisburger, J. H., Weisburger, E. K., Rice, J. M., and Cypher, R. (1972). Thyroid cancer in rats from ethylenethiourea intake. *J. Nat. Cancer Inst.*, **49**, 583–584.

Van den Brenk, H. A. S., Kelly, H., and Stone, M. G. (1976). Innate and drug-induced resistance to acute lung damage caused in rats by α-naphthylthiourea (ANTU) and related compounds. *Br. J. Exp. Path.*, **57**, 621–636.

Visser, T. J., Van Overmeeren, E., Fekkes, D., Doctor, R., and Hennemann, G. (1979). Inhibition of iodothyronine-5'-deiodinase by thioureylenes: structure–activity relationship. *FEBS Lett.*, **103**, 314–318.

Vonk, L. J. (1971). Ethylenethiourea, a systemic decomposition product of nabem. *Meded. Fac. Landbouwwetensch. Rijksuniv. Gent*, **36**, 109–112 (*CA* **76** 21724y).

Von Voigtlander, P. F. and Moore, K. E. (1970). Behavioural and brain catecholamine depleting actions of U-14,624, an inhibitor of dopamine β-hydrolase. *Proc. Soc. Exp. Biol. Med.*, **133**, 817–820.

Walter, W. and Randau, G. (1969a). Oxidation products of thioamides XIX. Thiourea-*S*-monoxides. *Liebigs. Ann. Chem.*, **722**, 52–79.

Walter, W. and Randau, G. (1969b). Oxidation products of thioamides XX. Thiourea-*S*-dioxides. *Liebigs. Ann. Chem.*, **722**, 80–97.

Walter, W. and Randau, G. (1969c). Oxidation products of thioamides XXI. Thiourea *S*-trioxides (guanylsulfonic acid betaines). *Liebigs. Ann. Chem.*, **722**, 98–109.

Ware, E. (1950). The chemistry of the hydantoins. *Chem. Rev.*, **46**, 403–470.

Wätjen, F., Buchardt, O. and Langvad, E. (1982). Affinity therapeutics. 1. Selective incorporation of 2-thiouracil derivatives in murine melanomas. Cystostatic activity of 2-thiouracil arotinoids, 2-thiouracil retinoids, arotinoids, and retinoids. *J. Med. Chem.*, **25**, 956–960.

Weisiger, R. A. and Jakoby, W. B. (1979). Thiol *S*-methyltransferase from rat liver. *Arch. Biochem. Biophys.*, **196**, 631–637.

Whittaker, J. R. (1971). Biosynthesis of the thiouracil pheomelanin in embryonic pigment cells exposed to thiouracil. *J. Biol. Chem.*, **246**, 6217–6226.

Williams, R. H. and Kay, G. A. (1945). Absorption, distribution and excretion of thiourea. *Amer. J. Physiol.*, **143**, 715–722.

Williams, R. T. (1961). The metabolism and toxicity of arylthioureas. *Biochem. J. (Proc.)*, **80**, 1–2P.

Wolf, G., Bergan, J. G., and Sunko, D. E. (1961). Metabolism studies with DL-[α-^{14}C]ergothioneine. *Biochim. Biophys. Acta*, **54**, 287–293.

Yamada, T. and Kaplowitz, N. (1980). Propylthiouracil: a substrate for the gluta-thione *S*-transferases that competes with glutathione. *J. Biol. Chem.*, **255**, 3508–3513.

Yip, C. and Klebanoff, S. J. (1963). Synthesis *in vitro* of thyroxine from diiodotyro-sine by myeloperoxidase and by a cell-free preparation of beef thyroid glands. 1. Glucose–glucose oxidase system. *Biochim. Biophys. Acta,* **74**, 747–755.

Yu, M. W., Sedlak, J., and Lindsay, R. H. (1973a). Metabolism of the nucleoside of 2-thiouracil (2-thiouridine) by rat liver slices. *Arch. Biochem. Biophys.*, **155**, 111–119.

Yu, M. W., Sedlak, J. and Lindsay, R. H. (1973b). Relative antithyroid effects of 2-thiouracil, 2-thiouridine and 2-thio-UMP. *Proc. Soc. Exp. Biol. Med.*, **143**, 672–676.

Zelman, S. J., Rapp, N. S., Zenser, T. V., Mattammal, M. B., and Davis, B. B. (1984). Antithyroid drugs interact with renal medullary prostaglandin H syn-thase. *J. Lab. Clin. Med.*, **104**, 185–192.

Zenser, T. V., Mattammal, M. B., Wise, R. W., Rice, J. R., and Davis, B. B. (1983). Prostaglandin H synthase-catalyzed activation of benzidine: a model to assess pharmacologic intervention of the initiation of chemical carcinogenesis. *J. Pharmacol. Exp. Ther.*, **227**, 545–550.

Zhislin, L. E. and Ovetskaya, N. M. (1972). Toxicological characteristics of thiourea and its dioxide. *Gig. Tr. Prof. Zabol.*, **16**, 51–52 (*CA* **77** 97443).

Ziegler, D. M. (1978). Intermediate metabolites of thiocarbamides, thioureylenes and thioamides: mechanism of formation and reactivity. *Biochem. Soc. Trans.*, **6**, 94–96.

Ziegler, D. M. (1980). Microsomal flavin-containing monnoxygenase: oxygenation of nucleophilic nitrogen and sulfur compoundfs. In W. B. Jakoby (ed.), *Enzymatic Basis of Detoxication*, Vol. I, Academic Press, New York, pp. 201–227.

Ziegler, D. M. (1984). Metabolic oxygenation of organic nitrogen and sulfur compounds. In J. R. Mitchell and M. G. Horning, (eds.), *Drug Metabolism and Drug Toxicity*, Raven Press, New York, pp. 33–53.

4

Carbamothioates and carbamodithioates

Yoffi Segall
Israel Institute for Biological Research, P.O. Box 19, Ness-Ziona 70450, Israel

SUMMARY

1. Carbamothioates and carbamodithioates are important agricultural chemicals, widely used in crop protection, mainly as herbicides, fungicides, bactericides and nematocides. Their broad spectrum of activity is determined by the type of substituents attached to the nitrogen and sulphur.
2. S-Alkyl and S-benzyl carbamothioates degrade rapidly in soils, plants and mammals and decompose in the environment. Metabolic oxidation activates carbamothioate proherbicides to potent and selective carbamoylating agents. Sulphoxidation, a major metabolic pathway in nearly all the species, further leads to conjugation with glutathione, cysteine and malonyl and acetylcysteine.
3. Peracid oxidation is a suitable biomimetic model for S-alkyl carbamothioates, including diallate, but not for carbamodithioates including sulphallate.
4. Carbamothioates and carbamodithioates are of moderate toxicity to mammals. The S-(2-chloropropenyl)-containing diallate, triallate and sulphallate, show mutagenic activity only after microsomal metabolic activation. Diallate and triallate are also neurotoxicants.

4.1 INTRODUCTION

4.1.1 Structural features and uses

Carbamothioates (CT),$RR'NC(O)SR''$, and carbamodithioates (CDT), $RR'NC(S)SR''$, often termed thiocarbamates and dithiocarbamates, are sulphur analogues of carbamic acid (H_2NCOOH) derivatives. These structural features are present in many soil application herbicides (CT and the CDT sulphallate), non-systemic and foliar fungicides (mostly CDT that, in terms of tonnage employed, are the most used organic fungicides), bactericides and nematocides (Büchel, 1983; Hassall, 1982). A few selected herbicides are listed in Table 1. Important fungicides and bactericides are the metallic complexes ziram (**13**) and ferbam (**14**) and their

Table 1 — Uses of some soil application herbicidal carbamothioates $R_1R_2 C(O)SR_3$[a]

	Common name	Trade name	R_1	R_2	R_3	Use	Control
(1)	Ethiolate	Prefox	C_2H_5	C_2H_5	C_2H_5	Corn, with cyprazine	Wild oat
(2)	EPTC	Eptam	$n\text{-}C_3H_7$	$n\text{-}C_3H_7$	C_2H_5	Potato, alfalfa, bean	Wild oat, green foxtail, cyperus
(3)	Butylate	Sutan	$(CH_3)_2CHCH_2$	$(CH_3)_2CHCH_2$	C_2H_5	Corn	Green foxtail, cyperus
(4)	Cycloate	Ro-Neet	C_2H_5	$c\text{-}C_6H_{11}$	C_2H_5	Sugar beet, spinach	Broad-leaved weeds
(5)	Molinate	Ordram	— $(CH_2)_6$ —		C_2H_5	Rice	Germinating weeds
(6)	Pebulate	Tillam	C_2H_5	$n\text{-}C_4H_9$	$n\text{-}C_3H_7$	Sugar beet, tomato, tobacco	Annual grasses, weeds
(7)	Vernolate	Vernam	$n\text{-}C_3H_7$	$n\text{-}C_3H_7$	$n\text{-}C_3H_7$	Soybean, tobacco, tomato, peanut	Annual grasses, nutsedges
(8)	Diallate	Avadex	$i\text{-}C_3H_7$	$i\text{-}C_3H_7$	$CH_2C(Cl){=}CHCl$	Beet, potato, corn, beans	Wild oat, black grass
(9)	Triallate	Avadex, BW Far-Go	$i\text{-}C_3H_7$	$i\text{-}C_3H_7$	$CH_2C(Cl){=}CCl_2$	Spring barley, spring wheat	Wild oat, black grass
(10)	Benthiocarb	Saturn, Bolero	C_2H_5	C_2H_5	$p\text{-}ClC_6H_4CH_2$	Rice, Soybean, peanut	Germinating weeds (paddy)
(11)	Orbencarb		C_2H_5	C_2H_5	$o\text{-}ClC_6H_4CH_2$	Wheat, soybean	Germinating weeds (upland)
(12)	Sulphallate[b]	Vegadex, CDEC	C_2H_5	C_2H_5	$CH_2C(Cl){=}CH_2$	Celery, spinach, ornamentals	Grass weeds

[a]Mostly from Büchel (1983), p. 348, by permission of John Wiley and Sons, Inc. [b] Carbamodithioate.

oxidation product thiram (**15**, TMTD), from N,N-dimethylcarbamodithioates, and zineb (**16**) and maneb (**17**) and their mixed complex, mancozeb, from ethylene bis (CDT), also used as foliar application and seed dressing.

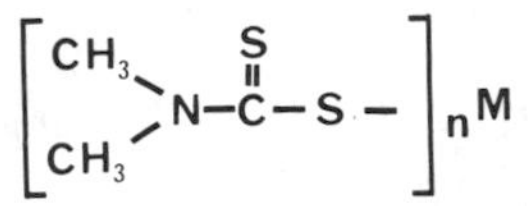

(**13**), n=2, M= Zn, Ziram   (**15**), Thiram (TMTD) 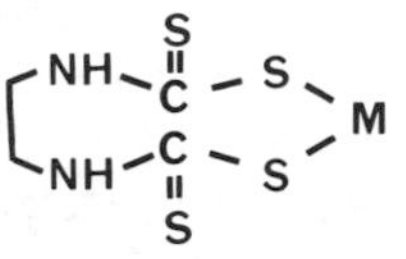   (**16**), M= Zn, Zineb

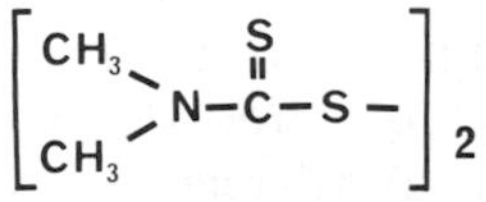

(**14**), n=3, M= Fe, Ferbam   (**17**), M= Mn, Maneb

4.1.2   CT and CDT in xenobiotics other than herbicides

The nature of the activity of CT and CDT is mainly dependent on the type of the substituents attached to the nitrogen and sulphur. O-ester CT, such as (**18**), are fungicides (Asahi Chemical Industry Co. Ltd., 1981). N-phenylcarbamothioate (**19**) effectively controls benzimidazole and thiophanate-resistant microorganisms (Haga *et al.*, 1985). S-Long alkyl-chain CT, such as (**20**), completely control scabies mites at 1% on rabbits without being harmful to the host (Kochansky and Wright, 1984). Maximum activity against scabies mites of cattle and sheep resides on those CT with a long-chain alkyl group on oxygen or sulphur and a monoalkyl group on nitrogen. Therefore, activity requires one and only one hydrogen attached to nitrogen (Kochansky and Wright, 1985). The tolerance to structural variations varied in the order

$$RNHC(S)OR' > RNHC(O)SR' > RNHC(O)OR'$$

Perhaps the most unusual activity was found with CT (**21**) that mimic insect juvenile hormones (Kisida *et al.*, 1984).

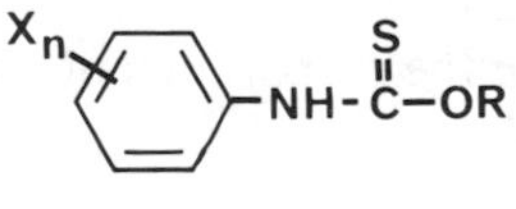

(**18**), X= halo., n= 0–3 ,   (**19**), R= $C_{1–8}$ alkyl,   (**20**)

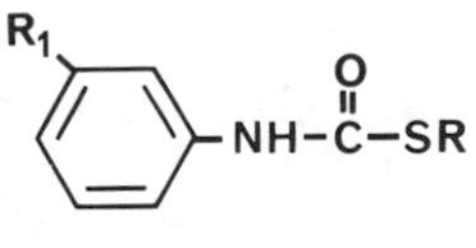

R = $C_{1–12}$ alkyl   R₁= H, halo., alkyl

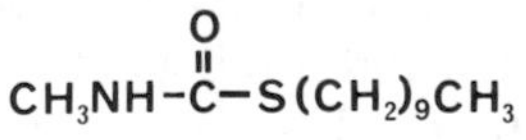

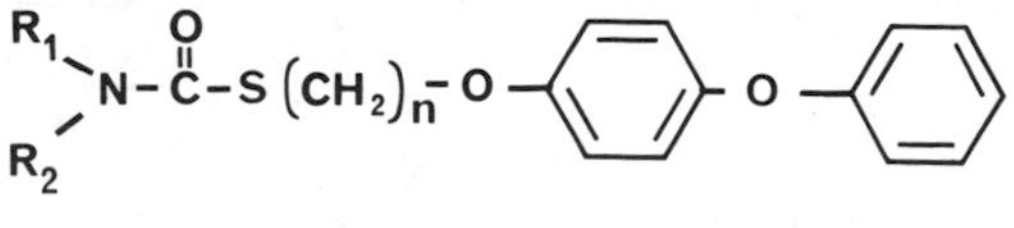

(**21**)

These compounds show outstanding juvenile hormone-like activity and high larvicidal activity against the larvae of the common mosquito and the yellow fever
mosquito. Based on its IC_{50} (0.00008 ppm for the common mosquito), **(21)**
($R_1=R_2=CH_3$, $n=2$) is twice as active as the commercial product, 'methoprene'
(Zoicon Corp.), a juvenile hormone mimic insecticide. Moreover, **(21)** has low
toxicity against mice (LD_{50} oral > 300 mg/kg).

N,N-disubstituted S-2,3-dibromopropyl CDT, $R_1R_2NC(S)SCH_2CH(Br)CHBr_2$,
[SIC], were suggested as plant growth regulators (Baicu *et al.*, 1981). However, for
the reasons discussed below, these should be treated as suspect promutagens.

The diethyl analogue of **(15)**, disulfiram (antabus), is used medically to treat
alcoholics (Physician's Desk Reference, 1987; Hald and Jacobson, 1948). Incubation
of this drug with human plasma resulted in a rapid reduction into CDT. Similar
results were obtained on incubation of disulfiram with bovine serum albumin,
indicating that the latter is the major component of human plasma involved in the
reduction of disulfiram (Agarwal *et al.*, 1983). This metabolite may be analysed by
high performance liquid chromatography on treatment of the plasma with methylio-
dide to form S-methyl-N,N-diethylcarbamodithioate (Giles *et al.*, 1982). Because of
their chelating properties, derivatives of diethylcarbamodithioates reduce organ
concentrations of heavy metals such as Pb (Gale *et al.*, 1986) and Cd (Gale *et al.*,
1985). In addition to the agricultural uses outlined above, DTC are also used in
industry as vulcanization accelerators and antioxidants in rubber (Kirk-Othmer,
1982).

4.1.3   Chemistry
The free acids $R_2NC(O)SH$ and $R_2NC(S)SH$ are not stable, reverting to dialkylamine and carbonylsulphide and to dialkylamine and carbon disulphide, respectively.
The parent CDT acid has been liberated from its ammonium salt at 0°C (Fourquet,
1969), but decomposed to CS_2 and $H_2NC(S)SNH_4$ on warming up to 20°C. Nevertheless, CDT acid is more stable than CT acid.

4.1.3.1   *Carbamothioates*
The following standard methods are still the most efficient for the preparation of CT
(eqns.(1)–(4)), e.g. the synthesis of diallate **(8)**:

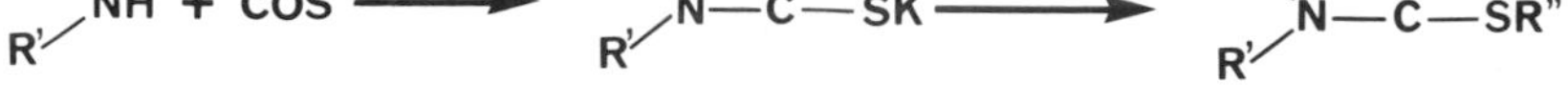

$$\underset{R'}{\overset{R}{>}}NH + COS \xrightarrow{KOH} \underset{R'}{\overset{R}{>}}N-\overset{\overset{O}{\|}}{C}-SK \xrightarrow{R''X} \underset{R'}{\overset{R}{>}}N-\overset{\overset{O}{\|}}{C}-SR''$$

(1)

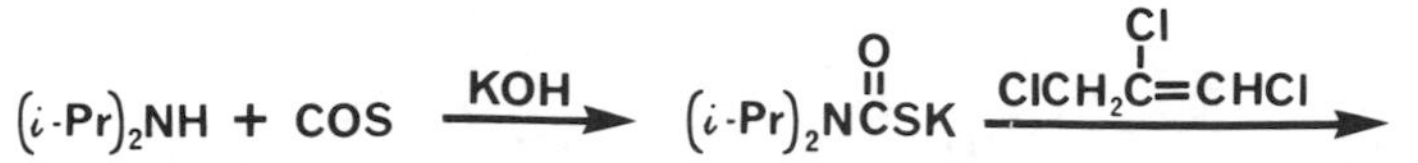

$$(i\text{-}Pr)_2NH + COS \xrightarrow{KOH} (i\text{-}Pr)_2N\overset{\overset{O}{\|}}{C}SK \xrightarrow{\overset{\overset{Cl}{|}}{ClCH_2C=CHCl}} \quad \textbf{(8)}$$

$$RNCO + NaSH \longrightarrow RNH\overset{\overset{O}{\|}}{C}SNa \qquad \qquad (2)$$

The synthesis of EPTC (**2**) involves phosgene:

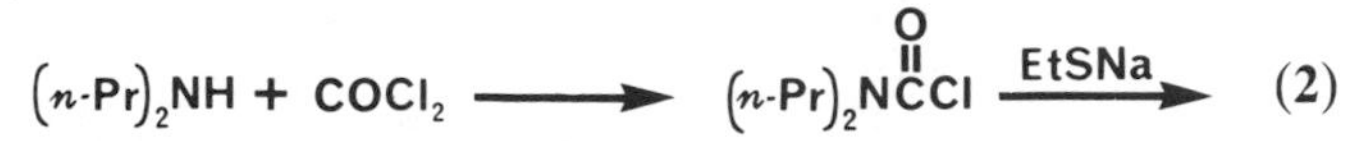

$$(n\text{-Pr})_2\text{NH} + \text{COCl}_2 \longrightarrow (n\text{-Pr})_2\overset{\overset{\text{O}}{\|}}{\text{N}}\text{CCl} \xrightarrow{\text{EtSNa}} (\textbf{2}) \qquad (3)$$

Alternatively, CT may be synthesized from thiochloroformic ester and amine:

$$\text{Ph}\overset{\overset{\text{O}}{\|}}{\text{S}}\text{CCl} + \text{PhNH}_2 \longrightarrow \text{PhNH}\overset{\overset{\text{O}}{\|}}{\text{C}}\text{SPh} \qquad (4)$$

The 'Newman Kwart rearrangement' (Newman and Karnes, 1966) utilizes rearrangement of *O*-ester to *S*-ester CT. *O*-thioesters of the structure $R_2NC(S)OR$ undergo rearrangement under the influence of heat or light to give *S*-thioesters (Scheithauer and Myer, 1979). This conversion is useful for the synthesis of *S*-aryl CT; these are otherwise difficult to obtain:

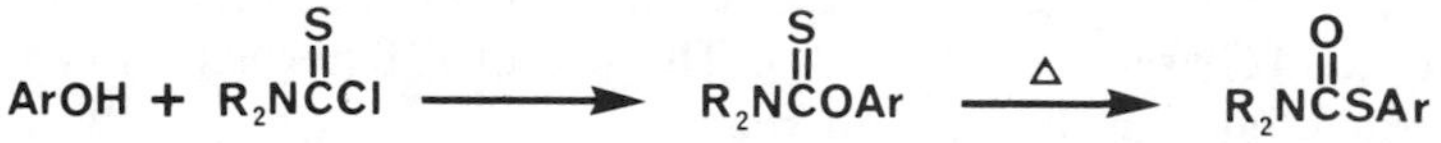

$$\text{ArOH} + R_2\overset{\overset{\text{S}}{\|}}{\text{N}}\text{CCl} \longrightarrow R_2\overset{\overset{\text{S}}{\|}}{\text{N}}\text{COAr} \xrightarrow{\Delta} R_2\overset{\overset{\text{O}}{\|}}{\text{N}}\text{CSAr} \qquad (5)$$

In simple esters the rearrangement gives the thermodynamically favoured CT only at high temperatures. However, when an intermediate carbonium ion is strongly stabilized, the reaction proceeds at lower temperatures (Smith, 1962):

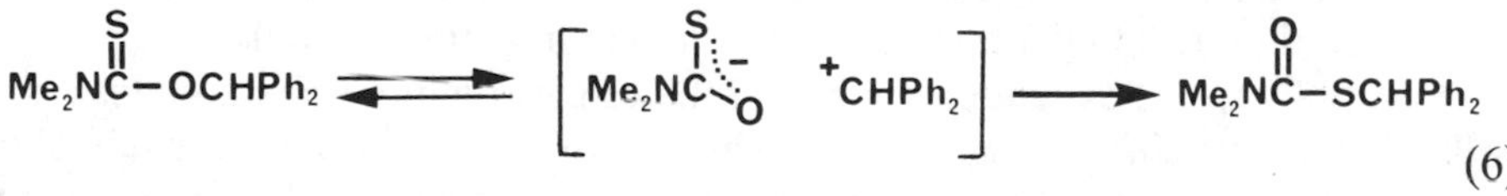

$$\text{Me}_2\overset{\overset{\text{S}}{\|}}{\text{N}}\text{C-OCHPh}_2 \rightleftharpoons \left[\text{Me}_2\text{N}\overset{\overset{\text{S}\colon^-}{\|}}{\text{C}\diagdown_{\text{O}}} \ ^+\text{CHPh}_2 \right] \longrightarrow \text{Me}_2\overset{\overset{\text{O}}{\|}}{\text{N}}\text{C-SCHPh}_2 \qquad (6)$$

In a similar way Kato *et al.* (1986) synthesized *S*-haloarylcarbamothioates:

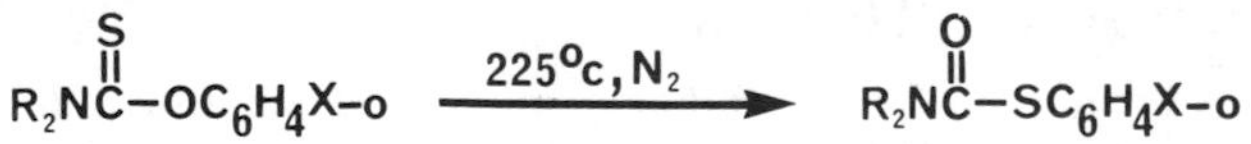

$$R_2\overset{\overset{\text{S}}{\|}}{\text{N}}\text{C-OC}_6\text{H}_4\text{X-o} \xrightarrow{225^\circ\text{c}, N_2} R_2\overset{\overset{\text{O}}{\|}}{\text{N}}\text{C-SC}_6\text{H}_4\text{X-o} \qquad (7)$$

4.1.3.2   Carbamodithioates

Analogous basic methods for the preparation of CT are also valid for the preparation of CDT:

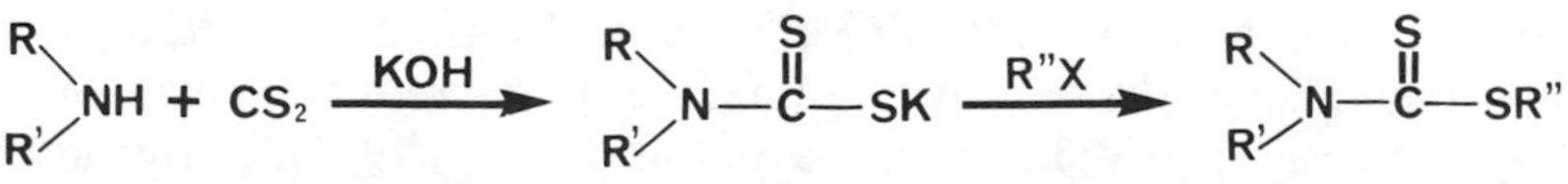

$$\overset{R}{\underset{R'}{>}}\text{NH} + CS_2 \xrightarrow{\text{KOH}} \overset{R}{\underset{R'}{>}}\text{N}-\overset{\overset{\text{S}}{\|}}{\text{C}}-\text{SK} \xrightarrow{R''X} \overset{R}{\underset{R'}{>}}\text{N}-\overset{\overset{\text{S}}{\|}}{\text{C}}-\text{SR}'' \qquad (8)$$

An alternative small-scale preparation of CDT utilizes a Grignard reagent with TMTD:

$$
\left[\underset{\text{TMTD}}{Me_2N\overset{\displaystyle S}{\overset{\|}{C}}S-} \right]_2 + RMgX \longrightarrow Me_2N\overset{\displaystyle S}{\overset{\|}{C}}SR + Me_2N\overset{\displaystyle S}{\overset{\|}{C}}SMgX \tag{9}
$$

The reaction proceeds smoothly with primary, secondary and tertiary alkyl, alkenyl or aryl halides, affording the desired CDT in good yield (Grunwell, 1970).

Thiuram disulphides are obtained on mild oxidation of an alkali metal salt of dimethyl CDT. Tetraethylthiuram disulphide was prepared by oxidation in aqueous solution of sodium diethyl CDT with atmospheric O_2 in the presence of mono- or dichlorocobalt phthalocyaninedisulphonic acid catalyst (Maizlish et al., 1986).

4.1.4 Potential exposure to man

Recently, Stauffer Chemical Co. suggested S-benzyl dipropylcarbamothioate as a novel CT herbicide, exhibiting excellent selective pre- and post-emergence control of many cereal weeds (Glasgow et al., 1987). The use of CT herbicides in weed control has been accelerated in the past few years.

Accumulated pesticides in the biota and environment might be hazardous to man; although bioaccumulation of zineb (16) and ziram (13) in rainbow trout is low (Leeuwen et al., 1986c), fish usually bioaccumulate pesticides in concentrations higher than those in water. Extensive work has been carried out in order to detect and identify hazardous CT (Akio et al., 1985; Lee and Chau, 1983; Ripley and Braun, 1983; Cook et al., 1982; Giles et al., 1982; Schairer et al., 1982) and to investigate their residual amounts in crops, rivers, mammals and fish.

Metabolism studies, understanding of the role of potential metabolites and the collection of data concerning acute and chronic toxicity of CT and the study of their fate in the environment are indispensible for the efficient and safe use of these pesticides.

4.1.5 Early studies on metabolism and toxicology

Compared with carbamates that have been used for almost half a century, CT are relatively late arrivals in weed control. The first member to reach a practical use was EPTC (2, Table 1), introduced by Stauffer Chemical Co. In subsequent years, this company developed a whole series of herbicidal CT for various applications.

As opposed to carbamates that are mostly acetylcholinesterase (AChE) inhibitors, CT are in fact propesticides, undergoing metabolic activation by reactions involving oxidation. These sulphoxidation reactions increase the reactivity of CT as carbamoylating agents (Unai and Tomizawa, 1986a; Casida, 1983; Schuphan and Casida, 1983; Schuphan and Casida, 1979a; Chen and Casida, 1978; Casida et al., 1975a,b) and may enhance the selectivity, sometimes with a reduction in mammalian toxicity (Lay and Casida, 1976; Lay et al., 1975; Casida et al., 1974). Early metabolic studies indicate that CT are rapidly degraded in mammals and plants. CT sulphoxides, which are major metabolites *in vivo* in mammals or with liver microsomal mixed function oxidase (MFO) systems, are carbamoylating agents for tissue thiols such as glutathione (GSH), cysteine and coenzyme A (CoA) and are more potent as herbicides than the parent CT. Other MFO metabolic pathways include hydroxylation at each carbon of EPTC (Chen and Casida, 1978).

In terms of acute toxicity, CT and CDT are considered slightly toxic to mammals (see Table 2). However, chronic tests indicate that *S*-chloroallyl CT herbicides diallate (8), triallate (9) and sulphallate (12) are carcinogens (National Cancer Institute, 1978) and promutagens (Schuphan and Casida 1979a,b). The *S*-alkyl CT herbicides did not show mutagenic activity (Moriga *et al.*, 1983). Diallate produces symptoms of severe ataxia in hens, whereas triallate is much less neurotoxic (Hansen *et al.*, 1985; Fisher and Metcalf, 1983).

4.1.6 Scope

This chapter comprises studies on the fate of CT herbicides in biota and the environment with the emphasis on metabolism in various organisms and chronic toxicity. Because of their toxicological significance and exceptional chemical behaviour, *S*-chloroallyl herbicides diallate, triallate and sulphallate are treated as a group separately from other carbamothioates. Where appropriate, toxicological data on CDT is also provided.

For additional information in relation to this review, the reader is advised to consider chapters 1 and 3 of Volume 1, Part A, chapters 1 and 5 of Volume 2, Part B, and chapter 5 of Volume 3, Part B of this series.

4.2 FATE OF CARBAMOTHIOATES IN THE ENVIRONMENT AND BIOTA

CT herbicides are rapidly degraded in soil, environment, plants and mammals with identifiable metabolites. Sulphoxidation is one of the initial steps in biodegradation under many environmental conditions. Carbamoylation reactions are most important in their metabolism and possibly determine their mode of action. Most of the CT are volatile, a property that greatly influences their agricultural use. They need to be incorporated into the soil soon after application.

4.2.1 Microbial degradation of carbamothioates in soil

Although chemically stable, CT degrade by the oxidative, reductive and hydrolytic enzymes of soil microorganisms, a major factor determining the rate of disappearance of CT from soils.

Alcoligenes, Micrococcus, Bacillus and *Pseudomonas* are bacteria isolated from soil that possibly degrade EPTC (Lee, 1984). Benthiocarb (10) was rapidly degraded by Corneyform bacteria (Abe and Kuwatsuka, 1979), but poorly in soil pre-treated with steam, indicating that soil microorganisms play an important role in the degradation process (Nakamura *et al.*, 1977a). The rates of degradation of both diallate (8) and triallate (9) in soil were directly related to the microbial biomass (Anderson, 1984).

Ishikawa ((1981) reported that degradation of ring ^{14}C benthiocarb and liberation of $^{14}CO_2$ were more rapid under non-flooded than flooded conditions. Small amounts of desethylbenthiocarb, 2-hydroxybenthiocarb, 4-chlorobenzoic acid, benthiocarb sulphoxide, 4-chlorobenzylmethylsulphoxide and sulphone were detected as degradation products, identical with the lipophilic metabolites in animals and plants. The benzene ring of these degradants was subsequently degraded to CO_2.

Orbencarb, a pre-emergence highly effective herbicide against weeds in upland crops, such as wheat and soy bean, degrades to 5 major metabolites as proposed in Fig. 1 (Ikeda et al., 1986a). The degradation products that contain the benzene ring of (11), partly bound to the soil particles, are decomposed finally to CO_2 by the soil microorganisms. Sterilization of the soil depressed the degradation of (11).

S-2-Chlorobenzyl-N-ethyl-N-vinylcarbamothioate (Fig. 1) and its 4-chlorobenzyl analogue are novel minor metabolites of orbencarb and benthiocarb respectively. They have recently been found in soil, plant and rat liver (Unai et al., 1986a). A possible biotransformation is N-and C-hydroxylation followed by dehydration (Fig. 1).

The persistence of bacterial degradation products of benthiocarb in different soils have been studied extensively (Ishikawa, 1981). As a result of a serious dwarfing of rice plants treated with benthiocarb in Takaska City, Japan, a very unusual soil degradative product, S-benzyl-N,N-diethylcarbamothioate (22), a reductive dechlorination product of benthiocarb, has been reported (Ishikawa et al., 1980). Repeated application of benthiocarb to the soil accelerated the dechlorination reaction (Moon and Kuwatsuka 1985a), which is correlated with soil phosphate content. It was greatly promoted by ascorbic acid and less so by ADP or ATP (Moon and Kuwatsuka, 1985b). Experiments with various antibiotics suggest that the activity is due to Gram-positive anaerobic bacteria. The reaction occurs with benthiocarb, its N,N-dipropyl and meta-chloro analogues, but the ortho-chloro analogue (orbencarb) is not affected. In continued studies, Moon and Kuwatsuka also report on population changes of benthiocarb-dechlorinating microorganisms (Moon and Kuwatsuka, 1987a) as well as nutritional conditions (Moon and Kuwatsuka, 1987b), factors influencing dechlorination (Moon and Kuwatsuka, 1985c) and properties and conditions of the soils causing this reaction (Moon and Kuwatsuka, 1984).

Recently, Unai and Tomizawa investigated the fate of fenothiocarb (23), a new type of acaricide for the control of citrus red mites (Unai and Tomizawa, 1986b), and fenothiocarb sulphoxide (Unai and Tomizawa, 1986a) in soil. Although (23) has a slightly different chemical structure from herbicidal CT, essentially similar types of degradative products have been found. These include S-oxidation, β-oxidation, ring hydroxylation, hydrolysis, S-methylation and N-demethylation reactions. The sulphoxide seems to be an important intermediate product in the metabolism of (23) and differs from its parent compound in that the quantity of the bound residues is much higher for the sulphoxide.

(22)

(23), Fenothiocarb

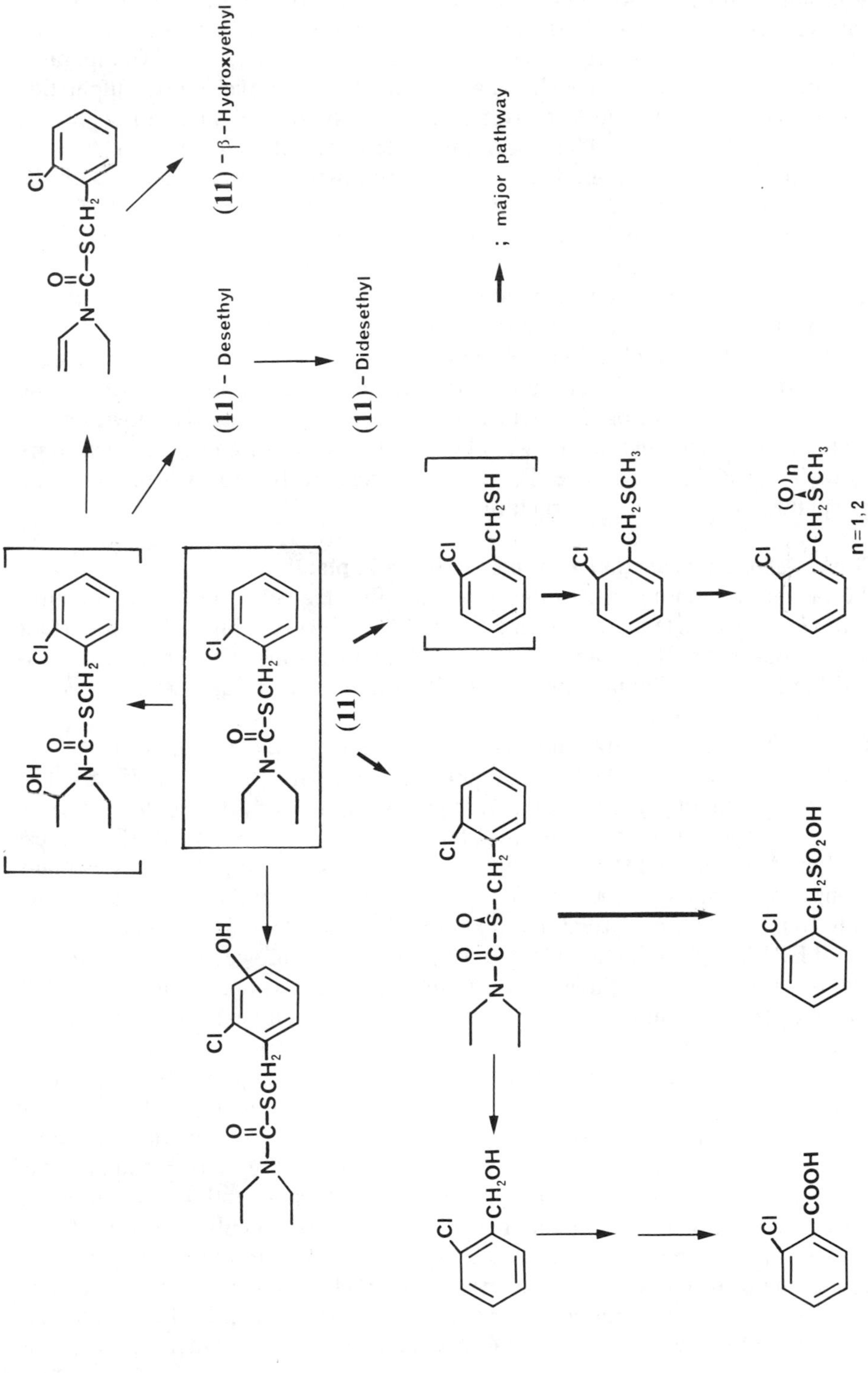

Fig. 1 — Proposed degradation pathways for orbencarb in soil (Ikeda *et al.*, 1986a).

4.2.2 Persistence of carbamothioates in soil

Factors influencing pesticide degradation in soil, such as volatility, soil moisture, soil and air temperature, depth of incorporation and wind velocity have been reviewed (Aharonson *et al.*, 1983; Obrigawitch *et al.*, 1982; Goring *et al.*, 1975). Comparable studies on the behaviour of EPTC and its major metabolic EPTC sulphoxide indicated that, when applied to moist and sandy soils, the sulphoxide lasts longer than EPTC in dry soil, or soil higher in clay content (Casida *et al.*, 1975b).

The rate of microbial degradation is influenced by previous treatment with the same or different herbicides. Obrigawitch *et al.* (1982) found that EPTC was degraded much more rapidly in the field on soil that had received eight previous successive annual applications of EPTC compared with soil with no history of CT application. Addition of the herbicide antidote R-33865 (*O, O*-diethyl-*O*-phenyl phosphorothioate) to butylate (**3**) or vernolate (**7**) extended their persistence with or without prior EPTC application (Obrigawitch *et al.*, 1983). Similar results were obtained from laboratory studies on the degradation rates of butylate, EPTC and vernolate in soil with prior CT herbicide exposure (Wilson, 1984). However, no cycloate (**4**) enhanced degradation was observed with prior three-year soil exposure with butylate, EPTC or vernolate. Therefore, soil sensitization to microbial degradation depends on the herbicide structure.

4.2.3 Metabolic degradation of carbamothioates in plants

CT herbicides are among many other pesticides that are converted to GSH conjugates *in vivo* in plants (Lamoureux and Frear, 1979). GSH conjugation is important in the metabolism of CT in plants because it frequently results in selective detoxification of the herbicide (Shimabukuro *et al.*, 1978; Lay and Casida, 1976; Lay *et al.*, 1975; Casida *et al.*, 1974).

One of the major metabolites of CT found by many researchers are CT sulphoxides (Ikeda *et al.*, 1986b,c,d; Wilkinson, 1983; Casida *et al.*, 1975a,b). The precise mechanism of sulphur oxidation in plants is not clear (Owen, 1987). Casida *et al.*, (1974), in a pioneering study of the fate of EPTC (**2**) and vernolate (**7**), suggest that sulphoxidation is one of the initial steps in biodegradation of CT under many environmental conditions. Detectable amounts of the corresponding sulphoxides are already found after 8 h of treatment. Therefore, this hypothesis on the mode of action of EPTC implies that it undergoes sulphoxidation followed by degradation or further reaction with plant thiols such as GSH and CoA (Lay and Casida, 1976). CT sulphoxides do not accumulate at high levels relative to their parent compounds and no sulphones are detected (Casida *et al.*, 1975b). Contrary to this, Horvath and Pulay (1980) report that CT sulphone, rather than the sulphoxide, is the true carbamoylating agent. The methanol extract from corn, germinated with addition of [^{14}C]EPTC, exhibited three components of which the main component was identified as the GSH conjugate of the *S*-(*N, N*-dipropylcarbamoyl) moiety. Comparison of the carbamoylating abilities of EPTC, EPTC sulphoxide and EPTC sulphone indicated that only the sulphone reacted in a 10-day incubation with GSH. Moreover, each of the ^{14}C-labelled compounds was added to corn shoot homogenates and the three mixtures were analysed after a one-day incubation period. This resulted in partial sulphoxidation of EPTC and no change with EPTC sulphoxide, while EPTC sulphone produced a metabolite similar to that found in the germination experiment. The authors, therefore, assume that oxidation of EPTC to the sulphone involves two enzyme systems, one for each oxidation state. The sulphoxide–sulphone conversion enzyme is destroyed when corn tissues are homogenized. Nevertheless, *in vitro* experiments with corn seedling homogenates showed effective carbamoylation reactions only when EPTC sulphone, rather than sulphoxide, was present.

The metabolic pathways of benthiocarb **(10)** and orbencarb **(11)** were studied in rice (Nakamura *et al.*, 1977b), soybean (Unai *et al.*, 1986a; Ikeda *et al.*, 1986c), wheat, corn and crabgrass seedling (Ikeda *et al.*, 1986b). The metabolic pattern of orbencarb in the cotton plant, which is qualitatively similar to that in soybean plant, is depicted in Fig. 2.

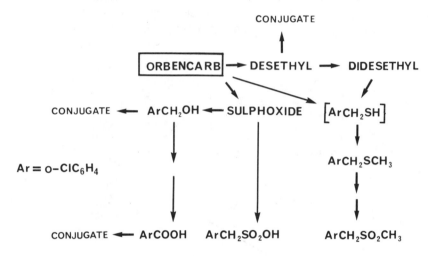

Fig. 2 — Proposed metabolic pathways of orbencarb in cotton plants (Ikeda *et al.*, 1986d).

Santi and Gozzo (1976) studied the fate of [^{14}C]drepamon **(24)**, a pre-emergence herbicide, that is applied to overspreading water in rice and barnyard grass. Aside from drepamon sulphoxide and sulphone, another major microbial oxidation product **(25)** was detected. The compound **(25)** was not isolated in the absence of soil. The sulphoxide had nonspecific herbicidal activity and phytotoxicity, remarkably higher than that of drepamon, when applied to water as the only growing medium for rice. Therefore, although CT sulphoxidation in mammalian sytems is the first detoxication stage, in plants formation of similar sulphoxides might represent increased toxicity.

(24) , Drepamon   **(25)**

As opposed to *S*-oxidation's being the major reaction detected in the metabolism of CT, the major pathway in the metabolism of fenothiocarb **(23)** in citrus seems to be

N-methyl oxidation to form N-hydroxymethyl fenothiocarb (Unai *et al.*, 1986b). Therefore, the metabolic patterns of fenothiocarb in citrus are different from those in soil (Unai and Tomizawa, 1986b) and closer to those of dimethylurea.

4.2.4   Role of glutathione in carbamothioate degradation in plants
A major metabolic pathway of CT degradation in Plants involves glutathione-S-transferase reactions. However, GSH can participate in reactions not involving these enzymes. Most GSH conjugations with CT take part only after the original substance has undergone oxidation or hydrolysis. Once formed, tripeptide conjugates are generally broken down in plant tissues.

Studies conducted by Unai *et al.* (1986c) indicated that soybean plants convert benthiocarb sulphoxide and orbencarb sulphoxide to cysteine-containing metabolites as proposed in Fig. 3. S-(2-chlorobenzyl)-L-cysteine, S-(2-chlorobenzyl)-N-

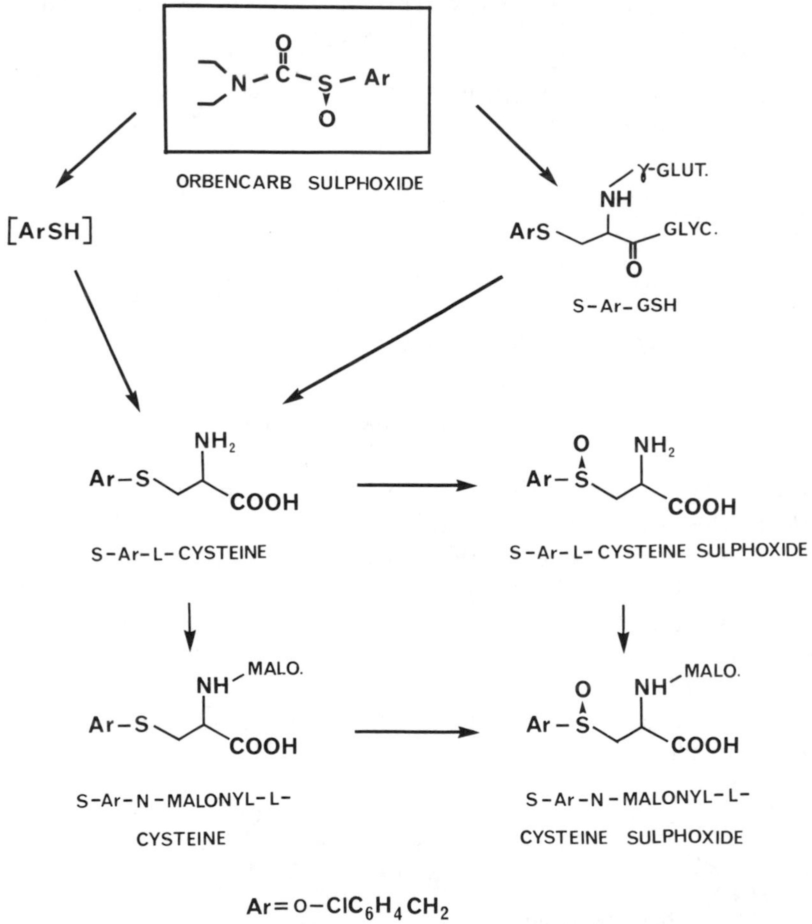

$$Ar = o - ClC_6H_4CH_2$$

Fig. 3 — Proposed metabolic fate of the 2-chlorobenzyl moiety of orbencarb sulphoxide in soybean plants, showing only conjugates (glut.=glutamic acid, glyc.=glycine, malo.=malonic acid, Ar=2-chlorobenzyl) (Unai *et al.*, 1986c).

malonyl-L-cysteine and their sulphoxides were the major metabolites from orbencarb. Previous studies indicated that CT sulphoxides were conjugated at the N,N-dialkylcarbamoyl moiety (Hubbell and Casida, 1977). The S-substituted N-malonyl cysteine conjugates (Fig. 3), also detected in the catabolism of GSH conjugates in higher plants (Lamoureux and Rusness, 1981), appear to be the end-product metabolites in the plant kingdom, in analogy to the mercapturic acid conjugates produced in the animal kingdom.

Komives *et al*., (1985) have reported that treatment of maize seedlings with EPTC, butylate, pebulate, vernolate, molinate and cycloate elevates the GSH-S-transferase levels. Treatment of maize seedlings with 2,2-dichloro-N,N-di-2-propenyl acetamide antidote, R-25788 (**26**), also increased the enzyme activity.

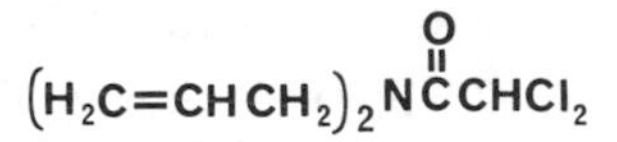

$$\left(H_2C{=}CHCH_2\right)_2 N\overset{\overset{\displaystyle O}{\|}}{C}CHCl_2$$

(**26**)

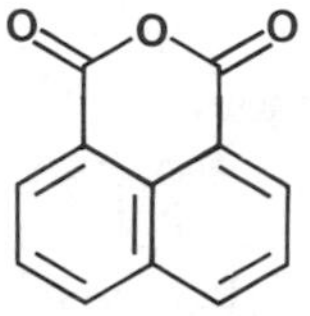

(**27**), Protect

Herbicide antidotes, such as R-25788 (**26**) (Gulf Oil, introduced in 1972) and protect-1,8-naphthalic anhydride (**27**) (Stauffer, introduced in 1973), extend the use of specific herbicides. The antidotes prevent damage to the crop plants from treatement with CT and improve selectivity, without affecting herbicidal potency. This can be achieved by coating the crop seed or by applying a mixture of the antidote with CT herbicides.

The antidotal action of R-25788 in corn is attributable to the elevation of GSH content and GSH-S-transferase activity in corn roots (Lay *et al*., 1975). Reaction of CT sulphoxide with GSH leads to rapid detoxification which prevents carbamoylation of physiologically important sites (Lay and Casida, 1978). This hypothesis is supported by studies on 32 dichloroacetamides and related compounds (Lay and Casida, 1976). Studies conducted by Stephenson *et al*. (1979) reveal that amides that are closely similar in structure to various CT herbicides are often effective antidotes to these herbicides in corn.

Other compounds that have been tested for antidotal activity are N-(benzenesulphonyl)carbamothioates (**28**) and phosphorodithioates (**29**).

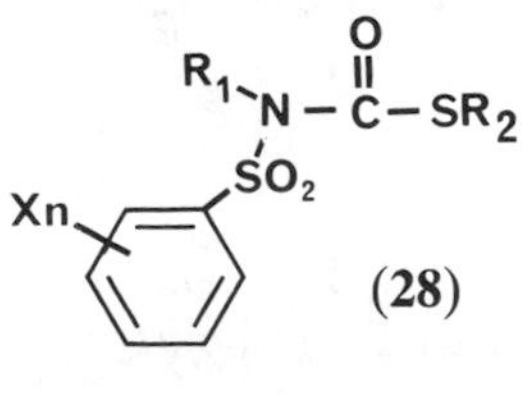

(**28**)

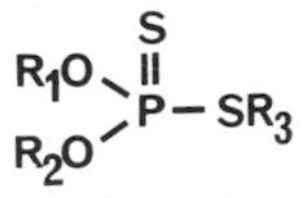

(**29**)

The compound (28) (R_1=H, R_2=C_2H_5, X=para-Cl) provided 50% protection from vernolate (7) to soybean (Gaughan and Kezerian, 1981). Compound (29) (R_1=R_2=C_2H_5, R_3=2,6-xylylcarbamoylmethyl) reduced herbicidal injury from triallate to wheat by 55% (Schafer and Czajkowski, 1983).

4.2.5 Metabolism of carbamothioates in animals

Studies suggest that the carbamoyl moiety in CT herbicides is essential for the important carbamoylation reaction in their metabolism (Unai and Tomizawa, 1986c; Hutson, 1981; Ishikawa, 1981; Lay and Casida, 1978; Hubbell and Casida, 1977; Lay and Casida, 1976; Casida et al., 1975a,b; Casida et al., 1974). Microsomally mediated S-oxygenation reactions allow GSH biotransformations, following splitting of the metabolite to fragments.

Radioactive EPTC and pebulate sulphoxides are rapidly metabolized in mice yielding the same amount of $^{14}CO_2$ as the parent CT. However, EPTC sulphone, which is also rapidly metabolized, gives very little $^{14}CO_2$. Incubation of benthiocarb, butylate, cycloate, EPTC, molinate, pebulate and vernolate with mouse liver microsomes and NADPH yields the corresponding sulphoxides with no detectable sulphone (Fig. 4). These findings indicate that the sulphoxide, rather than the sulphone, is the reactive intermediate.

CT $\xrightarrow[\text{NADPH}]{\text{microsomal enzyme}}$ CT-SULPHOXIDE $\xrightarrow[\text{GSH}]{\text{soluble enzyme}}$ **CLEAVAGE PRODUCTS**

Fig. 4 — General scheme for metabolic cleavage of CT in mammals.

4.2.5.1 In vitro studies

A detailed study conducted by Chen and Casida (1978) revealed that the microsomal oxygenase metabolism of [^{14}C]EPTC involves oxidative attack, in increasing preference, at sulphur, α-C of S-C_2H_5, α-, β- and γ-C of N-C_3H_7 groups and β-C of S-C_2H_5. The metabolites hydroxylated at the N-α-C further decompose to yield N-despropyl EPTC, and carbonylsulphide, dipropylamine and acetaldehyde from S-α-C hydroxylation. [^{14}C]EPTC sulphoxide is further oxidized to the sulphone (Fig. 5).

Sulphoxidation is the major microsomal oxygenase (MO) mediated metabolite of EPTC. This MO system lacks GSH and GSH-S-transferase necessary for EPTC sulphoxide detoxification. Although the sulphoxide undergoes MO metabolism to the sulphone, this is not likely to occur in vivo because of the GSH rapid sulphoxide detoxification.

4.2.5.2 In vivo studies

Early metabolic studies that identified radiolabelled metabolites of S-[^{14}C]methylene CT were as follows: CO_2 and urea from EPTC in living rats (Ong and Fang,

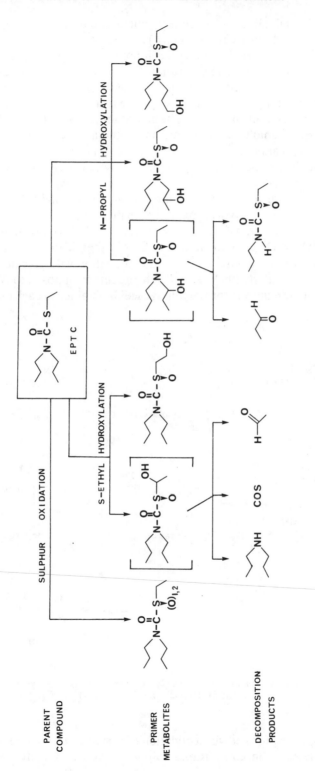

Fig. 5 — Metabolic pathways for EPTC in mouse liver microsome–NADPH system (Chen and Casida, 1978).

1970); CO_2, urea and amino acids from pebulate in living rats (Fang *et al.*, 1964); *N*-desethylbenthiocarb, bis(*p*-chlorobenzyl)sulphide or disulphide, *p*-chlorobenzylalcohol, *p*-chlorobenzoic acid and *p*-chlorohippuric acid from benthiocarb in living mice (Ishikawa *et al.*, 1973). The expiration of labelled [^{14}C]EPTC and [^{14}C]pebulate and their sulphoxides accounts for 35–40% of the administered dose of 100 mg/ kg orally in mice but only 6% for EPTC sulphone (Casida *et al.*, 1975a). Among the identified metabolites, ethyl methyl sulphone was the major one. The corresponding sulphoxides of EPTC, molinate, pebulate and vernolate are detected in the liver of mice 10 min after treatment, but they are transient, not being identifiable after 60 min. The sulphoxides of benthiocarb, butylate and cycloate are not detected at any time in the liver of CT-treated mice.

In a comprehensive work, Hubbell and Casida (1977) established that metabolism of EPTC and pebulate involves oxidation to their sulphoxides, conjugation with GSH assisted by GSH-*S*-transferase, cleavage of the GSH conjugate to the cysteine derivative, acetylation to the mercapturic acid or further degradation to the mercaptoacetic acid, a portion of which undergoes conjugation with glycine (Fig. 6). The metabolic pathways differ with *N*-*n*-alkyl substituents as opposed to *N*-branched or *N*-cyclic substituents mainly in the percentage yields obtained in each metabolic step.

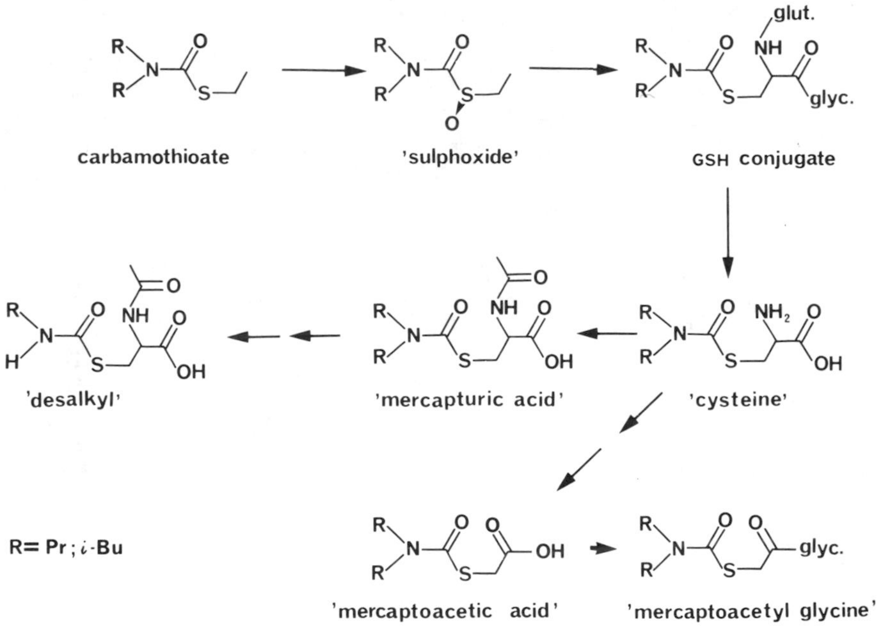

Fig. 6 — Partial metabolic pathway for EPTC(**2**) and butylate(**3**). (glyc.=glycine, glut.=glutamic acid) (Hubbell and Casida, 1977).

Mercapturic acids, formed via acetylation of cysteine-conjugated metabolites in rats, are not detected in corn. Rapid sulphoxidation of CT, followed by GSH conjugation in rats, is a significant detoxification pathway. In plants this process might represent an activation pathway (Santi and Gozzo, 1976).

Recently, Unai *et al.* (1986a) detected the new unexpected metabolites *S*-2-chlorobenzyl- and *S*-4-chlorobenzyl-*N*-ethyl-*N*-vinylcarbamothioates from orbencarb and benthiocarb respectively. These minor metabolites are found in rat liver microsomes, in soils under upland conditions and shoots of soybean plant.

Fenothiocarb (23) (an acaricide) penetrates rapidly in mites. Surprisingly, the primary site of fenothiocarb metabolism in citrus red mites seems to be oxidation at the *N*-methyl moiety, rather than *S*-oxidation or ring hydroxylation. However, metabolites identified in mites were essentially similar to those detected in other animals (Unai and Tomizawa, 1986c).

Molinate (5) has been found to undergo analogous sulphoxidation followed by GSH conjugation and cleavage reactions in rats (De Baum *et al.*, 1978) and in fish (Lay and Menn, 1979). Metabolites of benthiocarb extracted from stock water of carp, after 96 h of treatment, were identified as desethylbenthiocarb,4-chlorobenzyl sulphoxide and the sulphone. The major metabolite was the sulphoxide that was excreted into the water and did not bioaccumulate in the fish (Ishikawa, 1981, and references cited therein).

4.2.6 Photodegradation of carbamothioate herbicides

Photodegradation directly affects the fate of herbicides in the environment. The photodecomposition of pesticides in an aquatic environment occurs in both direct and indirect photochemical processes. The direct photochemical transformation in pure water proceeds by absorbtion of radiation directly. In indirect photolysis, a chromophore is needed to facilitate transfer of absorbed energy or to generate an oxidizing species.

Laboratory experiments indicated that dilute aqueous solutions of molinate (5) (Soderquist *et al.*, 1977) and solid molinate (Casida *et al.*, 1975a) are stable to sunlight. However, tryptophan, a naturally occurring photosensitizer in field water, promoted the photodecomposition of molinate in water with a half life of ≈10 days (Soderquist *et al.*, 1977). Examination by GLC and TLC indicated products shown in Fig. 7, which are formed by a combination of sensitized photolysis and dark hydrolysis.

Major products from tryptophan-sensitized photolysis of molinate are 2-oxo-hexahydroazepine (30), molinate sulphoxide and hexahydroazepine (Fig. 7). Whereas the sulphoxide and the sulphone were readily produced on *m*-chloro-peroxybenzoic acid (MCPBA) or hypochloride oxidation, ring oxidation to give (30) apparently requires radical mediation. Caprolactam was only detected in minor quantities and is probably formed in part by hydrolysis of (30). Draper and Crosby (1981) investigated the relative importance of singlet molecular oxygen and hydroxyl radical in indirect photooxidation of benthiocarb in water. Benthiocarb photodecomposes slowly to the sulphoxide in aqueous solutions exposed to sunlight, but rapidly in the presence of hydrogen peroxide, with a half-life of 41 h. On a preparative scale, four extractable metabolites have been isolated: benthiocarb sulphoxide, desethylbenthiocarb, and two derivatives with ring hydroxylation at

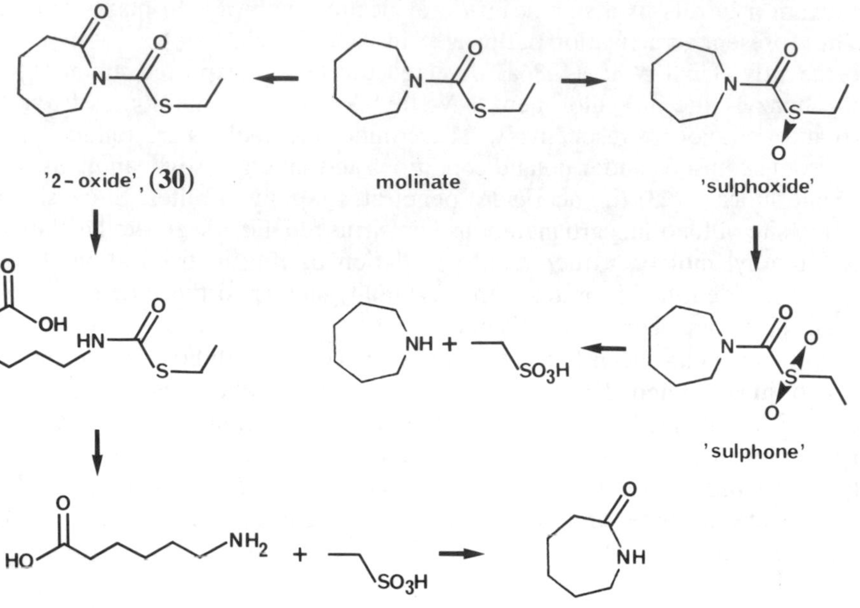

Fig. 7 — Proposed degradation pathway for the tryptophan–sensitized photolysis of molinate
(Soderquist *et al.*, 1977).

positions 2 and 3. The latter are identified by mass spectrometry by the characteristic hydroxychlorotropylium ion with $m/z=141$. The phenolic products of benthiocarb result from hydroxyl radical attack at the aromatic nucleus, which seems to be predominant over formation of the desethyl analogue.

Ishikawa *et al.* (1977) reported that about 30 degradation products are detected on aqueous benthiocarb exposure to a low pressure mercury lamp. Some of these are benthiocarb sulphoxide, desethyl benthiocarb, the 4-chloro analogues of benzyl alcohol, benzaldehyde and benzoic acid, and the analogues of benzyl alcohol, benzaldehyde and benzoic acid, and the analogous derivatives in which the 4-Cl is substituted with 4-OH.

Draper and Crosby (1984), on further investigation of the photooxidation of molinate and benthiocarb, found that traces of H_2O_2 significantly increased the photolysis rate of each pesticide in sunlight. Thus, 97–100% of molinate and benthiocarb remained on direct photolysis after 245 h, but 46% and 15% respectively, on photolysis in solutions containing trace amounts of H_2O_2. Evidence was obtained for radical addition, hydrogen abstraction and electron transfer (processes associated with hydroxyl radical reactions in water) in the photooxidation products determined for molinate and benthiocarb (Fig. 8). Hydrogens in the N-α position of molinate are readily abstracted, giving a high yield of the 2-oxoderivative (**30**). Formation of compound (**30**) involves combination of a carbon-centred radical with ground state diradical oxygen, to give peroxy radical intermediates. Hydrogen abstraction probably initiates a process in the formation of desethyl benthiocarb whereas electron transfer is proposed to initiate sulphur oxidation in molinate and benthiocarb. Sulphoxide formation may result from radical coupling with dissolved oxygen. The photo-induced reactions of peroxide occur at oxidant concentrations equivalent to those found in the sunlight-irradiated natural water.

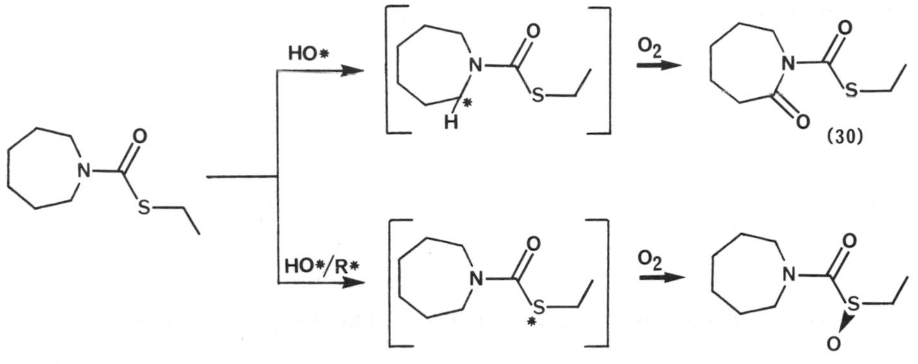

Fig. 8 — Proposed reaction scheme for the free radical photooxidation of molinate (Draper and Crosby, 1984). Reprinted with permission from *J. Agr. Food Chem.* Copyright (1984) American Chemical Society.

Ruzo and Casida (1985) studied extensively the oxidative and free radical processes associated with the photodecomposition of benthiocarb, diallate, triallate and the CDT sulphallate. Products were mostly formed by oxidation reactions but also by abstraction and recombination processes. Formation of the *O*-(4-chlorobenzyl) ester (the carbamothiono analogue of benthiocarb), previously not reported, might be a result of radical dissociation and recombination. From the many photodegradation studies on benthiocarb, it seems that this herbicide is highly degraded on exposure to sunlight compared with its degradation in soil, an important factor affecting its dissipation rate from paddy fields.

Irradiation of *cis*-diallate results primarily in isomerization to *trans*-diallate. The rate is increased 10-fold in the presence of the sensitizer benzophenone. Diallate also undergoes oxidation and radical reactions at both the *N*-alkyl and *S*-dichloroallyl substituents (Fig. 9).

Photooxidation of diallate, triallate and sulphallate degrades these compounds to mutagenic 2-chloroacrolein derivatives (see subsection on mutagenic activity of CT). The highest mutagenic activity was found with sulphallate.

Photodegradation of fenothiocarb (**23**) on silica gel plates exposed to sunlight (Unai and Tomizawa, 1986d), gave *N*-formylfenothiocarb, desmethylfenothiocarb, bis(4-phenoxybutyl)thiosulphinate and thiosulphonate, fenothiocarb sulphoxide and 4-phenoxybutylsulphonic acid, among the 12 compounds detected. Primary and secondary photochemical reactions seem to be sulphoxide formation and oxidation of the *N*-methyl moiety.

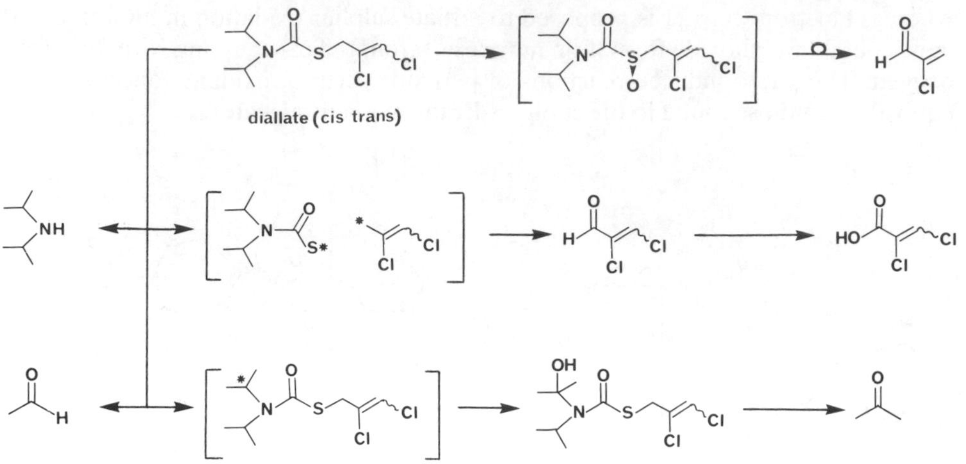

diallate (cis trans)

Fig. 9 — Photoreactions of diallate in oxygenated chloroform or water solutions (Ruzo and Casida, 1985).

4.2.7   Selected syntheses of carbamothioate metabolites

Although physical techniques, such as NMR, IR, UV, etc., are useful tools for identification and structure elucidation of compounds, comparison of unknowns with authentic, independently synthesized derivatives, is unequivocally a final proof to their structure or their presence in a mixture.

CT sulphoxides and sulphones are readily prepared on equimolar or excess MCPBA oxidation of the parent CT in chloroform or methylenechloride solution (Casida *et al.*, 1974; Schuphan *et al.*, 1981). Soderquist *et al.* (1977) indicate that sodium hypochlorite, rather than MCPBA, gave satisfactory results on attempted preparation of molinate sulphoxide.

Hubbell and Casida (1977) provide synthetic routes for products associated with the metabolism of EPTC and butylate in rats and corn (Fig. 6). These general procedures are also useful for the synthesis of analogous CT metabolites.

CT sulphones were utilized to prepare most of the CT derivatives. The sulphones are more stable on storage and much more reactive with thiols than the corresponding sulphoxides. Morover, N,N-dialkylcarbamoyl chlorides did not react under comparable conditions.

S-(N,N-dipropylcarbamoyl) GSH (Fig. 6), an initial GSH cleavage product of EPTC sulphoxide, is synthesized from sulphone (**31**) according to

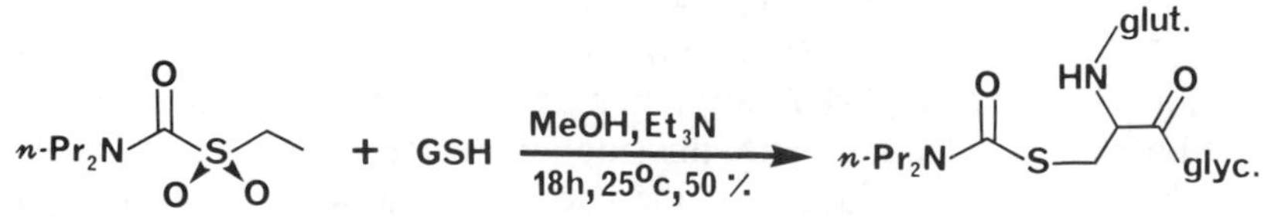

$$n\text{-Pr}_2\text{N} \quad + \quad \text{GSH} \xrightarrow[18\text{h},\,25^\circ\text{c},\,50\,\%]{\text{MeOH},\text{Et}_3\text{N}} \quad n\text{-Pr}_2\text{N} \qquad\qquad (10)$$

(**31**)

glut=glutamic acid; glyc=glycine

The cysteine analogue was synthesized according to

$$(31) \quad + \quad \underset{\text{HS}}{\overset{\text{H}_2\text{N}}{\bigvee}}\text{OH} \quad \xrightarrow[\text{24h, 25°c, 50\%}]{\text{MeOH, Et}_3\text{N,N}_2} \quad n\text{-Pr}_2\text{N}\overset{\text{O}}{\bigvee}\text{S}\overset{\text{H}_2\text{N}}{\bigvee}\overset{\text{O}}{}\text{OH} \qquad (11)$$

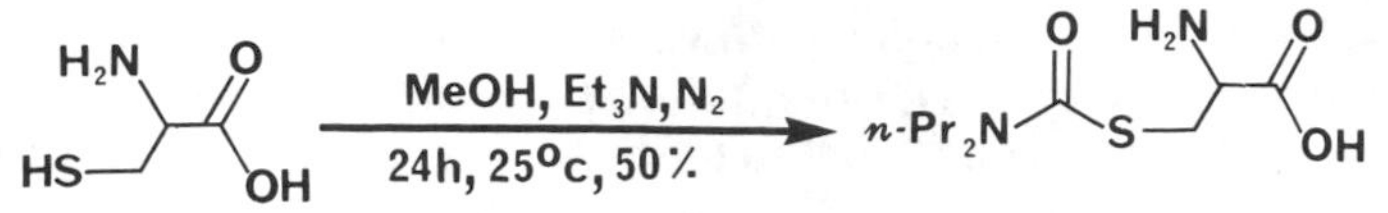

$S–(N$-Propylcarbamoyl)cysteine was prepared from propyl isocyanate and cysteine:

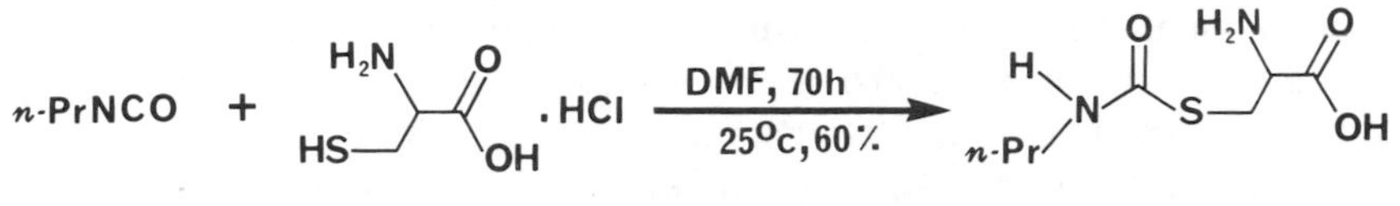

$$n\text{-PrNCO} \quad + \quad \underset{\text{HS}}{\overset{\text{H}_2\text{N}}{\bigvee}}\text{OH} \cdot \text{HCl} \quad \xrightarrow[\text{25°c, 60\%}]{\text{DMF, 70h}} \quad \underset{n\text{-Pr}}{\overset{\text{H}}{\bigvee}}\text{N}\overset{\text{O}}{\bigvee}\text{S}\overset{\text{H}_2\text{N}}{\bigvee}\overset{\text{O}}{}\text{OH} \qquad (12)$$

Mercapturic acid derivatives of EPTC and butylate were obtained on reacting the appropriate sulphones with N-acetylcysteine. β-hydroxyethyl-EPTC, a microsomal oxygenase metabolite of EPTC (Fig. 5), is obtained using β-mercaptoethanol (Chen and Casida, 1978):

$$(31) \quad + \quad \text{HS}\diagup\diagdown\text{OH} \quad \xrightarrow[\text{15h, 25°c, 45\%}]{\text{MeOH, Et}_3\text{N}} \quad n\text{-Pr}_2\text{N}\overset{\text{O}}{\bigvee}\text{S}\diagdown\diagup\text{OH} \qquad (13)$$

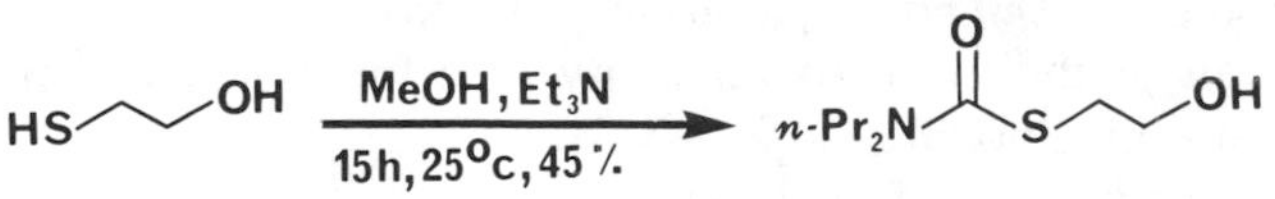

The novel metabolite N-vinyl-benthiocarb, from benthiocarb metabolism in soil, plant and rat liver microsomes (Fig. 1), is prepared by the following series of reactions (Unai *et al.*, 1986a):

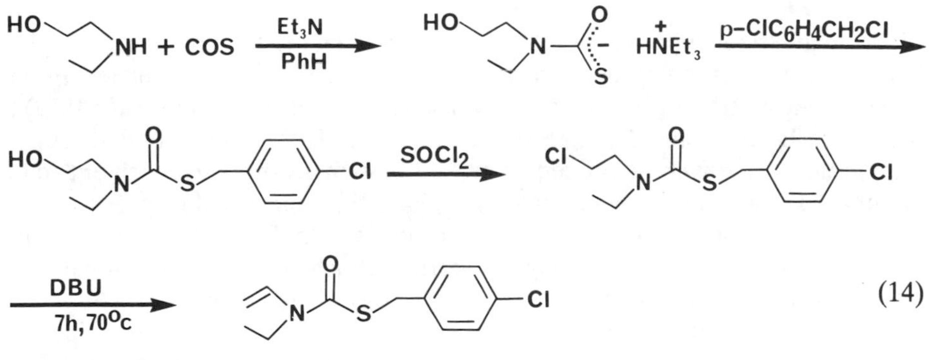

$$\text{HO}\diagdown\diagup\text{NH} + \text{COS} \quad \xrightarrow[\text{PhH}]{\text{Et}_3\text{N}} \quad \text{HO}\diagdown\diagup\text{N}\overset{\text{O}}{\overset{\parallel}{\text{C}}}\overset{-}{\text{S}} \quad \overset{+}{\text{HNEt}_3} \quad \xrightarrow{\text{p-ClC}_6\text{H}_4\text{CH}_2\text{Cl}}$$

$$\text{HO}\diagdown\diagup\text{N}\overset{\text{O}}{\bigvee}\text{S}\diagdown\text{C}_6\text{H}_4\text{-Cl} \quad \xrightarrow{\text{SOCl}_2} \quad \text{Cl}\diagdown\diagup\text{N}\overset{\text{O}}{\bigvee}\text{S}\diagdown\text{C}_6\text{H}_4\text{-Cl}$$

$$\xrightarrow[\text{7h, 70°c}]{\text{DBU}} \quad \diagup\diagdown\text{N}\overset{\text{O}}{\bigvee}\text{S}\diagdown\text{C}_6\text{H}_4\text{-Cl} \qquad (14)$$

4.3 TOXICOLOGICAL SIGNIFICANCE OF BIOMIMETIC SULPHOXIDATION STUDIES WITH CARBAMOTHIOATE AND RELATED COMPOUNDS

Sulphur-containing functional groups are present in about one-third of all organic pesticides (Schuphan and Casida, 1983). Biooxidation, an essential activation process for sulphur-containing pesticides, contributes importantly to their potency, selective toxicity, metabolism, type of reaction at target site and mode of action (section 4.2). Major metabolic pathways with CT involve sulphoxidation and S-methylene hydroxylation (Distlerath *et al.*, 1985; Schuphan *et al.*, 1981; Rosen *et al.*, 1980a; Schuphan *et al.*, 1979; Hubbell and Casida, 1977; Casida *et al.*,1974, 1975a). With sulphallate, only the latter mechanism is applicable (Rosen *et al.*, 1980a). The herbicidal activity of CT sulphoxides is probably a result of carbamoylation of critical thiols. In mammals, however, this increased reactivity may lead to a reduced toxicity.

The mechanisms and products involved in biological sulphoxidations are often similar to those of organic peracid oxidations, such as with MCPBA. Many reactive intermediates formed on epoxidation and *N*- and *S*-oxidation reactions proved difficult to study without the use of biomimetic models. Such models have several advantages:

(a) isolation in relatively large quantities of reactive intermediates and ultimate products;
(b) monitoring reaction propagation using multiple spectroscopic methods (NMR,UV,IR,MS);
(c) performing reactions in non-aqueous media;
(d) determination of the order of reactivity of a functional group in pesticides containing multifunctional groups;
(e) reduction of the number of reagents involved;
(f) working at low or high temperatures;
(g) ideal for reagent quantification;
(h) usually, low cost and non-time-consuming experiments.

Selected examples of biomimetic models, considering these advantages, are summarized below.

In many cases, the results obtained from biomimetic experiments might be helpful in understanding mechanisms and products involved in related research areas. Using NMR and IR spectroscopy, it was revealed (Schuphan *et al.*, 1979) that the thermally unstable sulphoxide obtained on MCPBA oxidation of diallate at low temperature, is converted by a sequence of [2,3] sigmatropic rearrangement and 1,2-elimination reactions to the ultimate mutagen 2-chloropropenal (Fig. 10).

A similar spontaneous rearrangement occurs with *S*-(2-chloropropenyl)-*N,N*-diethylcarbamothiate but, lacking a chlorine attached to the *S*-terminal C, the ultimate product is the sulphenate ester (in analogy to that in Fig. 10) (Schuphan and Casida, 1979b; Schuphan *et al.*, 1979).

As a result of these findings, a series of halopropenals were tested for their mutagenic activity (Segall *et al.*, 1985; Rosen *et al.*, 1980b). It was established that 2-bromopropenal is a powerful direct mutagen. That led Marsden and Casida (1982) to assume that an identical ultimate mutagen might be involved with the carcinogenic nematocide DBCP and the flame retardant tris-BP. Indeed, urinary bromopropenoic acids were excreted from rats treated with these carcinogens. The results support the assumption that part of the metabolic pathway involves 2-bromopropenal which is further oxidized to 2-bromopropenoic acid.

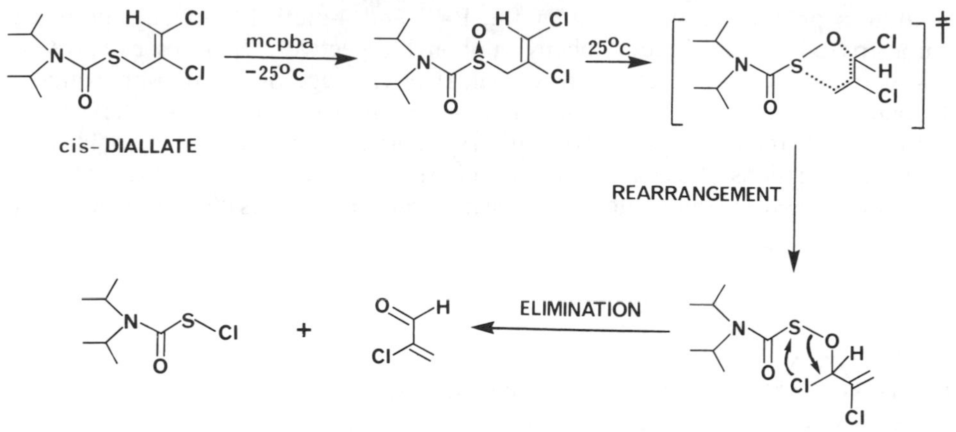

Fig. 10 — Spontaneous [2,3] sigmatropic rearrangement followed by 1,2-elimination reaction of *cis*-diallate sulphoxide yielding the potent bacterial mutagen 2-chloropropenal (Schuphan *et al.*, 1979).

The MCPBA system is also a suitable biomimetic model for sulphur- containing pesticides other than CT (Casida and Ruzo, 1986).

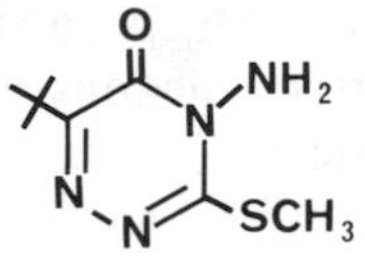

(32), Metribuzin 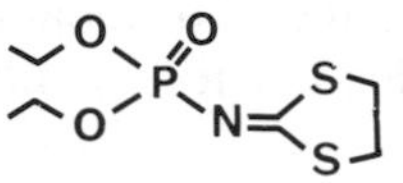   (33), Phosfolan 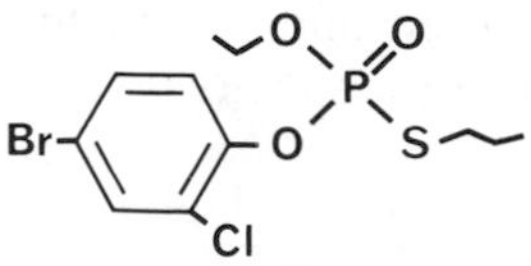   (34), Profenofos

Metribuzin (32) (a herbicide) is metabolized in rats and mice to a diketo derivative but sulphoxides were not detected, although they are likely reactive intermediates (Bleeke *et al.*, 1985). Metribuzin sulphoxide, prepared by oxidation with MCPBA, readily reacted with GSH and *N*-acetylcysteine.

Phosfolan (33) is a poor *in vitro* AChE inhibitor, yet acts as a potent inhibitor *in vivo*. A coupled MFO activation–AChE inhibition assay verified that phosfolan undergoes a bioactivation process prior to its AChE inhibition. Activation of a similar magnitude occurs on treatment of (33) with MFO system or equimolar MCPBA (Gorder *et al.*, 1985).

Several phosphorothioate and phosphorodithioate insecticides, such as profeno-fos (34) and sulprofos (35), are active on pest strains resistant to other organophos-phorus pesticides, but their acute mammalian toxicity is relatively low (Wing *et al.*, 1984; Hirashima *et al.*, 1984; Segall and Casida, 1982). These toxicants are AChE inhibitors only after conversion of the P=S and 4-methylthiophenyl groups of sulprofos to P=O and 4-methylphenylsulphonyl respectively. In the case of profeno-fos, activation occurs after extensive oxidative cleavage of the phosphorothioate linkage, and formation of an unstable inhibitor was detected in the coupled MFO–AChE assay. The peracid oxidation system was helpful in understanding these biotransformations (Casida and Ruzo, 1986). As monitored by NMR, MCPBA oxidation of (35) gives in sequence the reaction rate preferences illustrated in Fig. 11.

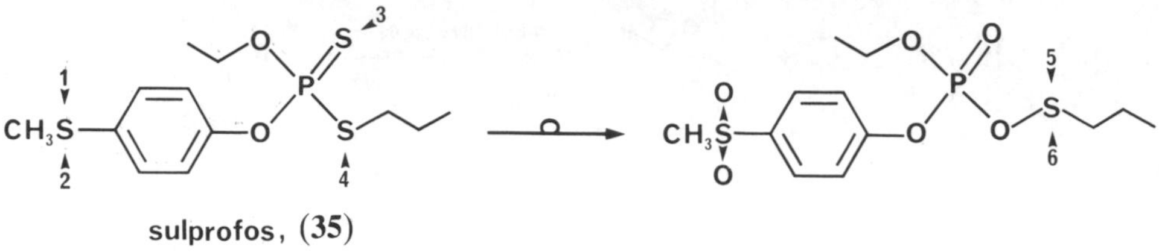

Fig. 11 — MCPBA sequential oxidation preferences for sulprofos (Segall and Casida, 1982).

A similar phosphorothioate rearrangement to phosphinyloxysulphonates was observed with profenofos and many other phosphorothioates (Segall and Casida, 1981; Segall and Casida, 1983a). Sulphoxidation of (36), a phosphorus analogue of diallate, leads exclusively to the oxysulphonate rather than products derived from rearrangement followed by elimination reactions, since no mutagenic activity was observed and no aldehyde detected. Presumably, the latter reactions are much slower than the rearrangement leading to oxysulphonate esters.

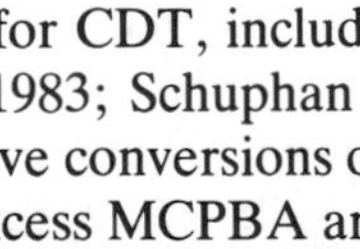

(36)

The MCPBA peracid system, a suitable model for activation of CT and many xenobiotics, is not appropriate for CDT, including sulphallate (Segall and Casida, 1983b; Schuphan and Casida, 1983; Schuphan et al., 1981; Rosen et al., 1980a). Possible mechanisms for oxidative conversions of alkyl dialkylcarbamodithioates in the presence of equivalent or excess MCPBA are depicted in Fig. 12.

NMR monitoring revealed that CDT, including sulphallate, react quickly with equivalent MCPBA to form sulphine analogues (Fig. 12A), as illustrated with S-methyl-N,N- diethylcarbamothioate (37), resulting in an upfield shift of S-α-C protons (Fig. 13A). On attempted product isolation, only unmodified CDT was obtained. Thus, sulphoxidation is exclusively preferred at the thion rather than thiol sulphur, in which case there would be a downfield shift of the same protons.

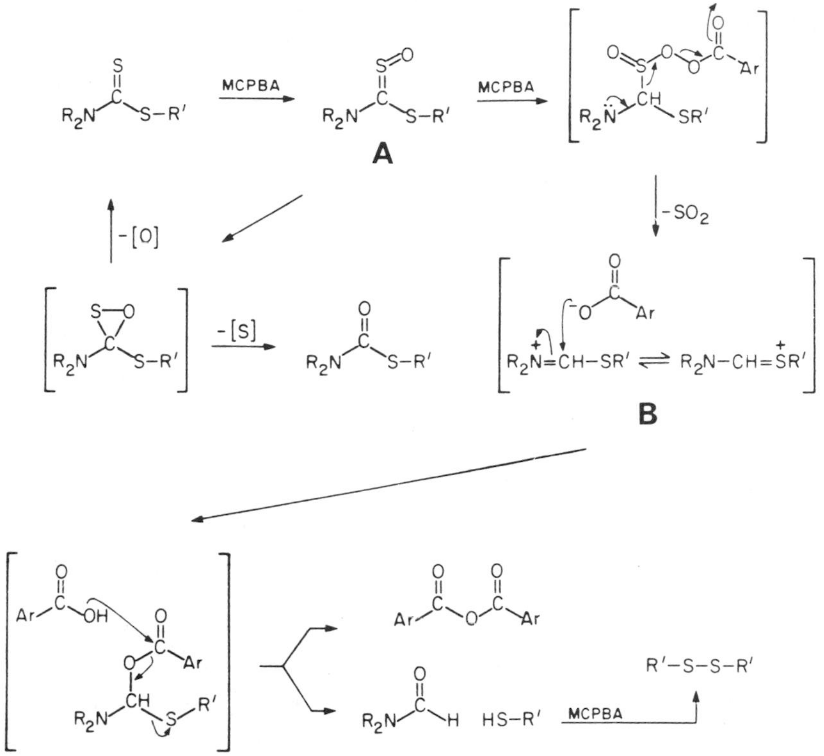

Fig. 12 — Possible mechanisms for oxidative conversion of sulphallate and dialkylcarbamodith-ioates to sulphine derivatives and further to dialkylformamides and dialkyldisulphides via trialkyliminium salt (Segall and Casida, 1983b).

Reaction of sulphallate with excess MCPBA and product isolation (Fig. 12) gave two major products, 2-chloropropenyl disulphide and diethylformamide. NMR spectra of the reaction mixture (Fig. 13) were not appropriate for those products ultimately isolated. A possible mechanism, accommodating the NMR spectral features of the intermediates, involves the iminium ion B (Figs. 12,13). The intermediate B has a very low field resonating proton. That this proton originates from MCPBA is verified on oxidation of sulphallate with MCPBA-d(ArCO$_3$-D). In this case, diethylformamide-d is isolated and a low field resonating proton is absent.

Monoequivalent MCPBA oxidation of sulphallate retains its mutagenic activity in the S-9 activated *Salmonella* mutagenicity test as if sulphallate alone was introduced. This result is in accordance with the chemical findings, since the sulphine readily deoxygenates to the starting material. However, on oxidation of sulphallate with excess MCPBA, there is a complete loss of mutagenic activity due to extensive breakdown to non-active fragments. Therefore, excess MCPBA oxidation is not a suitable biomimetic system for understanding the activation of sulphallate to a potent mutagen.

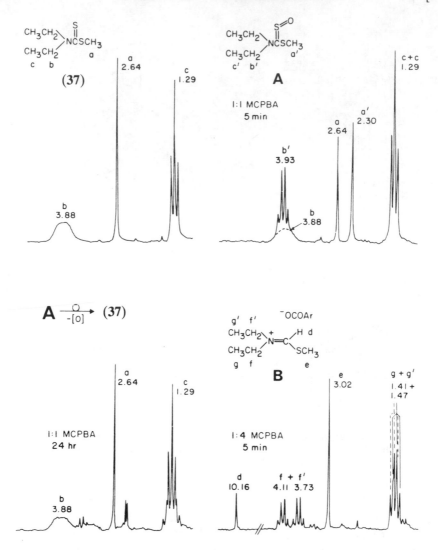

Fig. 13 — ¹H NMR spectra for **(37)**, for sulphine derivative A from oxidation of **(37)** with equivalent MCPBA, for reversion of A to **(37)** and for iminium ion intermediate B from oxidation with 4 equivalents MCPBA (Segall and Casida, 1983b).

4.4   ACUTE TOXICITY AND CHRONIC EFFECTS RELATED TO CARBAMOTHIOATES AND CARBAMODITHIOATES

Acute toxic effects normally appear rapidly and are readily quantifiable. The information obtained from acute toxicity studies is vital for the user in handling pesticides safely. In many cases, either through food or by impact, man is exposed to repeated subacute doses of pesticides that, in the long term, may produce chronic hazardous effects. Sometimes, chronic effects may appear after years of exposure.

It is difficult to predict a certain level of pesticide below which no chronic effects occur. In practice, a threshold daily dose (mg/kg/day) is determined from chronic studies in animals; this figure is then divided by a safety factor and used as a practical guideline for man.

4.4.1 Acute toxicity studies

An empirical scale was proposed for the assessment of toxicity levels of xenobiotics. Thus, a substance is extremely toxic with LD_{50} less than 1 mg/kg, highly toxic with LD_{50} between 1–50 mg/kg, moderately with 50–500 mg/kg and slightly toxic with 500–5000 mg/kg.

In recent years, acute toxicity data for many pesticides, including CT and CDT, have been provided (Gaines and Linder, 1986; Unai and Tomizawa, 1986c; Kochansky and Wright, 1985; Schafer and Bowles, 1985; Leeuwen et al., 1985a; Fisher and Metcalf, 1983; Schafer et al., 1983; Ted, 1983; Sanders and Hunn, 1982). Acute toxicity data on selected CT and CDT is given in Table 2.

Table 2 — LD_{50}, acute, oral, dermal and LC_{50} for selected carbamothioates and carbamodithioates

Pesticide	LD_{50}, oral, acute (mg/kg)	LD_{50}, percutaneous (mg/kg)	LC_{50}, 96 h (mg/l)
EPTC	2550, male albino rat (A) 1630, rat (B) 100, bird (F)	>5000, rabbit (A)	19, rainbow trout (A) 27, bluegill (A)
Butylate	3500, male albino rat (A) 4660, rat (B)	>5000, rabbit (A)	
Cycloate	3600, rat (B) 2710, rat (A) 1838, male weanling rat (E)	>4640, rabbit (A) 2467, male weanling rat (E)	4.5, rainbow trout (A)
Molinate	500–720, rat (B) 450, female rat (A) 369, male rat (A)	>4640, rabbit (A)	1.3, rainbow trout (A) 30, goldfish (A)
Pebulate	1120, rat (B) 100, bird (F)	4640, rabbit (F)	7.4, rainbow trout (A)
Vernolate	1780, rat (B) 100, bird (F)	>5000, rabbit (A)	
Drepamon	>10 000, rat (B) 8000, mice (A)	1200, no mortality (A)	≥8, fish (A) >60, mollusc (A)

Table 2 (continued)

Pesticide	LD$_{50}$, oral, acute (mg/kg)	LD$_{50}$, percutaneous (mg/kg)	LC$_{50}$, 96 h (mg/l)
Diallate	395, rat (B)	2000–2500, rabbit (A)	7.9, rainbow trout (A)
	684, male weanling rat (E)	2175, male weanling rat (E)	5.9, bluegill (A)
	510, dog (A)		7.2, culex larvae (G)
	37.2, *Musca domestica* (G)[a]		7,9 rainbow trout (A)
Triallate	1675–2165, rat (B)	8200, rabbit (A)	1.2, rainbow trout (A)
	>2251, bobwhite quail (A)		1.3, bluegill (A)
	68.1, *Musca domestica* (G)[a]		46.4, culex larvae (G)
Benthiocarb	1300, rat (B)	2900, rat (A)	1.2, rainbow trout (C)
	560, mice (A)		3.6, carp, 48 h (A)
			0.47, crawfish (N)
			0.33, estaurine mysid (D)
Sulphallate	850, rat (B)	2200–2800, rabbit (P)	6.5, culex larvae (G)
	10.8, *Musca domestica* (G)[a]		
Ziram	1400, rat (B)	>2000, rat (P)	0.5, Carassius auratus, 48 h (P)
	1070, deer mouse (H)		
	100, bird (F)		
Zineb	>5200, rat (A)	>10 000, rat (P)	
	1000, male rat (M)		
	100, bird (F)		
Thiram	780, rat (B)	>1000, rat (P)	4, carp, 48 h (A)
	210, rabbit (A)		0.13, trout, 48 h (A)
	1500–2000, mice (A)		
	200, deer mouse (M)		
	300, bird (F)		

[a]Applied to the pronota of the flies.
(A) Worthing and Walker (1987); (B), Büchel (1983), p. 348; (C) Sanders and Hunn (1982); (D) Schimmel *et al.* (1983); (E) Gaines and Linder (1986); (F) Schafer *et al.* (1983); (G) Fisher and Metcalf (1983); (H) Schafer and Bowles (1985); (M) Hassall (1982), p. 11; (N) Ted (1983). (P) Hartley and Kidd (1985).

The data from Table 2 suggest that CT and CDT are very slightly toxic to mammals by oral or dermal application but moderately toxic to birds. Triallate is of low toxicity to flies (*Musca domestica*), yet diallate and sulphallate are moderately toxic. The LC_{50} values are used as a basis for the determination of a threshold dose of the pesticide permitted in the aquatic environment.

Sanders and Hunn (1982) studied the acute toxicity and bioconcentration of benthiocarb herbicide in species of aquatic invertebrates and freshwater fish. Benthiocarb is moderately toxic, the LC_{50} ranging between 1 and 6.5 mg/l. Residues of [^{14}C]benthiocarb that accumulated in the organisms were rapidly eliminated when the species were placed in fresh water. Therefore, the use of benthiocarb will probably result in little environmental hazard to freshwater organisms.

Gaines and Linder (1986) report oral LD_{50} values for cycloate and diallate (Table 2) with a male:female ratio of 1.1 and 0.9 respectively.

EPTC, EPTC sulphoxide and EPTC sulphone are more toxic to i.p.-treated mice than those in the butylate series (Hubbell and Casida, 1977). The sulphones are the most toxic derivatives (Table 3).

Table 3 — Mouse intraperitoneal LD_{50} values for EPTC, butylate and their metabolites (Hubbell and Casida, 1977)

Compound	LD_{50} (mg/kg)
EPTC	>500
EPTC sulphoxide	325
EPTC sulphone	130
Butylate	>500
Butylate sulphoxide	440
Butylate sulphone	200

4.4.2 Chronic effects on exposure to carbamothioates and carbamodithioates

4.4.2.1 Mutagenicity and carcinogenicity

Cancer studies revealed that diallate is an active carcinogen in mice (Innes *et al.*, 1969) and sulphallate in mice and rats (National Cancer Institute, 1978).

Sikka and Florczyk (1978) and De Lorenzo *et al.* (1978) independently indicated that the pre-emergence herbicides diallate, triallate and sulphallate induced mutation in TA-100 and TA-1535 strains of *Salmonella typhimurium* only in the presence of liver microsomal preparation S-9 (Ames *et al.*, 1975), i.e. they require metabolic activation for their conversion to active mutagens. McCann and Ames (1976) found that most carcinogens are also mutagens.

The mutagenic activity of these herbicides (Table 4) is primarily attributed to the presence of a 2-chloropropenyl group in each. The figures in Table 4 should be considered minimum values since small but appreciable percentages of the mutagenic metabolites of diallate, triallate and sulphallate escaped detection in this conventional assay (Distlerath *et al.*, 1984).

Butylate (Plewa *et al.*, 1984), EPTC, molinate and the CDT ziram (**13**), ferbam (**14**), zineb (**16**), maneb (**17**) and mancozeb, which lacks a 2-chloropropenyl group,

Table 4 — Mutagenic activity (revertants/nmol) of carbamothioate herbicides in the presence and absence of the liver microsomal S-9 fraction

	Without S-9 (A), (D)	With S-9
cis-diallate	0	40 (D)
trans-diallate	0	25 (D)
Diallate (cis–trans mix)	0	27 (C); 8 (B)
Triallate	0	13 (A); 1.5 (B)
Sulphallate	0	16 (A); 2.2 (B)
cis-diallate sulphoxide	113	68 (A)

(A) Rosen et al. (1980a); (B) De Lorenzo et al. (1978); (C) Sikka and Florczyk (1978); (D) Schuphan et al. (1979).

are not mutagenic although they still show very strong herbicidal and fungicidal activity (De Lorenzo et al., 1978). Yasuhiko et al. (1984) also found that EPTC, cycloate, molinate, zineb, maneb and mancozeb are not mutagenic; however, ferbam, ziram and thiram (15), in strain TA-100, were direct mutagens with 1.33, 1.24 and 0.84 revertants/nmol, respectively.

Zdzienicka et al. (1979) indicate that thiram induced mutations in strains TA-1535 and TA-100 without metabolic activation, but required activation for the expression of its mutagenic activity in TA-1538 and TA-98 strains. Cysteine and GSH abolished mutagenic activity of thiram in all four strains.

In contrast to studies indicating sulphallate to be an indirect mutagen, Principe et al. (1981) report that sulphallate is a direct mutagen mainly of the base-pair substitution mutants (TA-100 and TA-1535).

Schuphan et al. (1979) provide a novel activation mechanism for promutagenic diallate (see section 4.3 and Fig. 10). The instability of S-2- chloropropenyl CT sulphoxides, in contrast to S-alkyl and S-benzyl CT sulphoxides, is due to their quantitative spontaneous conversion to the sulphenate ester via a [2,3] sigmatropic rearrangement reaction. In the case of diallate, it is followed by a 1,2-elimination reaction yielding 2-chloropropenal.

Under biological conditions, diallate sulphoxide undergoes either detoxification via GSH conjugation (Fig. 6) or activation through rearrangement–elimination reactions (Fig. 10) to liberate 2-chloropropenal, revealed by mutagenesis studies to be the ultimate mutagen. Thus, diallate is not mutagenic without the S-9 fraction; however, diallate sulphoxide is a direct mutagen, with 113 revertants/nmol, the same value obtained with 2-chloropropernal (Table 4). Therefore, the herbicidal activity of diallate may be due to the sulphoxide's acting as a carbamoylating agent with critical tissue thiols or to liberation of 2-chloropropenal, similar in potency to diallate in inhibiting root growth.

Rosen et al. (1980a) provide a mechanism for the mutagenic activation of sulphallate. 2-Chloropropenal was established as a microsomal oxidation metabolite of sulphallate by trapping this aldehyde as its 2,4-dinitrophenylhydrazone. This evidence strongly indicates that 2-chloropropenal is also likely to be the ultimate mutagen formed on sulphallate metabolism. The rearrangement–elimination sequence reactions associated with diallate sulphoxide are not applicable to sulphal-late because it lacks a 3-chloro substituent necessary for the final elimination reaction (see also Fig. 12 for MCPBA oxidation of sulphallate).

A possible and highly efficient mechanism by which 2-chloropropenal might be formed is α-hydroxylation of sulphallate followed by immediate decomposition of the α-hydroxy compound to 2-chloropropenal (Fig. 14). Similar α-hydroxylation accounts for 7–17% of the overall MO metabolism of EPTC, leading to acetaldehyde and carbonyl sulphide (Fig. 5). α-Hydroxylation may also be applicable in part to the metabolic activation of diallate and triallate forming 2,3-dichloro- and 2,3,3-trichlor-opropenal respectively, each a highly potent mutagen (>100 revertants/nmol) (Rosen *et al.*, 1980b).

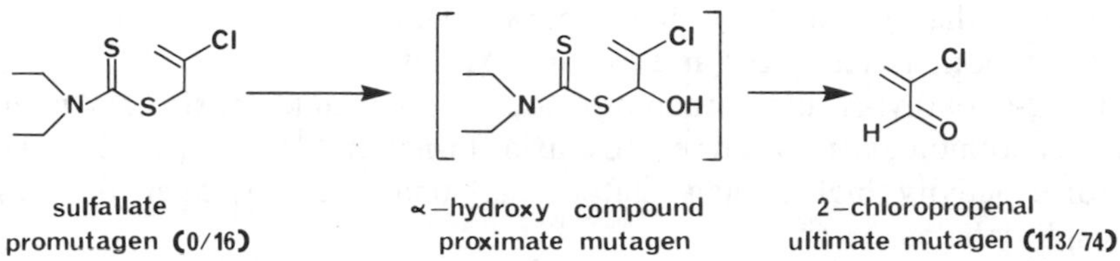

Fig. 14 — Oxidation of sulphallate indicating mutagenic activities in *Salmonella typhimurium* TA-100 assay (revertants/nmole; values without activation and with activation are given) (Rosen *et al.*, 1980b).

Diallate yields the most potent mutagens (Table 4), possibly because it gives highly active mono– and dichloropropenals, both on the preferred sulphoxidation–elimination sequence and on α-hydroxylation.

The structure–mutagenic activity relationships for compounds related to 2-chloropropenals reflect the need for two reactive functional groups, such as a double bond and a carbonyl group, to facilitate cross-linking with DNA (Rosen *et al.*, 1980b). Chemicals that are metabolized to 2,3-dihalopropanals and butanals, or 2-halopropanals and butanals, should be considered as candidate promutagens (Segall *et al.*, 1985). Marsden and Casida (1982) demonstrated that rats treated orally with several promutagens and carcinogens containing halopropyl or halopro-penyl substituents excrete small amounts of urinary 2-halpropenoic acids. Thus, 2-bromopropenoic acid was excreted after treatment with DBCP (a nematocide) and the flame retardant tris-BP (see section 4.3), 2-chloropropenoic acid after treatment with sulphallate, diallate and triallate, 2,3-dichloropropenoic acid after treatment with diallate and trichloropropenoic acid after treatment with triallate. Most of these conversions involve halo- or dihalopropenal intermediates.

Dihalopropanal formation may be most efficiently accomplished by initial hydroxylation at $-CH_2Cl$, $-CH_2Br$ or $-CH_2OP$, following direct aldehyde release. They undergo rapid physiological dehydrohalogenation to halopropenals, presum-ably ultimate mutagens and carcinogens.

Distlerath *et al*. (1985) also propose α-hydroxylation of 3-(2-chloroethoxy)-1,2-dichloropropene (CP) (a *Salmonella* promutagen isolated from the residual organics of drinking water) at both sides of the ether linkage to explain the production of multiple mutagenic products from metabolic activation.

4.4.2.2   *Delayed neurotoxicity*
Fisher and Metcalf (1983) and Hansen *et al*. (1985, and references cited therein) support reports that diallate and triallate are delayed neurotoxicants.

Diallate, given orally at 312 mg/kg, consistently produced ataxia in hens. This results in leg weaknesses, a propensity to sit or stand with closed eyes and periods of convalescence. Some hens had severe ataxia, suffering drastic weight loss and muscular atrophy, and died within a few days of becoming paralysed. Triallate showed only mild ataxia and sulphallate lacked an appreciable neurotoxic activity. It is noteworthy that, of the three halopropenyl-containing herbicides, diallate is the most active, both as a mutagen and a neurotoxicant.

Some 2-naphthyl-*N*-alkylcarbamothioates showed mild ataxia, but phenyl-*N*-alkylcarbamothioates produced severe ataxia (Fisher and Metcalf, 1983). CTs show neurotoxic activity higher than that of carbamates but weaker than that of organosphosphates.

Relatively large topical doses of diallate (40 mg/kg day) and triallate (300 mg/kg day) did not show adverse effects in hens. Diallate is a cumulative neurotoxicant but apparently does not share similar mechanisms associated with organophosphorus-induced delayed neurotoxicants (Hansen, *et al*., 1985).

Chronic effects such as mutagenicity, teratogenicity and carcinogenicity observed with CDT fungicides have led some countries to restrict their use. A good deal is known about the effects of CDT in warm-blooded animals. Diethylcarbamodithioate, a major metabolite of disulfiram, decrease hepatic cytochrome P-450 concentrations in rats (Miller *et al*., 1983). Its methyl ester produces symptoms of toxicity preceding death of rats, resembling those observed with disulfiram toxicity (Fairman *et al*., 1983). Acute central nervous system changes due to intoxication by manzidan (combination of maneb and zireb) in a single human subject were observed (Israeli *et al*., 1983). These were followed by loss of consciousness, convulsions and night hemiparesis. The symptoms disappeared after a few days.

Leeuwen *et al*. (1985b,c; 1986a,b) carried out a few studies on cold-blooded animals. It was established that some dialkylcarbamodithioates and ethylene bis(carbamodithioates) have harmful effects on aquatic life, the latter derivatives being less toxic. CDT are toxic to bacteria and fish, after a short-term exposure, and induced embryo toxicity and teratogenicity in rainbow trout at concentrations below ppb levels. In contrast, mancozeb was not teratogenic to rats following inhalation exposure (Lu and Kennedy, 1986). Based on the mechanisms of toxicity, inhibition of metabolism with CDT must occur at many sites and in many processes. Therefore, CDT are regarded as broad-spectrum biocides.

4.5   SUMMARIZING REMARKS
CT and CDT are widely used in industry and crop protection. Their specific activity is mainly determined by the substituents attached to the sulphur and nitrogen, i.e. they

may be used as vulcanization accelerators and antioxidants in the rubber industry, in crop protection, or medically to treat alcoholics (Table 5).

Table 5 — Structure–activity relationship for RR′NC(X)YR″ carbamothioates and carbamodithioates

R	R′	X	Y	R″	Activity
Alkyl	Alkyl	O	S	Alkyl, benzyl	Herbicides
Alkyl	Alkyl	O or S	S	Halopropenyl	Herbicides, procarcinogens promutagens, neuro-toxicants
H	Arylsulphonyl	O	S	Alkyl	Herbicidal antidotes
Alkyl	Alkyl	S	S	Metal	Fungicides, some neuro-toxicants
Aryl	H	S	O	C_{1-12} alkyl	Fungicides
Alkyl	Hydroxymethyl	S	S	Alkyl	Fungicides, bactericides, nematocides
Alkyl	H	O or S	S or O	Long chain alkyl	Nematocides
Alkyl	Alkyl	S	S	2,3,3-tribromopro-pyl	Plant growth regulators
Alkyl	Alkyl	O	S	Alkoxyaryl	Juvenile hormone mimics
Alkyl, benzyl	Alkyl or H	S	S	Benzothiazyl, Bi, Cd, Cu, Pb, Se	Vulcaniztion accelerators, antioxidants

CT herbicides degrade rapidly in the biota and environment. Sulphoxidation and/ or S-methylene hydroxylation are two of the initial steps in biodegradation of CT in many species and under various environmental conditions. The herbicidal action of CT sulphoxides probably results from their carbamoylating action for sensitive plant thiols, although the specific target site is not defined. GSH conjugation is an important process in certain plants and mammals because it might lead to selective detoxification of the herbicide. Many CT are, in fact, proherbicides.

The various metabolic pathway reactions for CT in soil, plants, animals and the environment are summarized in Table 6. N-Oxidation reactions in glyphosate and fluvalinate, leading ultimately to many of the same products formed on their metabolism and environmental degradation, have been established by utilizing MCPBA oxidation reactions (Gohre et al., 1987; Holden et al., 1982). No evidence is available to suggest whether a similar N-oxidation might be involved with CT-metabolism. N-Oxidation followed by rearrangement to alkoxyamine, might explain important metabolic pathways such as N-dealkylation. It is, therefore, interesting to review early studies with CT in order to establish that prediction.

CT and CDT are slightly toxic to mammals by oral or dermal application.

S-(2-Chloropropenyl) CT and CDT herbicides, such as diallate, triallate and sulphallate, in contrast to S-alkyl and S-benzyl CT, show mutagenic activity in the *Salmonella typhimurium* assay only on metabolic activation with microsomal S-9 mix. They are promutagens, since ultimate mutagens are formed on oxidation with liver microsomal MFO system.

Organic biomimetic oxidations of CT with MCPBA provide a convenient model for studying certain bioactivation pathways and identifying transient metabolites and ultimate products. 2-Chloropropenal was isolated from MCPBA oxidation of

Table 6 — Metabolic reactions of carbamothioate herbicides in the biota and environment (some reactions are only applicable to certain carbamothioates)

Metabolic reaction	Soil (microorganisms)	Plant (enzymes)	Animal (enzymes)	Environment (photodegradation)
S-oxidation	+	+	+	+
S-methylene hydroxylation	−	+	+	−
GSH and cysteine conjugation	−	+	+	−
Malonyl cysteine conjugation	−	+	−	−
Mercapturic acid conjugation	−	−	+	−
N-dealkylation	+	+	+	+
N-alkyloxidation	+	+	+	+
Ring hydroxylation	+	+	+	+
N-alkyldehydrogenation	+	+	+	−
S-methylation	+	+	+	−
Hydrolysis	+	+	+	+
Reductive dechlorination	+	−	−	−
Oxidative dechlorination	−	−	−	+
Thiol–thion and cis-trans rearrangement	−	−	−	+
CO_2 formation	+	+	+	−

diallate and its mutagenic activity, as well as that of a whole series of substituted halopropenals, was investigated. It was established that diallate, triallate and sulphallate are mutagens and carcinogens. Diallate and triallate are also delayed neurotoxicants, producing ataxia in hens. Accumulation of these herbicides in fish, vegetables or the environment might be a potential hazard to man.

Pesticides including CT herbicides and CDT fungicides have helped to increase crop productivity. However, as indicated by many studies referred to in this review, some of these have severe chronic hazardous effects. Investigation of these effects and collection of data related to chronic and acute toxicity are indispensable for the efficient and safe use of pesticides.

REFERENCES

Abe, H. and Kuwatsuka, S. (1979). Degradation of benthiocarb and its related compounds by soil bacteria. *Abst. 4th Annual Meeting of the Pesticide Science Society of Japan,* p. 207.

Agarwal, R. P., McPherson, R. A., and Phillips, M. (1983). Rapid degradation of disulfiram by serum albumin. *Res. Commun. Chem. Pathol. Pharmacol.,* **42**, 293–310.

Aharonson, N., Rubin, B., Katan, J., and Benjamin, A. (1983). *Proc. 5th Int. Congress on Pesticide Chemistry,* Vol. *4*, 189–194.

Akio, T., Atsuo, U., Hisako, K., and Minoru, M. (1985). IV. Vertebral curvature of the tropical fish, guppy, from the carbamate and dinitroaniline pesticides and the acute toxicity of their decomposition products after ultraviolet irradiation. *CA*, **103**, 18087.

Ames, B. N., McCann, J. and Yamasaki, E. (1975). Methods for detecting carcinogens and mutagens with the *Salmonella*/mammalian microsome mutagenicity test. *Mutat. Res.*, **31**, 347–363.

Anderson, J. P. E. (1984). Herbicide degradation in soil: influence of microbial biomass. *Soil Biol. Biochem.*, **16**, 483–489.

Asahi Chemical Industry Co. Ltd. (1981). Alkyl phenylthiocarbamate fungicides. *Japanese Patent 81,115,704* (*CA*, **96**, 16067).

Baicu, T., Bucur, E., Constantin, D., and Staicu, S. (1981). *N,N*-disubstituted *S*-2,3-dibromopropyl dithiocarbamate esters, a new class of growth regulators. *An. Inst. Cercet. Prot. Plant. Acad. Stiinte Agric. Silvice*, **16**, 277–283 (*CA*, **99**, 135449).

Bleeke, M. S., Smith, M. T., and Casida, J. E. (1985). Metabolism and toxicity of metribuzin in mouse liver. *Pest. Biochem. Physiol.*, **23**, 123–130.

Büchel, K. H. (1983). *Chemistry of Pesticides*, Wiley, New York, pp. 273–282.

Casida, J. E. (1983). Propesticides: bioactivation in pesticide design and toxicological evaluation. In J. Miyamoto (ed.), *IUPAC Pesticide Chemistry: Human Welfare and Environment*, Pergamon, New York.

Casida, J. E. and Ruzo, L. O. (1986). Reactive intermediates in pesticide metabolism: peracid oxidations as possible biomimetic models. *Xenobiotica*, **16**, 1003–1015.

Casida, J. E., Gray, R. A., and Tilles, H. (1974). Thiocarbamate sulphoxides: potent, selective and biodegradable herbicides. *Science*, **184**, 573–574.

Casida, J. E., Kimmel, E. C., Ohkawa, H., and Ohkawa, R. (1975a). Sulphoxidation of thiocarbamate herbicides and metabolism of thiocarbamate sulphoxides in living mice and liver enzyme systems. *Pest. Biochem. Physiol.*, **5**, 1–11.

Casida, J. E., Kimmel, E. C., Lay, M., Ohkawa, H., Rodebush, J. E., Gray, R. A., Tseng, C. K., and Tilles, H. (1975b). Thiocarbamate sulphoxide herbicides *Environ. Qual. Safety, Suppl.*, **3**, 675–679.

Chen, S. Y. and Casida, J. E. (1978). Thiocarbamate herbicide metabolism: microsomal oxygenase metabolism of EPTC involving mono- and dioxygenation at the sulphur and hydroxylation at each alkyl carbon. *J. Agric. Food Chem.*, **26**, 263–267.

Cook, L. W., Zach, F. W., and Fleeker, J. R. (1982). Steam distillation and gas–liquid chromatographic determination of triallate and diallate in milk and plant tissue. *J. Assoc. Off. Anal. Chem.*, **65**, 215–217.

De Baum, J. R., Bova, D. L., Tseng, C. K., and Menn, J. J. (1978). Metabolism of [ring-^{14}C] ordram (molinate) in the rat.2. Urinary metabolite identification. *J. Agric. Food Chem.*, **26**, 1098–1104.

De Lorenzo, F., Staiano, N., Silengo, L., and Cortese, R. (1978). Mutagenicity of diallate, sulfallate and triallate and relationship between structure and mutagenic effects of carbamates used widely in agriculture. *Cancer Res.*, **38**, 13–15.

Distlerath, L. M., Loper, J. C. and Dey, C. R. (1984). Aliphatic halogenated hydrocarbons produce volatile *Salmonella* mutagens. *Mutat. Res.*, **136**, 55–64.

Distlerath, L. M., Loper, J. C. and Tabor, M. W. (1985). Metabolic activation of 3-(2-chloroethoxy)-1,2-dichloropropene: a mutagen structurally related to diallate, triallate, and sulfallate. *Environ. Mutagen,* **7**, 303–312.

Draper, W. M. and Crosby, D. G. (1981). Hydrogen peroxide and hydroxyl radical: intermediates in indirect photolysis reactions in water. *J. Agric. Food Chem.,* **29**, 699–702.

Draper, W. M. and Crosby, D. G. (1984). Solar photooxidation of pesticides in dilute hydrogen peroxide. *J. Agric. Food Chem.,* **32**, 231–237.

Fairman, M. D., Artman, L., and Maziasz, T. (1983). Diethyldithiocarbamic acid methyl ester distribution, elimination, and LD_{50} in the rat after intraperitoneal administration. *Alcoholism. Clin. Exp. Res.,* **7**, 307–311.

Fang, S. C., George, M., and Freed, V. H. (1964). The metabolism of S-propyl-1-[14]C n-butylethylthiocarbamate (Tillam-^{14}C) in rats. *J. Agric. Food Chem.,* **12**, 37–40.

Fisher, S. W. and Metcalf, R. L. (1983). Production of delayed ataxia by carbamate acid esters. *Pestic. Biochem. Physiol.,* **19**, 243–253.

Fourquet, J. L. (1969). Preparation of dithiocarbamic acid and pyridinium dithiocarbamate. *Bull. Soc. Chim. Fr.,* **9**, 3001–3002.

Gaines, T. B. and Linder, R. E. (1986). Acute toxicity of pesticides in adult weanling rats. *Fundam. Appl. Toxicol.,* **7**, 299–308.

Gale, G. R., Atkins, L. M., Smith, A. B., and Jones, M. M. (1985). Effects of diethyldithiocarbamate and selected analogs on cadmium metabolism following chronic cadmium ingestion. *Res. Commun. Chem. Pathol. Pharmacol.,* **47**, 107–114.

Gale, G. R., Atkins, L. M., Smith, A. B. and Jones, M. M. (1986). Effect of diethyldithiocarbamate and selected analogs on lead metabolism in mice. *Res. Commun. Chem. Pathol. Pharmacol.,* **52**, 29–44.

Gaughan, E. J. and Kezerian, C. (1981). N-(Benzenesulphonyl) thiocarbamates as herbicidal antidotes. *Canadaian Patent 1,110,081* (*CA*, **96**, 29975).

Giles, H. G., Au, J. and Sellers, E. M. (1982). Analysis of plasma diethyldithiocarbamate, a metabolite of disulfiram. *J. Liq. Chromatogr.,* **5**, 945–951.

Glasgow, J. L., Mojica, E., Baker, D. R., Tillis, H., Gore, N. R., and Kurtz, P. J. (1987). In *British Crop Protection Conference, Weeds,* Vol. 1, BCPC, Brighton, pp. 27–33.

Gohre, K., Casida, J. E., and Ruzo, L. O. (1987). N-Oxidation and cleavage of the amino acid derived herbicide glyphosate and anilino acid of the insecticide fluvalinate. *J. Agric. Food Chem.,* **35**, 388–392.

Gorder, G. W., Holden, I., Ruzo, L. O., and Casida, J. E. (1985). Phosphinyliminodithiolane insecticides: oxidative bioactivation of phosfolan and mephosfolan. *Bioorg. Chem.,* **13**, 344–352.

Goring, C. A. I., Laskowski, D. A., Hamaker, J. W., and Meikle, R. W. (1975). *Environmental Dynamics of Pesticides,* Plenum, New York, p. 135.

Grunwell, J. R. (1970). Reaction of Grignard reagents with tetramethylthiuram disulfide. *J. Org. Chem.,* **35**, 1500–1501.

Haga, T., Komyoji, T., Nakajima, T., Oshima, T., and Suzuki, K. (1985). Phenylthiolcarbamates as microbicides. *Japanese Patent 60,185,704* (*CA*, **104**, 143963).

Hald, J. and Jacobson, E. (1948). A drug sensitising the organism of ethyl alcohol. *Lancet,* **225**, 1001–1004.

Hansen, L. G., Francis, B. M., Metcalf, R. L., and Reinders, J. H, (1985). Neurotoxicity of diallate and triallate when administered orally or topically to hens. *Environ. Sci. Health., Part B,* **20**, 97–111.

Hartley, D. and Kidd, H. (1985). The Agrochemicals Handbook. The Royal Society of Chemistry, Nottingham, UK.

Hassall, K. A. (1982). *The Chemistry of Pesticides,* Verlag Chemie, Deerfield Beach, FL, pp. 318–323.

Hirashima, A., Leader, H., Holden, I., and Casida, J. E. (1984). Resolution and stereoselective action of sulprofos and related S-Propyl phosphorothiolates. *J. Agric. Food Chem.,* **32**, 1302–1307.

Holden, I., Segall, Y., Kimmel, E. C., and Casida, J. E. (1982). Peracid-mediated N-oxidation and rearrangement of dimethylphosphoroamides: plausible model for oxidative bioactivation of the carcinogen hexamethylphosphoramide (HMPA). *Tetrahedron Lett.,* **23**, 5107–5110.

Horvath, L. and Pulay, A. (1980). Metabolism of EPTC in germinating corn: sulfone as the true carbamoylating agent. *Pest. Biochem. Physiol.,* **14**, 265–270.

Hubbell, J. P. and Casida, J. E. (1977). Metabolic fate of the N,N-dialkylcarbamoyl moiety of thiocarbamate herbicides. *J. Agric. Food Chem.,* **25**, 404–413.

Hutson, D. H. (1981). S-Oxygenation in herbicide metabolism in mammals. In J. D. Rosen, P. S. Magee, and J. E. Casida (eds.), *Sulfur in Pesticide Action and Metabolism, ACS Symposium Series,* Vol. 158, Washington, DC, pp. 53–63.

Ikeda, M., Unai, T., and Tomizawa, C. (1986a). Degradation of the herbicide orbencarb in soils. *J. Pest. Sci.,* **11**, 85–96.

Ikeda, M., Unai, T., and Tomizawa, C. (1986b). Metabolism of the herbicide orbencarb in soybean, wheat, corn and crabgrass seedlings. *Weed. Res. Japan.,* **31**, 238–243.

Ikeda, M., Unai, T., and Tomizawa, C. (1986c). Absorption, translocation and metabolism of orbencarb in soybean plants. *J. Pest. Sci.,* **11**, 97–110.

Ikeda, M., Unai, T., and Tomizawa, C. (1986d). Absorption, translocation and metabolism of the herbicide orbencarb in cotton plant. *J. Pest. Sci.,* **11**, 379–385.

Innes, J. R. M., Ulland, B. M., Valerio, M. G., Petrucelli, L., Fishbein, L., Hart, E. R., Pallotta, A. J., Bates, R. R., Falk, H. L., Gart, J. J., Klein, M., Mitchell, I., and Peters, J. J. (1969). Bioassay of pesticides and industrial chemicals for tumorigenicity in mice: a preliminary note. *J. Nat. Cancer. Inst.,* **42**, 1101–1114.

Ishikawa, K. (1981). Fate and behaviour of benthiocarb (thiobencarb) herbicide in biota and the environment. *Rev. Plant. Protec. Res.,* **14**, 149–168.

Ishikawa, K., Okuda, I., and Kuwatsuka, S. (1973). Metabolism of benthiocarb in mice. *Agric. Biol. Chem.,* **37**, 165–173.

Ishikawa, K., Nakamura, Y., Niki, Y., and Kuwatsuka, S. (1977). Photodegradation of benthiocarb herbicide. *J. Pest. Sci.,* **2**, 17–25.

Ishikawa, K., Shinohara, R., Yagi, A., Shigematsu, S., and Kimura, I. (1980). Identification of S-benzyl N,N-diethylthiocarbamate in paddy field soil applied with benthiocarb herbicide. *J. Pest. Sci.,* **5**, 107–109.

Israeli, R., Sculsky, M., and Tiberin, P. (1983). Acute central nervous system changes due to intoxication by manzidan (a combined dithiocarbamate of maneb and zineb). *Arch. Toxicol. Suppl.,* **6**, 238–243.

Kato, K., Kawamura, M., Itsuda, H., and Sato, M. (1986). *S*-o-Halophenyl dialkylthiocarbamates. Japanese Patent 61 63,651 (*CA*, **105**, 152741).

Kirk-Othmer (1982). *Encyclopedia of Chemical Technology*, Vol. 20, Wiley, New York, pp. 337–348.

Kisida, H., Hatakoshi, M., Itaya, N., and Nakayama, I. (1984). New insect juvenile hormone mimics: thiolcarbamates. *Agric. Biol. Chem.*, **48**, 2889–2891.

Kochansky, J. P. and Wright, F. C. (1984). Control of parasitic mites with alkyl carbamates. *US Patent 4,464,390 (CA*, **101**, 165592).

Kochansky, J. P. and Wright, F. C. (1985). Synthesis and structure–activity studies on aliphatic carbamates and thiocarbamates toxic to scabies mites, *Psoroptes* spp. (Acari: Psoroptidae). *J. Econ. Entomol.*, **78**, 599–606.

Komives, A. V., Komives, T., and Dutka, F. (1985). Effects of thiocarbamate herbicides on the activity of gluthathione *S*-transferase in maize. *Cereal Res. Commun.*, **13**, 253–257.

Lamoureux, G. L. and Frear, D. S. (1979). In G. D. Paulson, D. S. Frear, and E. P. Marks (eds.), *Xenobiotic Metabolism: in vitro Methods*, *ACS Symposium Series*, Vol. 97, Washington, DC, p. 102.

Lamoureux, G. L. and Rusness, D. G. (1981). Catabolism of GSH conjugates of pesticides in higher plants. In J. D. Rosen, P. S. Magee, and J. E. Casida, (Eds.), *Sulfur in Pesticide Action and Metabolism,. ACS Symposium Series*, Vol. 158, Washington, DC, pp. 133–164.

Lay, M. M. and Casida, J. E. (1976). Dichloroacetamide antidotes enhance thiocarbamate sulphoxide detoxification by elevating corn root GSH content and GSH *S*-transferase activity. *Pest. Biochem. Physiol.*, **6**, 442–456.

Lay, M. M. and Casida, J. E. (1978). Involvement of GSH and GSH-transferases in the action of dichloroacetamide antidotes for thiocarbamate herbicides. In F. M. Pallos and J. E. Casida (eds.), Chemistry and Action of Herbicide Antidotes, Academic Press, New York, pp. 151–160.

Lay, M. M., Hubbell, J. P., and Casida, J. E. (1975). Dichloroacetamide antidotes for thiocarbamate herbicides: mode of action. *Science*, **189**, 287–288.

Lay, M. M. and Menn, J. J. (1979). Mercapturic acid occurrence in fish bile. A terminal product of metabolism of the herbicide molinate. *Xenobiotica*, **9**, 669–673.

Lee, A. (1984). EPTC (*S*-ethyl-*N,N*-dipropylthiocarbamate)-degrading microorganisms isolated from a soil previously exposed to EPTC. *Soil. Biol. Biochem.*, **16**, 529–531.

Lee, H. B. and Chau, A. S. Y. (1983). Determination of trifluvalin, diallate, triallate, atrazine, barban, diclofop-methyl and benzolyprop-ethl in natural waters at parts per trillion levels. *J. Assoc. Off. Anal. Chem.*, **66**, 651–658.

Leeuwen, C. J. V., Griffioen, P. S., Vergouw, W. H. A., and Maas-Diepeveen, J. L. (1985a). Differences in susceptibility of early stages of rainbow trout to environmental pollutants. *Aquat. Toxicol.*, **7**, 59–78.

Leeuwen, C. J. V., Moberts, F., and Niebeck, G. (1985b). Aquatic toxicological aspects of dithiocarbamates and related compounds. II. Effects of survival, reproduction and growth of *Daphnia magna. Aquat. Toxicol.*, **7**, 165–175.

Leeuwen, C. J. V., Maas-Diepeveen, J. L., Niebeek, G., Vergouw, W. H. A., Griffioen, P. S., and Luijken, M. W. (1985c). Aquatic toxicological aspects of dithiocarbamates and related compounds. I. Short term toxicity tests. *Aquat. Toxicol.*, **7**, 145–164.

Leeuwen, C. J. V., Helder, T., and Seinen, W. (1986a). Aquatic toxicological aspects of dithiocarbamates and related compounds. IV. Teratogenicity and histopathology in rainbow trout. *Aquat. Toxicol.*, **9**, 147–159.

Leeuwen, C. J. V., Espeldoorn, A., and Mol, F. (1986b). Aquatic toxicological aspects of dithiocarbamates and related compounds. III. Embryolarval studies with rainbow trout. *Aquat. Toxicol.*, **9**, 129–145.

Leeuwen, C. J. V., Hameren, P. V., Bogers, M., and Griffioen, P. S. (1986c). Uptake, distribution and retention of zineb and ziram in rainbow trout. *Toxicology*, **42**, 33–46.

Lu, M. H. and Kennedy, G. L. (1986). Teratogenic evaluation of mancozeb in the rat following inhalation exposure. *Toxicol. Appl. Pharmacol.*, **84**, 355–368.

Maizlish, V. E., Titova, G. F., Nikulina, T. A., and Smirnov, R. P. (1986). Tetraethylthiuram disulfide. *Russian Patent 1,227,626 (CA*, **105**, 174196).

Marsden, P. J. and Casida, J. E. (1982). 2-Haloacrylic acids as indicators of mutagenic 2-haloacrolein intermediates in mammalian metabolism of selected promutagens and carcinogens. *J. Agric. Food. Chem.*, **30**, 627–631.

McCann, J. and Ames, B. N. (1976). Detection of carcinogens as mutagens in the *Salmonella*/microsome test: assay of 300 chemicals: discussion. II. *Proc. Natl. Acad. Sci. USA*, **73**, 950–954.

Miller, G. E., Zemaitis, M. A., and Greene, F. E. (1983). Mechanism of diethyldithiocarbamate-induced loss of cytochrome P450 from rat liver. *Biochem. Pharmacol.*, **32**, 2433–2442.

Moon, Y. H. and Kuwatsuka, S. (1984). Properties and conditions of soils causing the dechlorination of the herbicide benthiocarb (thiobencarb) in flooded soils. *J. Pest. Sci.*, **9**, 745–754.

Moon, Y. H. and Kuwatsuka, S. (1985a). Dechlorination of benthiocarb in soil II. Microbial aspects of dechlorination of the herbicide benthiocarb in soil. *J. Pest. Sci.*, **10**, 513–521.

Moon, Y. H. and Kuwatsuka, S. (1985b). Dechlorination of benthiocarb in soil IV. Characterisation of microbes causing dechlorination of benthiocarb in a diluted soil suspension. *J. Pest. Sci.*, **10**, 541–547.

Moon, Y. H. and Kuwatsuka, S. (1985c). Factors influencing dechlorination of benthiocarb in the soil suspension. *J. Pest. Sci.*, **10**, 523–528.

Moon, Y. H. and Kuwatsuka, S. (1987a). Population changes of benthiocarb dechlorinating microorganisms in soil. *J. Pest. Sci.*, **12**, 11–16.

Moon, Y. H. and Kuwatsuka, S. (1987b). Nutritional conditions for microorganisms to dechlorinate benthiocarb. *J. Pest. Sci.*, **12**, 3–10.

Moriga, M., Ohta, T., Watanabe, K., Miyazawa, T., Kato, K., and Shirasu, Y. (1983). Further mutagenicity studies on pesticides in bacterial reversion assay systems. *Mutat. Res.*, **116**, 185–216.

Nakamura, Y., Ishikawa, K., and Kuwatsuka, S. (1977a). Degradation of benthiocarb herbicide in soil as affected by soil conditions. *J. Pest. Sci.*, **2**, 7–16.

Nakamura, Y., Ishikawa, K., and Kuwatsuka, S. (1977b). Metabolic fate of benthiocarb herbicide in plants. *Agric. Biol. Chem.*, **41**, 1613–1620.

National Cancer Institute (1978). Bioassay of sulfallate for possible carcinogenicity. In *Carcinogens, Tech. Rep. Ser. 115*, DHEW, p. 78–1370.

Newman, M. S. and Karnes, H. A. (1966). The conversion of phenols to thiophenols via dialkylthiocarbamates. *J. Org. Chem.*, **31**, 3980–3984.

Obrigawitch, T., Wilson, R. G., Martin, A. R., and Roeth, F. W. (1982). The influence of temperature, moisture and prior EPTC application on the degradation of EPTC in soils. *Weed. Sci.*, **30**, 175–181.

Obrigawitch, T., Martin, A. R., and Roeth, F. W. (1983). Degradation of thiocarbamate herbicides in soils exhibiting rapid EPTC breakdown. *Weed Sci.*, **31**, 187–192.

Ong, V. Y. and Fang, S. C. (1970). In vivo metabolism of ethyl-1-^{14}C-*N,N*-di-n-propylthiocarbamate in rats. *Toxicol. Appl. Pharmacol.*, **17**, 418–425.

Owen, W. J. (1987). Herbicide detoxification and selectivity. In *British Crop Protection Conference Weeds*, Vol. 1, BCPC, Brighton, p. 312.

Physician's Desk Reference (1987). Medical Economics Inc., Oradell, NJ, 41st edn., pp. 632–633.

Plewa, M. J., Wagner, E. D., Gentile, G. J., and Gentile, J. M. (1984). An evaluation of the genotoxic properties of herbicides following plant and animal activation. *Mutat. Res.*, **136**, 233–245.

Principe, P., Dogliotti, E., Bignami, M., Crebelli, R., Falcone, E., Fabrizi, M., Conti, G., and Comba, P. (1981). Mutagenicity of chemicals of industrial and agricultural relevance in *Salmonella, Streptomyces and Aspergillus. J. Sci. Food Agric.*, **32**, 826–832.

Ripley, B. D. and Braun, H. E. (1983). Retention time data for organochlorine, organophosphorus and organonitrogen pesticides on SE-30 capillary column and application of capillary gas chromatography to pesticide residue analysis. *J. Assoc. Off. Anal. Chem.*, **66**, 1084–1095.

Rosen, J. D., Schuphan, I., Segall, Y., and Casida, J. E. (1980a). Mechanism for the mutagenic activation of the herbicide sulfallate. *J. Agric. Food. Chem.*, **28**, 880–881.

Rosen, J. D., Segall, Y. and Casida, J. E. (1980b). Mutagenic potency of haloacroleins and related compounds. *Mutat. Res.*, **78**, 113–119.

Ruzo, L. O. and Casida, J. E. (1985). Photochemistry of thiocarbamate herbicides: oxidative and free radical processes of thiobencarb and diallate. *J. Agric. Food Chem.*, **33**, 272–276.

Sanders, H. O. and Hunn, J. B. (1982). Toxicity, bioconcentration, and depuration of the herbicide bolero 8EC in freshwater invertebrates and fish. *Bull. Japan. Soc. Sci. Fish.*, **48**, 1139–1143.

Santi, R. and Gozzo, F. (1976). Degradation and metabolism of drepamon in rice and barnyard grass. *J. Agric. Food Chem.*, **24**, 1229–1235.

Schafer, D. E. and Czajkowski, A. J. (1983). Compositions and methods for reducing herbicidal injury. *US Patent 4,379,716* (*CA*, **99**, 1830).

Schafer, E. W. and Bowles, W. A. (1985). Acute oral toxicity and repelling of 933 chemicals to house and deer mice. *Arch. Environm. Contam. Toxicol.*, **14**, 111–129.

Schafer, E. W., Bowles, W. A., and Hurlbut, J. (1983). The acute oral toxicity, repellency, and hazard potential of 998 chemicals to one or more species of wild and domestic birds. *Arch. Environm. Cantam. Toxicol.*, **12**, 355–383.

Schairer, L. A., Sautkulis, R. C., and Tempel, N. R. (1982). Monitoring ambient air for mutagenicity using the higher plant Tradescantia. In R. R. Tice, D. L. Costa, and K. M. Scaich (eds.), *Genotoxic Effects of Airborne Agents*, Plenum, New York, pp. 123–139.

Scheithauer, S. and Myer, R. (1979). *Topics in Sulphur Chemistry*, Vol. 4, Georg Thiem, Stuttgart, pp. 311–316.

Schimmel, S. C., Garnas, R. L., Patrick, J. M., and Moore, J. C. (1983). Acute toxicity, bioconcentration and persistence of AC 222,705, benthiocarb, chlorpyrifos, fenvalerate, methyl parathion and permethrin in the estuarine environment. *J. Agric. Food Chem.*, **31**, 104–113.

Schuphan, I. and Casida, J. E. (1979a). *S*-Chloroallyl thiocarbamate herbicides: chemical and biological formation and rearrangement of diallate and triallate sulphoxides. *J. Agric. Food Chem.*, **27**, 1060–1067.

Schuphan, I. and Casida, J. E. (1979b). [2,3] Sigmatropic rearrangement of *S*-(3-chloroallyl)thiocarbamate sulphoxides followed by a 1,2-elimination reaction yielding unsaturated aldehydes and acid chlorides. *Tetrahedron Lett.*, (10), 841–844.

Schupan, I. and Casida, J. E. (1983). Metabolism and degradation of pesticides and xenobiotics: bioactivations involving sulfur-containing substituents. In J. Miyamoto (ed.), *IUPAC Pesticide Chemistry: Human Welfare and Environment*, Pergamon, New York, pp. 287–294.

Schupan, I., Rosen, J. D., and Casida, J. E. (1979). Novel activation mechanism for the promutagenic herbicide diallate. *Science*, **205**, 1013–1015.

Schuphan, I., Segall, Y., Rosen, J. D., and Casida, J. E. (1981). Toxicological significance of oxidation and rearrangement reactions of *S*-chloroallyl thio- and dithiocarbamate herbicides. In J. D. Rosen, P. S. Magee, and J. E. Casida (eds.), *Sulfur in Pesticide Action and Metabolism, Vol 171 ACS Symposium Series*, Vol. 158, Washington, DC, pp. 65–82.

Segall, Y. and Casida, J. E. (1981). Products of peracid oxidation of *S*-alkyl phosphorothiolate pesticides. In L. D. Quin and J. Verkade (eds.), *Phosphorus Chemistry, ACS Symposium Series*, Washington, DC, pp. 337–340.

Segall, Y. and Casida, J. E. (1982). Oxidative conversion of phosphorothiolates to phosphinyloxysulfonates probably via phosphorothiolate *S*-oxides. *Tetrahedron Lett.*, **23**, 139–142.

Segall, Y. and Casida, J. E. (1983a). Reaction of proposed phosphorothiolate *S*-oxide intermediate with alcohols. *Phosphorus and Sulphur*, **18**, 209–212.

Segall, Y. and Casida, J. E. (1983b). Oxidation of sulfallate and related alkyl dialkyldithiocarbamates to dialkylformamides via sulfine and iminium ion intermediates. *J. Agric. Food Chem.*, **31**, 242–246.

Segall, Y., Kimmel, E. C., Dohn, D. R. and Casida, J. E. (1985). 3-substituted 2-halopropenals: mutagenicity, detoxification and formation from 3-substituted 2,3-dihalopropanal promutagens. *Mutat. Res.*, **158**, 61–68.

Shimabukuro, R. H., Lamoureux, G. L., and Frear, D. S. (1978). In F. M. Pallos and J. E. Casida (eds.), *Chemistry and Action of Herbicide Antidotes*, Academic Press, New York, p. 133.

Sikka, H. C. and Florczyk, P. (1978). Mutagenic activity of thiocarbamate herbicides in *Salmonella typhimurium*. *J. Agric. Food Chem.*, **26**, 146–148.

Smith, S. G. (1962). Solvolysis of benzhydril thionobenzoates, a criterion for ionization. *Tetrahedron Lett.*, 979–982.

Soderquist, C. J., Bowers, J. B., and Crosby, D. G. (1977). Dissipation of molinate in rice field. *J. Agric. Food Chem.*, **25**, 940–945.

Stephenson, G. R., Bunce, N. J., Makowski, R. I., Bergsma, M. D., and Curry, J. C. (1979). Structure–activity relationships to thiocarbamate herbicides in corn. *J. Agric. Food Chem.*, **27**, 543–547.

Ted, S. R. (1983). Laboratory and field studies on the toxic effects of thiobencarb to the crawfish *Procambarus clarkii*. *J. World Maric. Soc.*, **14**, 434–440.

Unai, T. and Tomizawa, C. (1986a). Metabolic fate of fenothiocarb sulphoxide in soils. *J. Pest. Sci.*, **11**, 357–361.

Unai, T. and Tomizawa, C. (1986b). Metabolic fate of the acaricide fenothiocarb in soils. *J. Pest. Sci.*, **11**, 335–345.

Unai, T. and Tomizawa, C. (1986c). Metabolism of fenothiocarb acaricide in the citrus red mites (Acarina:Tetranychidae). *Appl. Ent. Zool.*, **21**, 283–288.

Unai, T. and Tomizawa, C. (1986d). Photodegradation of fenothiocarb on silica gel plate exposed to sunlight. *J. Pest. Sci.*, **11**, 363–367.

Unai, T., Ikeda, M., Doi, M. and Tomizawa, C. (1986a). Novel metabolites of orbencarb and benthiocarb herbicides in soil, plant and rat liver microsomes: S-2-chlorobenzyl-N-ethyl-N-vinylthiocarbamate and its 4-chloro isomer. *J. Pest. Sci.*, **11**, 527–532.

Unai, T., Tamaru, M., and Tomizawa, C. (1986b). Translocation and metabolism of the acaricide fenothiocarb in citrus. *J. Pest. Sci.*, **11**, 347–356.

Unai, T., Ikeda, M., and Tomizawa, C. (1986c). Metabolic fate of the chlorobenzyl moiety of orbencarb sulphoxide and benthiocarb sulphoxide in soybean seedlings. *Weed. Res. Japan*, **31**, 228–237.

Wilkinson, R. E. (1983). Metabolism of [^{14}C]vernolate in soybean. *Pest. Biochem. Physiol.*, **20**, 347–353.

Wilson, R. G. (1984). Accelerated degradation of thiocarbamate herbicides in soil with prior thiocarbamate herbicide exposure. *Weed Sci.*, **32**, 264–268.

Wing, K. D., Glickman, A. H., and Casida, J. E. (1984). Phosphorothiolate pesticides and related compounds:oxidative bioactivation and aging of the inhibited acetylcholinesterase. *Pest. Biochem. Physiol.*, **21**, 22–30.

Worthing, C. R., and Walker, S. B. (1987). The Pesticide Manual, BCPC, Thornton Heath, UK.

Yasuhiko, S., Moriya, M., Tezuka, H., Teramoto, S., Ohta, T., and Inoue, T. (1984). Mutagenicity of pesticides. *Environm. Sci. Res.*, **31**, 617–624.

Zdzienicka, M., Zielenska, M., Tudek, B., and Szymczyk, T. (1979). Mutagenic activity of thiram in Ames tester strains of *Salmonella typhimurium*. *Mutat. Res.*, **68**, 9–13.

5

Sulphoxides and sulphones

A. G. Renwick
Clinical Pharmacology Group, University of Southampton, Medical & Biological Sciences Building, Bassett Crescent East, Southampton SO9 3TU, UK

SUMMARY

1. Sulphoxide groups present in foreign compounds may undergo oxidation to the corresponding sulphone and/or reduction to a thioether or sulphide. The reduction is a reversible process.
2. Reduction of sulphoxides may be effected by mammalian tissues and by the intestinal microflora. The cytosolic enzyme aldehyde oxidase is largely responsible for sulphoxide reduction by tissues under anaerobic conditions. Thioredoxin is important under aerobic conditions. Few data are available concerning the nature of the microbial enzymes.
3. The tissue and microbial enzyme systems responsible for sulphoxide reduction show different substrate specificities since sulindac is reduced by both tissue and intestinal microorganisms *in vitro*, but sulphinpyrazone undergoes extensive microbial but not tissue reduction. These differences *in vitro* correlate with the site of reduction *in vivo*.
4. Oxidation of sulphoxides to sulphones, which can be an irreversible process, is catalysed by cytochrome P-450, although other enzymes may be involved.
5. Pharmacological and toxicological activity may reside in one or more of the redox states. Therefore separate investigations are necessary for each redox state during the development of novel sulphoxide-containing pharmaceutical and agricultural chemicals.

5.1 INTRODUCTION

Sulphoxides and sulphones represent two of three interconvertible redox states (Fig. 1). Sulphoxide groups occur in xenobiotics either as part of the parent molecule

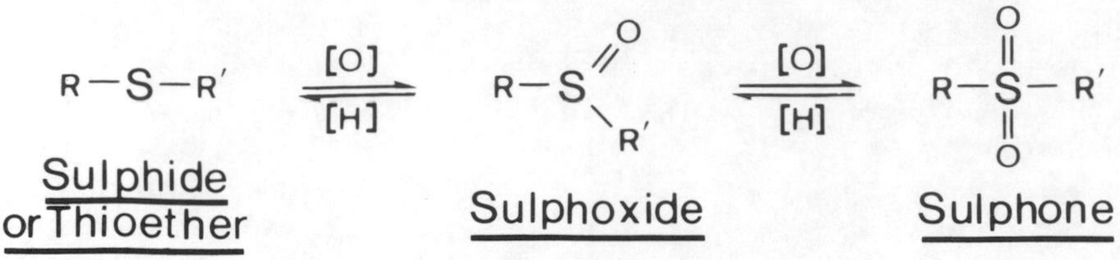

$$R-S-R' \;\underset{[H]}{\overset{[O]}{\rightleftharpoons}}\; R-S\overset{\displaystyle O}{\underset{R'}{\diagdown}} \;\underset{[H]}{\overset{[O]}{\rightleftharpoons}}\; R-\overset{\displaystyle O}{\underset{\displaystyle O}{S}}-R'$$

| Sulphide
or Thioether | Sulphoxide | Sulphone |

Fig. 1 — The interconversion of sulphur redox states.

(drug, insecticide, etc.) or following *S*-oxidation of a thioether. The latter reaction is important when the parent compound contains a thioether linkage (Chapter 6, this volume, Part A) or when such a bond is introduced following conjugation with glutathione, followed by the action of C–S lyase and methylation (see Chapters 5 and 6, Volume 2, Part B of this series).

Sulphoxides and sulphones exist as metabolites of a number of thioether-containing chemicals and the initial oxidation to a sulphoxide is considered in other chapters in this volume. Essentially there are two microsomal sulphoxidation systems; one involving cytochrome P-450 and the other a flavin-containing monooxygenase. The K_m values of thioether substrates may be vastly different in these systems, such that their relative contributions are often dose dependent (Souhaili El Amri *et al.*, 1987). The present chapter concentrates on the other interconversions of

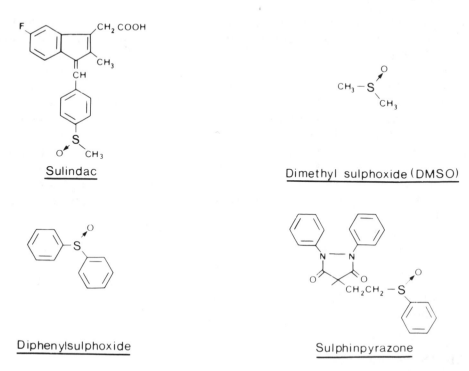

Sulindac

Dimethyl sulphoxide (DMSO)

Diphenylsulphoxide

Sulphinpyrazone

Fig. 2 — Xenobiotic substrates used for *in vitro* studies on the oxidation and reduction of sulphoxides.

the redox states, i.e. reduction of the sulphoxide to sulphide (or thioether) and interconversion of the sulphone and sulphoxide. Considerably more data are available concerning the reduction of the sulphoxide than its oxidation to the sulphone or the possibility of reduction of the state back to the sulphoxide. Although the sulphoxide moiety is present in a large number of xenobiotics, *in vitro* studies on the mechanisms involved in its reduction have used a relatively small number of substrates (Fig. 2).

5.2   THE REDUCTION OF SULPHOXIDES TO SULPHIDES

5.2.1   Sulphoxide reduction by mammalian tissues

The reduction of sulphoxides by mammalian tissues is a complex process which may involve both soluble and membrane-bound enzyme systems. One of the first reports on the hepatic reduction of sulphoxides was that of 4,4'-diaminodiphenylsulphoxide which was readily reduced by 10000g supernatant fractions of rat liver in the presence of NADPH but not NADH (Mazel *et al.*, 1969). Interestingly, neither microsomes nor supernatant were active alone but microsomal reducing activity could be restored by the addition of soluble fraction. The activity was inhibited by SKF 525A but not by carbon monoxide and was stimulated by flavin mononucleotide (FMN).

Table 1 — The interconversions of sulphoxides and sulphides by rabbit liver fractions

			Product formed (μmol/g liver/h)		
			10 000 supernatant	Microsomes (unwashed)	Cytosol
Sulindac					
SO→S	Aerobic		0.04	ND	0.22
	Anaerobic		0.20	ND	0.31
S→SO	Aerobic		2.10	1.23	0.40
	Anaerobic		2.11	1.15	0.24
SO→SO$_2$	Aerobic		0.78	0.25	0.02
	Anaerobic		0.52	0.24	0.01
Sulphinpyrazone					
SO→S	Aerobic		<0.01	ND	ND
	Anaerobic		0.08	ND	ND
S→SO	Aerobic		0.05	0.03	ND
	Anaerobic		0.03	0.02	ND
SO→SO$_2$	Aerobic		ND	ND	ND
	Anaerobic		ND	ND	ND

The data are the mean values for the interconversions of the different redox states of the sulphoxide drugs sulindac and sulphinpyrazone (Fig. 2).
ND, not detected.
Adapted from Strong *et al.* (1984b).

The nature of the tissue enzymes involved in sulphoxide reduction has since been studied extensively (see below), but a major problem is that nearly all of these studies have ignored the possibility of reoxidation of the resulting sulphide (thioether) (see chapter 6, this volume, Part A). Thus it must be realized that, in the following account, the 'reductase activity' represents a balance between two opposing processes. While it appears reasonable to assume that oxidation does not occur in some of these preparations, e.g. purified aldehyde oxidase under reducing conditions, it should be appreciated that this is an assumption. The potential importance of the reverse reaction has been demonstrated by the *in vitro* data of Strong *et al.* (1984b) who showed extensive oxidation of sulindac sulphide by rabbit liver 10 000g supernatant under aerobic and anaerobic conditions largely due to the microsomal fraction. The reduction of sulindac to its sulphide was detected at a considerably lower rate than oxidation. The highest reductase activity was found in the cytoplasmic fraction (Table 1). Data for sulphinpyrazone were similar to those described above for 4,4'-diaminodiphenylsulphoxide in that the very low levels of sulphoxide reduction detected were found in the 10 000g supernatant only. As in the case of sulindac, the rate of oxidation of 'sulphinpyrazone sulphide' to the sulphoxide (i.e. sulphinpyrazone) exceeded that of the corresponding reduction under aerobic conditions.

5.2.1.1   *Cytosolic enzymes involved in sulphoxide reduction*

An important early finding was the role played by thioredoxin, a protein of molecular weight approximately 12 000, in the reduction of sulindac by rat hepatic cytosolic enzymes under aerobic conditions (Anders *et al.*, 1980). The cytosolic enzymes, after precipitation by ammonium sulphate, were further purified by passing through Sephadex G25 and G50 columns. The latter column removed 94% of the *in vitro* reducing activity, but this could be restored by the addition of pure thioredoxin isolated from *E. coli* (which did not itself reduce sulindac). Dithiothreitol could be used in the place of thioredoxin to restore the reducing activity. These data demonstrated that thioredoxin was involved in sulindac reduction, possibly via maintainance of a soluble 'sulphoxide reductase' in a reduced state (Fig. 3). The *in vitro* activity of the cytosol was inhibited by the thioredoxin inhibitors insulin, glutathione disulphide, L-cystine, 5,5'-dithiobis(2-nitrobenzoic acid) and sodium arsenite (Anders *et al.*, 1980). Similar findings were reported subsequently for the kidney (Anders *et al.*, 1981), which is also an important site for the reduction of

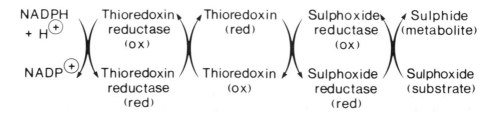

Fig. 3 — The sulphoxide reductase system present in rat hepatic and renal cytosol, operating under aerobic conditions.

sulindac (Duggan *et al.*, 1980; Ratnayake *et al.*, 1981). The cytosolic sulindac reducing activity was not induced by classic microsomal cytochrome P-450 inducers, i.e. 3-methylcholanthrene or phenobarbitone (Ratnayake *et al.*, 1981).

Thioredoxin contains two thiol groups which are readily oxidizable to form a disulphide bond but which can be reduced by the enzyme thioredoxin reductase (Laurent *et al.*, 1964). The inhibitors described above act either on the reductase (arsenite) or as competitive substrates (disulphides). Thioredoxin was first described as a hydrogen donor involved in the synthesis of deoxyribonucleotides (Laurent *et al.*, 1964). The protein isolated from *E. coli* is an acidic peptide made up of 108 amino acids, which has a redox potential of -0.26 V (Holmgren, 1968). The redox function resides in the two thiol groups on residues 32 and 35 and does not involve a transition metal. The peptide structure varies between species since the thioredoxin in calf liver contains 103 amino acids and four cysteine residues, only two of which are involved in the redox functioning (Engstrom *et al.*, 1974). An antibody raised against bovine hepatic thioredoxin did not cross-react with the protein from *E. coli*, yeast, rat liver or human platelets (Holmgren and Luthman, 1978). Using the antibody, thioredoxin was found in all organs studied, and was not restricted to the cytosol, but was found in microsomal fractions as well. Thioredoxin could therefore contribute to sulphoxide reduction by microsomal fractions under aerobic and anaerobic conditions (but see later).

The enzyme thioredoxin reductase which catalyses the reaction

$$\text{thioredoxin-S}_2 + \text{NADPH} + \text{H}^+ \rightarrow \text{thioredoxin-(SH)}_2 + \text{NADP}^+ \qquad (1)$$

contains bound flavin adenine dinucleotide (FAD). The enzyme does not reduce oxidized glutathione directly but can effect the reduction of a number of disulphides in the presence of catalytic amounts of thioredoxin plus NADPH (Moore *et al.*, 1964; Holmgren, 1977).

Recent studies have focused on the nature and the role of other electron transfer systems linked to the 'sulphoxide reductase' enzyme in the scheme given in Fig. 3. The 'reductase activity' in the $105\,000g$ supernatant of guinea pig liver under anaerobic conditions was partially purified by ammonium sulphate precipitation (at 30–45% saturation) (Kitamura *et al.*, 1981). High reducing activity was found on co-incubation of the purified fraction with the microsomal NADPH–cytochrome-*C* reductase (see later), and with the flavoenzymes xanthine oxidase (from milk) and lipoamide dehydrogenase (from pig heart). Weaker activity was also found with NADH–cytochrome-b_5 reductase and NADH dehydrogenase. The potential role of these flavoenzymes is depicted in Fig. 4. Subsequent studies on the ammonium sulphate precipitated, partially purified material from guinea pig liver (Tatsumi *et al.*, 1982) showed that compounds which can act as electron donors to aldehyde oxidase, e.g. aldehydes, *N*-methylnicotinamide and 2-hydroxypyrimidine, resulted in greatly enhanced sulphoxide reductase activity (using sulindac as the substrate) compared with NADPH or NADH. Inhibition of sulphoxide reductase activity by a range of inhibitors was mirrored closely by the inhibition of aldehyde oxidase activity, while purification on a DEAE cellulose column resulted in the co-elution of both enzyme activities (Tatsumi *et al.*, 1982). Reduction of a wide range of substrates was found when acetaldehyde was used as an electron donor, e.g. dimethylsulphoxide, diphenylsulphoxide, dibenzylsulphoxide, sulphinpyrazone, biotin sulphoxide methyl ester and methionine sulphoxide. Thus it appears that the sulphoxide reductase in Fig. 4 is the enzyme aldehyde oxidase, a molybdenum-containing enzyme closely related to xanthine oxidase.

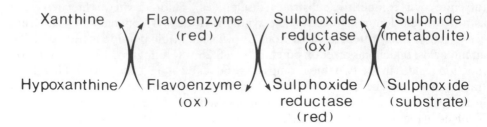

Fig. 4 — The contribution of flavoproteins to sulphoxide reduction under anaerobic conditions by cytosolic enzymes. (1) The flavoprotein in the above system would be xanthine oxidase, although other flavoenzymes and substrates could be involved (see text). (2) The 'sulphoxide reductase' has been identified as aldehyde oxidase which can be reduced directly by cosubstrates such as aldehydes and 2-hydrooxypyrimidine (see text). (3) FAD and methyl viologen can increase the electron transfer between the flavoprotein and aldehyde oxidase (see text).

Details of the separation of milk xanthine oxidase from the co-eluting aldehyde oxidase–sulphoxide reductase in guinea pig liver were reported in a subsequent paper (Kitamura and Tatsumi, 1983), which also showed that the reduction of sulindac could be accomplished by purified rabbit aldehyde oxidase plus acetaldehyde. Acetaldehyde could be replaced by xanthine + xanthine oxidase but not by either of these factors alone. The subsequent paper from these authors (Tatsumi *et al.*, 1983) showed that the purified rabbit aldehyde oxidase could also reduce a range of sulphoxide substrates (Table 2). The reducing activity occurred in the

Table 2 — Sulphoxide and related substrates reduced by purified aldehyde oxidase

Substrate	Rabbit liver[a]	Guinea pig liver[b]
Sulindac	1.14	—
Diphenylsulphoxide	1.00	1.00
Dibenzylsulphoxide	0.46	0.40
Phenothiazine sulphoxide	1.20	0.62
(+)Biotin (+)sulphoxide methyl ester	0.13	0.42
L-methionine sulphoxide	0.01	0.00
Diphenylsulphone	—	0.00
Quinoline *N*-oxide	—	0.51

The values are the rate of reduction calculated relative to diphenylsulphoxide.
[a] Adapted from Tatsumi *et al.* (1983). Determined by direct measurement of the thioether metabolite, using acetaldehyde as co-substrate.
[b] Adapted from Yoshihara and Tatsumi (1985b). Determined by the oxidation of 2-hydroxypyrimidine to uracil in the presence of the sulphoxide substrate.

Table 3 — Inhibitors of mammalian tissue sulphoxide reductase activity

Inhibitor	Sulphoxide reductase activity (% control)					
	Rat liver[a]	Rat kidney[b]	Guinea pig liver[c]	Guinea pig liver[d]	Guinea pig liver[e]	Guinea pig aldehyde oxidase[f]
Insulin	13	4	—	—	—	—
Glutathione disulphide	3	4	—	—	—	—
L-cystine	8	6	—	—	—	—
5,5'-dithiobis (2-nitrobenzoic acid)	6	1	—	—	—	11
Sodium arsenite	2	2	11	0	18	65
Sodium azide	42	52	—	—	—	27
Sodium cyanide	101	—	112	3	9	12
Mercuric chloride	<1	<1	—	—	—	66
Iodoacetate or its amide	2	3	8	9	—	100
Menadione	—	—	—	—	0	102
Dimethyl sulphoxide	48	71	66	78	—	93
Methionine sulphoxide	35	48	—	—	—	—
Cupric ions	—	—	—	9	—	4
Substrate Conditions	Sulindac, Aerobic	Sulindac, Aerobic	Sulindac, Aanaerobic	Sulindac, Anaerobic	Sulindac, Anaerobic	Diphenylsulphoxide, anaerobic

[a]Thioredoxin-dependent system; inhibitors at 10^{-3} M; Ratnayake et al. (1981).
[b]Thioredoxin-dependent system; inhibitors at 10^{-3} M; Anders et al. (1981).
[c]105000g supernatant; inhibitors at 10^{-4}–10^{-3} M; Kitamura et al. (1980).
[d]Unwashed microsomes; inhibitors at 10^{-4}–10^{-3} M; Kitamura et al. (1980).
[e]105000g supernatant; ammonium sulphate fractionated; in the presence of acetaldehyde; inhibitors at 10^{-5}–2.5×10^{-4} M; Tatsumi et al. (1983).
[f]Purified aldehyde oxidase linked with either hypoxanthine–xanthine oxidase plus methyl viologen or methyl viologen semiquinone; inhibitors at 5×10^{-4} or 10^{-3} M; Yoshihara and Tatsumi (1985a).

absence of acetaldehyde if xanthine + xanthine oxidase were added, and this activity was enhanced by the addition of FAD (2-fold) or methyl viologen (paraquat) (10-fold). Similar results were obtained with purified guinea pig aldehyde oxidase (Yoshihara and Tatsumi, 1985a) with 6-fold (FAD) and 100-fold (methyl viologen) increases over basal activity of aldehyde oxidase + xanthine oxidase–hypoxanthine system. Whereas sulphoxide reduction by acetaldehyde and xanthine oxidase linked aldehyde oxidase was inhibited by the aldehyde oxidase inhibitor menadione, methyl viologen semiquinone linked reduction was not inhibited by menadione, but by cyanide and arsenite. This suggested that methyl viologen semiquinone (and FAD) acts on the molybdenum site of aldehyde oxidase, rather than via xanthine oxidase (Yoshihara and Tatsumi, 1985a).

The importance of these various enzymes in the activity of less purified systems was shown by the fact that the sulphoxide redutase activity in crude 9000g supernatant from guinea pig liver was inhibited by menadione (an inhibitor of aldehyde oxidase) but not by allopurinol (an inhibitor of xanthine oxidase). Addition of acetaldehyde, but not NADPH or NADH, resulted in markedly enhanced activity (Tatsumi *et al.*, 1983), suggesting that aldehyde oxidase is also of potential importance.

It is possible that the contributions of the components thioredoxin, xanthine oxidase, aldehyde oxidase, etc., could be studied by the use of enzyme-specific inhibitors. Despite a large number of substrates' being tested for inhibitory activity in the various studies discussed above, there has been little uniformity (Table 3). The findings are generally consistent across different studies in that the arsenite and iodoacetate inhibit most systems, while cyanide and cupric ions appear to be highly active against the purified anaerobic system. The absence of inhibition by cyanide in unfractionated guinea pig cytosol suggests that the thioredoxin system may play an important role in cytosolic reduction of sulphoxides.

5.2.1.2 *Microsomal enzymes involved in sulphoxide reduction*

Studies on sulphoxide reduction by guniea pig liver (Kitamura *et al.*, 1980) demonstrated activity in both microsomal and supernatant fractions. Washing the microsomes removed the microsomal activity, but this could be restored by the addition of a small amount of supernatant owing to the presence of a non-dialysable, heat labile, acetone precipitable factor. The unwashed microsomal sulindac reducing activity was NADPH dependent, strongly inhibited by iodoacetate, arsenite, cupric ions and cyanide (Table 2) but not inhibited by carbon monoxide. Partial purification (63-fold) of the reducing activity resulted in similar increase (60-fold) in NADPH–cytochrome-C reductase. Inhibition of the NADPH dependent microsomal activity by addition of cytochrome-c led the authors (Kitamura *et al.*, 1980) to conclude that the contribution of the microsomes was to cause the reduction of the soluble factor which then effected the sulphoxide reduction (Fig. 5). The soluble factor which was reduced by the microsomes was not characterized further and it could be thioredoxin, its reductase or the sulphoxide reductase (aldehyde oxidase). The strong inhibition of microsomal activity by arsenite indicates that it could be linked via thioredoxin reductase.

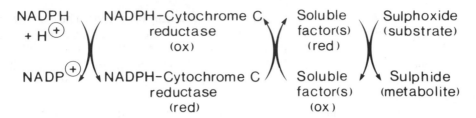

Fig. 5 — The contribution of microsomal cytochrome-*C* reductase to sulphoxide reduction by guinea pig liver preparations under anaerobic conditions. Adapted from Kitamura *et al.* (1980).

Interspecies differences in the contribution of microsomal cytochrome-*C* reductase to reductive activity may occur. Addition of 100 000*g* supernatant to washed microsomes from rat liver did not produce measurable reduction in one study (Ratnayake *et al.*, 1981) and produced low levels of activity in a different study (Kitamura *et al.*, 1981). In rabbit livers the activity in 10 000*g* supernatants could be accounted for by that present in the cytosol fraction, with no detectable reduction by unwashed microsomes (Strong *et al.*, 1984b). Species differences may reside in the supernatant fraction, since the partially purified NADPH–cytochrome-*C* reductase from rat liver produced considerable reduction of sulindac when supplemented with the 100 000*g* supernatant from guinea pig liver (Kitamura *et al.*, 1981). Considerable interspecies differences were found in the activity in the soluble fraction using diphenylsulphoxide as a substrate (Kitamura *et al.*, 1981), with the guinea pig about 10 times more active (expressed per gram of liver) than the rat.

5.2.2   Sulphoxide reduction by microorganisms

Early studies on the microbial reduction of sulphoxide groups concerned the sulphoxides of normal cell constituents, i.e. methionine sulphoxide (Sourkes and Trano, 1953) and biotin sulphoxide (Cleary and Dykhuizen, 1974). An early report on the *in vitro* enzymic reduction of the sulphoxide moiety was the reduction of L(−)methionine sulphoxide by a system isolated from yeast (Black *et al.*, 1960). The system comprised three different enzymes, required NADPH as a cofactor, and was inhibited by arsenite and iodoacetate. Reduction of disulphides to thiols could also be carried out by two of the three component enzymes and it was suggested that the three components acted as an electron transport system with the terminal enzyme causing the reduction of sulphoxides. Methionine sulphoxide and biotin sulphoxide are probably reduced by different enzymes since mutants of *E. coli* unable to grow on biotin sulphoxide could use methionine sulphoxide as a sulphur source (Zinder and Brock, 1978). The reduction of dimethylsulphoxide (DMSO) by *E. coli* was inhibited by methionine sulphoxide and was decreased but not abolished in the biotin sulphoxide reductase deficient mutants, indicating that this xenobiotic substrate was probably reduced by both enzyme systems (Zinder and Brock, 1978). DMSO reduction by *E. coli* was dependent on NADH or NADPH and was inhibited by oxygen, nitrate and other sulphoxides, e.g. methionine sulphoxide and tetramethylene sulphoxide (tetrahydrothiophene sulphoxide), but not by dimethylsulphone, which was not reduced by *E. coli* or other DMSO reducing organisms. The enzyme in *E. coli* responsible for the reduction of methionine sulphoxide was maximal during logarithmic growth, utilized NADPH in preference to NADH and was unable to reduce methionine sulphone (Ejiri *et al.*, 1979). The reduction of methionine sulphoxide moieties present within peptides is catalysed by a different reductase enzyme (Brot *et al.*, 1981).

Extensive reduction has been reported on incubation of sulphoxides with faeces or intestinal contents. For example sulindac is reduced *in vitro* by faeces from humans (Duggan *et al.*, 1977a; Strong *et al.*, 1985, 1987); sulphinpyrazone is reduced by caecal contents of rats (Renwick *et al.*, 1982) and rabbits (Strong *et al.*, 1984b) and by human faeces (Strong *et al.*, 1984a); cimetidine sulphoxide is reduced by human faeces (Mitchell *et al.*, 1982).

In a study on the reduction of sulindac by over 200 isolated strains of human intenstinal bacteria *in vitro*, Strong *et al.* (1987) found extensive reduction of sulindac by most aerobes. Under anaerobic conditions reduction occurred with all strains of *E. coli* ($n=47$), *Enterobacter* ($n=13$), *Proteus* ($n=7$), *Klebsiella* ($n=14$) and 2 out of 14 strains of *Enterococci*. In contrast, anaerobes such as *Bacteriodes* and *Bifidobacteria* showed much lower or no detectable enzyme activity; an exception were *Clostridia* in which considerable activity was found in all strains tested ($n=14$). Sulphinpyrazone was also reduced *in vitro* by the same strains of aerobes, but to a much lesser extent, while all the anaerobes showed either marginal or no activity. Thus the microbial reduction of sulphoxides may show both species and substrate specificity.

Both *S*-oxidation and reduction of sulindac was found with pure cultures of the soil organisms, *Arthrobacter, Sporobolomyces pararoseus, Aspergillus alliaceus* and *Nocardia corallina* (Davis and Guenthner, 1985). Only *Nocardia* performed both reactions and *Aspergillus* produced only the sulphone. The authors proposed that these organisms could provide useful models of mammalian metabolism and provide methods for the biosynthesis of metabolites. However, the influence of potential cofactors (e.g. acetaldehyde) or inhibitors (e.g. allopurinol) was not studied, and the enzymes involved were not characterized.

5.2.3   The role of mammalian tissues and intestinal flora in the reduction of sulphoxides *in vivo*

Both the tissues and intestinal flora are potential sites for sulphoxide reduction *in vivo* but they are likely to show different substrate specificities and drug interactions. Thus definition of the contribution of each is important in our understanding of the overall fate of the compound and in the rational therapeutic use of sulphoxide containing drugs.

An important study was that of Renwick *et al.* (1982) which showed that about 3 times more sulphinpyrazone reduction occurred following oral than following parenteral administration to rats. The formation of sulphide *in vitro* by the caecal contents was approximately 200 times greater than that by the liver, indicating that the gut flora were probably the main site of reduction. The essential role of the gut flora was confirmed by the absence of sulphinpyrazone reduction in rats treated with antibiotics and in germ-free rats. A subsequent study in rabbits (Strong *et al.*, 1984b) showed a 10-fold greater sulphoxide reduction following oral dosing compared with intravenous administration and suppression but not abolition of sulphide formation following antibiotics. The activity of the caecal contents *in vitro* was about 50 times that in liver, indicating that the gut flora are also the main site for sulphoxide reduction in this species. Route of administration did not affect the extent of reduction of sulphinpyrazone in man (Strong *et al.*, 1984a), although antibiotics suppressed sulphide formation *in vivo* (Strong *et al.*, 1984a, 1986) and *in vitro* (Strong *et al.*, 1986, 1987). The definitive proof of the essential role of the gut flora in sulphoxide reduction in man was the 30-fold lower amounts of sulphide formed by patients with ileostomies compared with normal subjects (Strong *et al.*, 1984a). These patients showed a similar extent of absorption of sulphinpyrazone based on the area under the concentration–time curve and the percentage excreted in urine unchanged, and therefore the potential for tissue reduction was the same as controls. The major difference in such patients is the absence of an anaerobic gut flora which therefore must be implicated in the reduction process. Complete suppression of sulphinpyrazone sulphide formation in a patient receiving oral lincomycin (Strong *et al.*, 1984a) also pointed to the anaerobes. The discovery that most enzyme activity resides in aerobic organisms such as *E. coli* (see above; Strong *et al.*, 1987) has led to the suggestion that the importance of the anaerobes is to provide the reductive environment in which the aerobes are able to express their 'sulphoxide reductase' activity. Evidence consistent with this hypothesis is the report that growth of *E. coli* on a minimal medium containing DMSO resulted in the induction of sulphoxide reductase activity under anaerobic but not under aerobic conditions (Bilous and Weiner, 1985a). The enzyme activity was in the membrane fraction and could utilize FMN or FAD as electron donors, in addition to the artificial donors benzyl and methyl viologen. The enzyme, which is probably part of an anaerobic electron transport chain (Bilous and Weiner, 1985b), could also reduce methionine sulphoxide, but its activity to other xenobiotics was not reported.

Sulindac, which has been used as the main substrate for *in vitro* studies, is reduced far more extensively (20–100 fold) by rabbit liver than is sulphinpyrazone (Strong *et al.*, 1984b). This suggests that the tissues may be more important in the *in vivo* reduction of this substrate. Following oral administration of sulindac to man the peak concentration of the sulphide metabolite almost coincided with that of the parent drug, at 1–2 h after the dose (Duggan *et al.*, 1977a; Strong *et al.*, 1985). This contrasts with the data for sulphinpyrazone where the peak concentration of the sulphide formed by the gut flora occurred at 15 h (4–31 h) after the dose (Strong *et al.*, 1984a). This temporal difference suggests that the liver and/or other tissues must be involved in the reduction of sulindac. This was confirmed by the study of Strong *et al.* (1985) who studied the fate of sulindac in patients with ileostomies. The absorption of sulindac was similar to that in controls based on the area under the plasma concentration–time curve. For the sulphide metabolite, the time to peak, peak concentration and area under the concentration–time curve up to 12 h were almost identical in ileostomy patients and controls, suggesting that the initial peak is formed by the liver. However, in the controls the sulphide showed a long terminal half-life (14 h) and the area under the curve from 12 h to infinity accounted for 56% of the total area under curve. In contrast, the slow late phase was totally absent from the ileostomy patients, the terminal half-life was only 2.5 h and the area under curve from 12 h to inifinity represented only 7% of the total. It was concluded (Strong *et al.*, 1985) that the late phase arose from the microbial reduction of sulindac which is excreted extensively in the bile (Dujovne *et al.*, 1982). Thus, in the case of sulindac, both the tissues and the intestinal flora are involved in reduction to the active sulphide metabolite. Their contributions are temporally separated following a single dose, and based on areas under the curve each site is responsible for about one-half of the total reduction. The longer terminal half-life associated with enterohepatic circulation and microbial reduction suggests that this site may be more important during chronic therapy.

5.3   THE OXIDATION OF SULPHOXIDES TO SULPHONES

Although sulphones are frequently detected as excreted metabolites of sulphoxides (Table 4), little is known concerning their site of formation, or the enzyme(s) involved. *In vivo* the oxidation of sulindac to the sulphone exceeds its reduction to the sulphide in rat, dog, guinea pig and rabbit, while in monkey and man approximately equal amounts undergo oxidation and reduction (Duggan *et al.*, 1978). Sulphinpyrazone is also oxidized to its sulphone in all animal species studied, i.e. rat, rabbit, guinea pig, dog, monkey and swine (Dieterle and Faigle, 1981), although the sulphone and hydroxysulphone represented minor constitutents (<10%) of the total metabolites in urine and plasma.

Formation of sulindac sulphone was found following the incubation of sulindac with rabbit liver fractions, especially the microsomes (Table 1; Strong *et al.*, 1984b), although this reaction occurred at a lower rate than the oxidation of the thioether to sulindac. Similarly, the oxidation of albendazole to its sulphoxide by ovine liver microsomes occurred at a rate approximately 100 times greater than the subsequent oxidation to the sulphone (Galtier *et al.*, 1986).

The enzymes involved have not been identified but it appears that the microsomal flavin-containing monooxygenase, which oxidizes thioethers to sulphoxides, is probably not responsible for the further oxidation to the sulphone (Hajjar and Hodgson, 1980). The sulphur heteroaromatic compounds (thiaarenes) such as benz[b]naphtho[2,3-d]thiophene (Table 4), are oxidized by rat liver microsomes to the corresponding sulphones by an NADPH-dependent system (Jacob *et al.*, 1986). The enzymic activity was increased by pretreatment with DDT or benzo[k]fluoranthene, which induces cytochrome P-450mc. Similarly, the conversion of didesmethylchlorpromazine sulphoxide to its sulphone by rat liver microsomes was enhanced by prior treatment with β-naphthoflavone and phenobarbitone and inhibited by SKF 525A and metyrapone (Kreft and Breyer-Pfaff, 1979), suggesting the involvement of cytochrome P-450. However, while the classic inhibitory ligand of P-450, carbon monoxide, blocked the additional induced enzyme activity, it did not inhibit the basic level of oxidation to the sulphone. In contrast, formation of the sulphone metabolite of the simple heterocyclic compound dibenzothiophene was almost completely blocked by carbon monoxide, largely because of inhibition of the further oxidation of the sulphoxide (Vignier *et al.*, 1985). Therefore it appears that cytochrome P-450 is largely responsible for the oxidation of sulphoxides to sulphones, but the involvement of other enzyme systems cannot be excluded.

Table 4 — Examples of thioether compounds metabolized to the sulphone via initial S-oxidation to the sulphoxide

Compound	Structure	Species	Reference
Albendazole	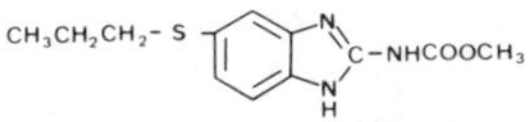	Sheep	Marriner and Bogan (1980)
Tiadenol (at either or both S atoms)	$HOCH_2CH_2S-(CH_2)_{10}-SCH_2CH_2OH$	Rat, man	Facino et al. (1986)
Benzo[b]naphtho[2,3-d]thiophene	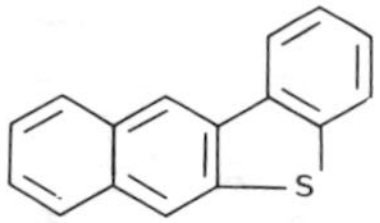	Rat (microsomes)	Jacob et al. (1986)
Thioridazine (at the side chain or at the ring S atoms)	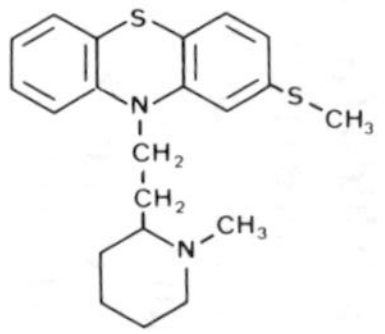	Cattle (liver) Man	Traficante et al. (1979) Papadopoulos and Crammer (1986)
Didesmethylchlorpromazine	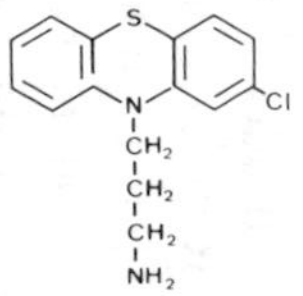	Rat	Kreft and Breyer-Pfaff (1979)
Methiocarb	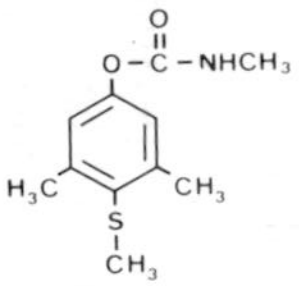	House fly	See Kulkarni and Hodgson (1980)
Methiochlor	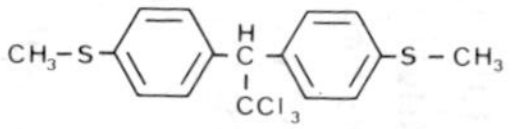	Mouse	See Kulkarni and Hodgson (1980)

Little research has focused on the formation of sulphones from their corresponding sulphoxides. This is probably because they are frequently less active than their reduced analogues, while their higher polarity results in a greater clearance and thus little accumulation and low blood levels during chronic administration.

5.4 THE REDUCTION OF SULPHONES TO SULPHOXIDES

The sulphone group is relatively metabolically stable. No evidence of conversion to the corresponding sulphoxide had been found for the antileprotic drug dapsone (4,4'-diaminodiphenylsulphone; Ellard, 1966) although this compound undergoes extensive metabolism at other sites via N-oxidation (Israili et al., 1973) and N-conjugations (Andoh et al., 1974). Administration of the sulphone metabolite of sulindac to rats and monkeys (Duggan et al., 1978) did not result in detectable formation of sulindac. Similarly, the metabolites of tolmesoxide sulphone did not include either sulphoxide or sulphide derivatives (Greenslade et al., 1981).

In contrast, administration of sulphinpyrazone sulphone to rats resulted in plasma levels of sulphinpyrazone equivalent to about 2–6% reduction (Renwick et al., 1982). Administration of the sulphone metabolite of pentachlorothioanisol (a metabolite of hexachlorobenzene) to rats gave a pattern of urinary and faecal metabolites similar to that of pentachlorothioanisol itself, suggesting extensive reduction back to the sulphoxide and sulphide (Koss et al., 1979). However, this result could also have been produced by replacement of the methylsulphonyl group by glutathione, followed by the action of C-S lyase and methylation of the resulting thiol (Bakke, J., personal communication).

It is possible that the sulphone moiety could be reduced by the same enzymes which convert sulphoxides to sulphides. However, sulphones did not act as inhibitors of sulphoxide reduction in various incubation systems, e.g. sulindac sulphone with sulindac reduction by rat liver cytosol (Ratnayake et al., 1981); diphenylsulphone with diphenylsulphoxide reduction by guinea pig liver aldehyde oxidase (Yoshihara and Tatsumi, 1985b); dimethyl sulphone with dimethylsulphoxide reduction by extracts of E. coli (Zinder and Brock, 1978). Thus it is unlikely that sulphones are competitive substrates for either tissue or microbial 'sulphoxide reductase' systems.

There is a relative paucity of information on the reduction of sulphones to their corresponding sulphoxides, but clearly this process could contribute to the overall balance of the redox states present in the body. The electrochemical redox potential for sulphones (−2 V; although to the sulphinic acid not the sulphoxide) is similar to that of the corresponding sulphoxide (Bowers and Russell, 1960). This suggests that the lack of reduction of some sulphones may be related to the substrate specificity of the reductase rather than the redox potential. The formation of a sulphone by the addition of a second electronegative oxygen atom greatly increases the positive charge on the sulphur atom. This in turn can affect adjacent methyl or methylene groups such that these can become acidic groups able to lose a proton. If the adjacent structure is an aromatic ring then this will be subjected to a stronger electron withdrawing effect (Price and Oae, 1962). Thus the higher polarity of sulphones may prevent their binding to the 'sulphoxide reductase' enzyme systems.

5.5 PHARMACOLOGICAL AND TOXICOLOGICAL CONSEQUENCES OF SULPHOXIDE OXIDATION AND REDUCTION

One of the first documented examples where the reduction of a sulphoxide group was implicated in overt toxicity was the severe haemolytic anaemia affecting ruminants fed large amounts of Brassica such as kale. The association was first recognized in 1943, but it was not until some 30 years later that the active component *S*-methylcysteine sulphoxide was identified. Administration of the sulphoxide caused haemolytic anaemia, but it required further microbial metabolism since it was inactive in germ-free lambs (Smith, 1980). The rumen microorganisms perform two reactions; an initial lyase reaction to produce dimethyldisulphide sulphoxide, followed by reduction to the active haemolytic agent dimethyldisulphide (Fig. 6).

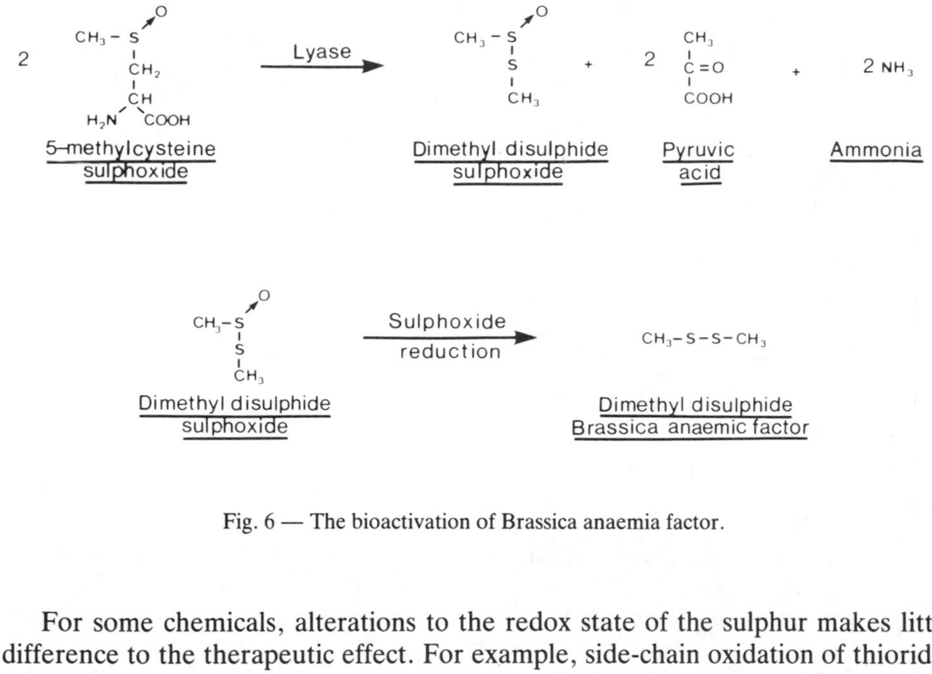

Fig. 6 — The bioactivation of Brassica anaemia factor.

For some chemicals, alterations to the redox state of the sulphur makes little difference to the therapeutic effect. For example, side-chain oxidation of thioridazine to the sulphoxide and sulphone produces clinically active compounds mesoridazine and sulphoridazine respectively (see Table 4; Papadopoulos and Crammer, 1986). In contrast, the cardiotoxicity of thioridazine is probably due to the 5-sulphoxide (ring sulphur) rather than the parent drug or the (side chain) 2-sulphoxide or 2-sulphone (Hale and Poklis, 1986).

However, for some compounds increasing oxidation status of the sulphur atom results in increasing polarity and reduced potency. Possibly the classic example of this is the anti-inflammatory drug sulindac (Table 5), since the parent drug is less active than the sulphide while the sulphone is essentially inactive (Duggan, 1981). Most of the activity *in vivo* is due to the sulphide metabolite (Duggan *et al.*, 1977b) which explains the intense interest in this drug and its selection as a substrate for *in vitro* mechanistic studies.

Table 5 — Examples of sulphoxide drugs undergoing *S*-oxidation and/or reduction reactions

Compound	Structure	Species	Sulphide(s)	Sulphone(s)	Reference
Sulfinalol	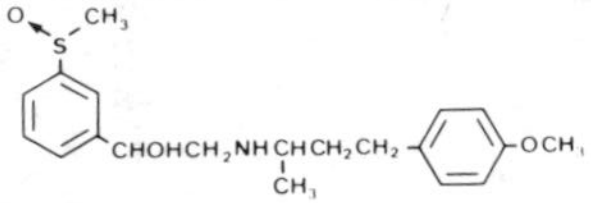	Dog Monkey	Urine, Faeces Urine, Faeces	Urine, Faeces Urine, Faeces	Benziger *et al.* (1981)
Tolmesoxide	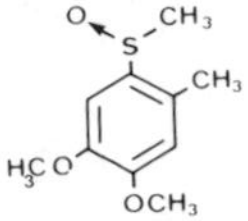	Rat Dog Man	ND Urine (trace) ND	Urine, bile Urine Urine	Greenslade *et al.* (1981)
		Man	—	Plasma	Silas *et al.* (1981)
AR-L 115BS	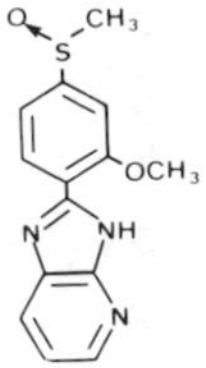	Rat Dog Rabbit Monkey Baboon	Plasma Plasma Plasma — —	Plasma, urine Plasma, urine Plasma, urine Urine Urine	Roth *et al.* (1981)
Omeprazole	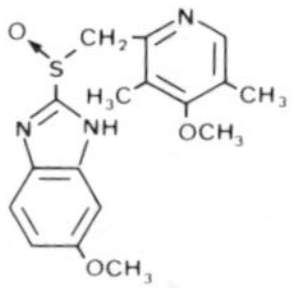	Man	Plasma	Plasma	Mihaly *et al.* (1983)
Sulindac	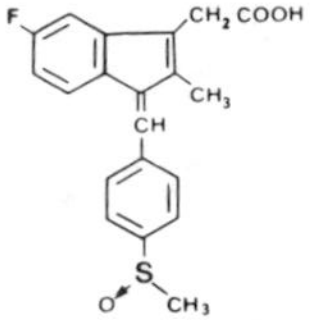	Various	Urine Faeces Plasma	Urine Faeces Plasma	Duggan *et al.* (1978)
Sulphinpyrazone	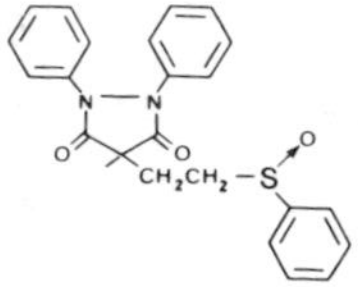	Various	Urine Plasma	Urine Plasma	Dieterle and Faigle (1981)

ND, not detected.

In other cases the sulphoxides and sulphones are more active than the thioethers. The anti-arthritic and analgesic activities of a number of 2-thioalkyl substituted 4,5-diarylimidazoles and the corresponding sulphoxides and sulphones have been studied in animal test systems (Sharpe *et al.*, 1985). Minor modifications to the aromatic and alkyl substitutents caused large differences in potency. Alteration to the redox state of the sulphur did not consistently affect the antiarthritic activity, but the sulphoxides and sulphones were generally more potent analgesics than their corresponding sulphides.

The inhibition of acid secretion caused by substituted benzimidazoles such as omeprazole (Table 5) is probably due to the sulphoxide since this is considerably more potent at inhibiting the H^+, K^+ ATP-ase than the corresponding sulphide (Frylund and Wallmark, 1986), which represents only a minor metabolite during chronic therapy (Mihaly *et al.*, 1983).

Sulphinpyrazone represents another interesting example since the drug initially under development was the thioether analogue (Brodie *et al.*, 1954). The discovery that the uricosuric activity resided in its sulphoxide metabolite (sulphinpyrazone) led to the development of this compound as the clinically used agent (Burns *et al.*, 1957). The therapeutic use of sulphinpyrazone decreased, following the introduction of allopurinol for gout, but renewed interest has been generated in recent years due to the discovery of the platelet antiaggregatory activity. This activity resides in the thioether metabolite of sulphinpyrazone (Kirstein Pedersen and Jakobsen, 1979) so that the interest in these compounds has actually revolved round a complete redox cycle.

For veterinary and agricultural chemicals often the sulphoxides or sulphones are as potent or more potent than the parent thioethers, e.g. benzimidazoles such as albendazole (Table 4; Averkin *et al.*, 1975; Marriner and Bogan, 1980), alkyl phosphodithioate (Bull *et al.*, 1976) and a number of structurally diverse insecticides contain a thioether linkage (Kulkarni and Hodgson, 1980). However, for other insecticides sulphoxidation represented a detoxication process such that no overall conclusion is possible (Kulkarni and Hodgson, 1980).

The sulphone metabolites of *m*-dichlorobenzene produced following glutathione conjugation, C–S lyase, methylation and *S*-oxidation have been implicated in the microsomal enzyme induction produced by this compound (Kato *et al.*, 1986). The absence of induction bile duct cannulated and antibiotic treated animals pointed to the importance of gut flora C–S lyase activity. However, microsomal enzyme induction following direct administration of the methylsulphone metabolite cannot be assigned unequivocally to that specific redox state, especially in view of the possible reduction of the corresponding sulphone metabolite of hexachlorobenzene (see above).

Thus overall conclusions and generalizations concerning the effects of changes in redox state on pharmacological or toxicological activity are probably best avoided. Investigations on novel thioethers or sulphoxides should define the relative activity for each oxidation state for that substrate but these must be combined with metabolism studies to define the extent of oxidation and/or reduction within the test system.

REFERENCES

Anders, M. W., Ratnayake, J. H., Hanna, P. E., and Fuchs, J. A. (1980). Involvement of thioredoxin in sulfoxide reduction by mammalian tissues. *Biochem. Biophys. Res. Commun.*, **97**, 846–851.

Anders, M. W., Ratnayake, J. H., Hanna, P. E., and Fuchs, J. A. (1981). Thioredoxin-dependent sulfoxide reduction by rat renal cytosol. *Drug Metab. Disp.*, **9**, 307–310.

Andoh, B. Y. A., Renwick, A. G., and Williams, R. T. (1974). The excretion of [^{35}S]dapsone and its metabolites in the urine, faeces and bile of the rat. *Xenobiotica*, **4**, 571–583.

Averkin, E. A., Beard, C. C., Dvorak, C. A., Edwards, J. A., Fried, J. H., Kilian, J. G., Schiltz, R. A., Kistner, T. P., Drudge, J. H., Lyons, E. T., Sharp, M. L., and Corwin, R. M. (1975). Methyl 5(b)-phenylsulfinyl-2-benzimidazolecarbamate, a new potent anthelmintic. *J. Med. Chem.*, **18**, 1164–1166.

Benziger, D. P., Fritz, A., and Edelson, J. (1981). Metabolism and disposition of sulfinalol in laboratory animals. *Drug Metab. Disp.*, **9**, 493–498.

Bilous, P. T. and Weiner, J. H. (1985a). Dimethyl sulfoxide in anaerobically grown *Escherichia coli* HB101. *J. Bacteriol.*, **162**, 1151–1155.

Bilous, P. T. and Weiner, J. H. (1985b). Proton translocation coupled to dimethyl sulfoxide reduction in anaerobically grown *Escherichia coli* HB101. *J. Bacteriol.*, **163**, 369–375.

Black, S., Harte, E. M., Hudson, B., and Wartofsky, L. (1960). A specific enzymatic reduction of L(−)methionine sulfoxide and a related nonspecific reduction of disulphides. *J. Biol. Chem.*, **235**, 2910–2916.

Bowers, R. C. and Russell, H. D. (1960). Polarographic behaviour of aryl sulfones and sulfoxides. *Anal. Chem.*, **32**, 405–407.

Brodie, B. B., Yu, T. F., Burns, J. J., Chenkin, T., Paton, B. C., Steele, J. M., and Gutman, A. B. (1954). Observations of G-25671, a phenylbutazone analogue (4-phenylthioethyl-1,2-diphenyl-3,5-pyrazolidinedione). *Proc. Soc. Exp. Biol. Med.*, **86**, 884–894.

Brot, N., Weissbach, L., Worth, J., and Weissbach, H. (1981). Enzymatic reduction of protein-bound methionine sulphoxide. *Proc. Natl. Acad. Sci. USA*, **78**, 2155–2158.

Bull, D. L., Whitten, C. J., and Ivie, G. W. (1976). Fate of O-ethyl O-[4-(methylthio)phenyl]S-propyl phosphorodithioate (BAY NTN 9306) in cotton plants and soil. *J. Agric. Food Chem.*, **24**, 601–605.

Burns, J. J., Yu, T. F., Ritterband, A., Perel, J. M., Gutman, B. and Brodie, B. B. (1957). A potent new uricosuric agent, the sulfoxide metabolite of the phenylbutazone analogue, G-25671. *J. Pharmacol. Exp. Therap.*, **119**, 418–426.

Cleary, P. P. and Dykhuizen, D. (1974). Enzymatic reduction of D-biotin-d-sulfoxide with cell-free extracts of *Escherichia coli*. *Biochem. Biophys. Res. Commun.*, **56**, 629–634.

Davis, P. J. and Guenthner, L. E. (1986). Sulindac oxidation/reduction by microbial cultures; microbial models for mammalian metabolism. *Xenobiotica*, **15**, 845–857.

Dieterle, W. and Faigle, J. W. (1981). Species differences in the disposition and metabolism of sulphinpyrazone. *Xenobiotica*, **11**, 559–568.

Duggan, D. E. (1981). Sulindac: therapeutic implications of the prodrug/pharmaco-phore equilibrium. *Drug Metab. Rev.*, **12**, 325–337

Duggan, D. E., Hare, L. E., Ditzler, C. A., Lei, B. W. and Kwan, K. C. (1977a). The disposition of sulindac. *Clin. Pharmacol. Therap.*, **21**, 326–335.

Duggan, D. E., Hooke, K. F., Risley, E. A., Shen, T. Y., and Van Arman, C. G. (1977b). Identification of the biologically active form of sulindac. *J. Pharmacol. Exp. Therap.*, **201**, 8–13.

Duggan, D. E., Hooke, K. F., Noll, R. M., Hucker, H. B., and Van Arman, C. G. (1978). Comparative disposition of sulindac and metabolites in five species. *Biochem. Pharmacol.*, **27**, 2311–2320.

Duggan, D. E., Hooke, K. F., and Hwang, S. S. (1980). Kinetics of the tissue distributions of sulindac and metabolites. Relevance of sites and rates of bioactivation. *Drug Metab. Disp.*, **8**, 241–246.

Dujovne, C. A., Pitterman, A., Vincek, W. C., Dobrinska, M. R., Davies, R. O., and Duggan, D. E. (1982). Enterohepatic circulation of sulindac and its metabolites. *Clin. Pharmacol. Ther.*, **33**, 172–177.

Ejiri, S. I., Weissbach, H. and Brot, N. (1979). Reduction of methionine sulfoxide to methionine by *Escherichia coli*. *J. Bacteriol.*, **139**, 161–164.

Ellard, G. A. (1966). Absorption, metabolism and excretion of di(p-aminophenyl)-sulphone (dapsone) and di(p-aminophenyl)sulphoxide in man. *Brit. J. Pharmacol.*, **26**, 212–217.

Engstrom, N. E., Holmgren, A., Larsson, A., and Soderhall, S. (1974). Isolation and chracterisation of calf liver thioredoxin. *J. Biol. Chem.*, **249**, 205–210.

Facino, R. M., Carini, M., and Tofanetti, O. (1986). Metabolism of the hypolipide-mic agent tiadenol in man and in the rat. *Arzneim. Forsch. Drug Res.*, **36**, 722–728.

Frylund, J. and Wallmark, B. (1986). Sulfide and sulfoxide derivatives of substituted benzimidazoles inhibit acid formation in isolated gastric glands by different mechanisms. *J. Pharmacol. Exp. Therap.*, **236**, 248–253.

Galtier, P., Alvinerie, M. and Delatour, P. (1986). *In vitro* sulfoxidation of albendazole by ovine liver microsomes; assay and frequency of various xenobio-tics. *Am. J. Vet. Res.*, **47**, 447–450.

Greenslade, D., Havler, M. E., Humphrey, M. J., Jordan, B. L., Lewis, C. J., and Rance, D. J. (1981). Biotransformation of tolmesoxide in animals and man. *Xenobiotica*, **11**, 89–96.

Hajjar, N. P. and Hodgson, E. (1980). Flavin adenine dinucleotide-dependent monooxygenase: its role in the sulfoxidation of pesticides in mammals. *Science*, **209**, 1134–1136.

Hale, P. W. and Poklis, A. (1986). Cardiotoxicity of thioridazine and two stereoiso-meric forms of thioridazine-5-sulfoxide in the isolated perfused rat heart. *Toxicol. Appl. Pharmacol.*, **86**, 44–55.

Holmgren, A. (1968). Thioredoxin. 6. The aminoacid sequence of the protein from *Escherichia coli* B. *Eur. J. Biochem.*, **6**, 475–484.

Holmgren, A. (1977). Bovine thioredoxin system. Purification of thioredoxin reductase from calf liver and thymus and studies of its function in disulphide reduction. *J. Biol. Chem.*, **252**, 4600–4606.

Holmgren, A. and Luthman, M. (1978). Tissue distribution and subcellular localisation of bovine thioredoxin determined by radio immunoassay. *Biochemistry*, **17**, 4071–4077.

Israili, Z. H., Cucinell, S. A., Vaught, J., Davis, E., Lesser, J. M., and Dayton, P. G. (1973). Studies on the metabolism of dapsone in man and experimental animals: formation of N-hydroxy metabolites. *J. Pharmacol. Exp. Therap.*, **187**, 138–151.

Jacob, J., Schmoldt, A., and Grimmer, GH. (1986). The predominant role of S-oxidation in rat liver metabolism of thiaarenes. *Cancer Lett.*, **32**, 107–116.

Kato, Y., Kogure, T., Sato, M., Murata, T. and Kimura, R. (1986). Evidence that methylsulfonyl metabolites of *m*-dichlorobenzene are causative substances of induction of hepatic microsomal drug-metabolising enzymes by the parent compound in rats. *Toxicol. Appl. Pharmacol.*, **82**, 505–511.

Kirstein Pedersen, A. and Jakobsen, P. (1979). Two new metabolites of sulfinpyrazone in the rabbit: a possible cause of the prolonged *in vivo* effect. *Thrombosis Res.*, **16**, 871–876.

Kitamura, S. and Tatsumi, K. (1983). A sulphoxide-reducing enzyme system consisting of aldehyde oxidase and xanthine oxidase — a new electron transfer system. *Chem. Pharm. Bull.*, **31**, 760–763.

Kitamura, S., Tatsumi, K. and Yoshimura, H. (1980). Metabolism *in vitro* of sulindac. Sulfoxide-reducing enzyme systems in guinea pig liver. *J. Pharm. Dyn.*, **3**, 290–298.

Kitamura, S., Tatsumi, K., Hirata, Y., and Yoshimura, H. (1981). Further studies on sulfoxide-reducing enzyme system. *J. Pharm. Dyn.*, **4**, 528–533.

Koss, G., Koransky, W., and Steinbach, K. (1979). Studies on the toxicology of hexachlorobenzene. IV. Sulphur-containing metabolites. *Arch. Toxicol.*, **42**, 19–31.

Kreft, H. and Breyer-Pfaff, U. (1979). Formation of the sulfone metabolite of didesmethylchloropromazine in the rat *in vivo* and *in vitro*. *Drug Metab. Disp.*, **7**, 404–410.

Kulkarni, A. P. and Hodgson, E. (1980). Metabolism of insecticides by mixed function oxidase. *Pharmac. Ther.*, **8**, 379–475.

Laurent, T. C., Moore, E. C. and Reichard, P. (1964). Enzymatic synthesis of deoxyribonucleotides. IV. Isolation and characterisation of thioredoxin, the hydrogen donor from *Escherichia coli* B. *J. Biol. Chem.*, **239**, 3436–3443.

Marriner, S. E. and Bogan, J. A. (1980). Pharmacokinetics of albendazole in sheep. *Am. J. Vet. Res.*, **41**, 1126–1129.

Mazel, P., Katzen, J., Skolnick, P., and Shargel, L. (1969). Reduction of sulfoxides by hepatic enzymes. *Fed. Proc. Fed. Am. Soc. Exp. Biol.*, **28**, 546.

Mihaly, G. W., Prichard, P. J., Smallwood, R. A., Yeomans, N. D., and Louis, W. J. (1983). Simultaneous high-performance liquid chromatographic analysis of omeprazole and its sulphone and sulphide metabolites in human plasma and urine. *J. Chromat.*, **278**, 311–319.

Mitchell, S. C., Idle, J. R., and Smith, R. L. (1982). Reductive metabolism of cimetidine sulfoxide in man. *Drug Metab. Disp.*, **10**, 289–290.

Moore, E. C., Reichard, P., and Thelander, L. (1964). Enzymatic synthesis of deoxyribonucleotides. V. Purification and properties of thioredoxin reductase from *Escherichia coli* B., *J. Biol. Chem.*, **239**, 3445–3452.

Papadopoulos, A. S. and Crammer, J. L. (1986). Sulphoxide metabolites of thioridazine in man. *Xenobiotica*, **16**, 1097–1107.

Price, C. C. and Oae, S. (1962). *Sulfur Bonding*, Ronald Press, New York, pp. 61–128.

Ratnayake,. J. H., Hanna, P. E., Anders, M. W., and Duggan, D. E. (1981). Sulphoxide reduction. *In vitro* reduction of sulindac by rat hepatic cytosolic enzymes. *Drug Metab. Disp.*, **9**, 85–87.

Renwick, A. G., Evans, S. P., Sweatman, T. W., Cumberland, J., and George, C. F. (1982). The role of the gut flora in the reduction of sulphinpyrazone in the rat. *Biochem. Pharmacol.*, **31**, 2649–2656.

Roth, V. W., Prox, A., Reuter, A., Schmid, J., Zimmer, A., and Zip, H. (1981). AR-L 115BS, Speziesvergleich der metabolitenmuster und biotransformation beim menschen. *Arzneim. Forsch. Drug Res.*, **31**, 232–235.

Sharpe, T. R., Cherkofsky, S. C., Hewes, W. E., Smith, D. H., Gregory, W. A., Haber, S. B., Leadbetter, M. R., and Whitney, J. G. (1985). Preparation and antiarthritic and analgesic activity of 4,5-diaryl-2-(substituted thio)-1H-imidazoles and their sulfoxide and sulfones. *J. Med. Chem.*, **28**, 1188–1194.

Silas, J. H., Phillips, F. C., Freestone, S., Tucker, G. T., and Ramsay, L. E. (1981). A clinical and pharmacokinetic evaluation of tolmesoxide in hypertensive patients. *Eur. J. Clin. Pharmacol.*, **19**, 133–118.

Smith, R. H. (1980). Kale poisoning: the brassica anaemia factor. *Veterinary Record*, 5 July, 12–15.

Souhaili El Amri, H., Fargetton, X., Delatour, P., and Batt, A. M. (1987). Sulphoxidation of albendazole by the FAD-containing and cytochrome P-450 dependent mono-oxygenases from pig liver microsomes. *Xenobiotica*, **17**, 1159–1168.

Sourkes, T. L. and Trano, Y. (1953). Reduction of methionine sulfoxides by *Escherichia coli*. *Arch. Biochem. Biophys.*, **42**, 321–326.

Strong, H. A., Oates, J., Sembi, J., Renwick, A. G., and George, C. F. (1984a). Role of the gut flora in the reduction of sulphinpyrazone in humans. *J. Pharmacol. Exp. Therap.*, **230**, 726–732.

Strong, H. A., Renwick, A. G., and George, C. F. (1984b). The site of reduction of sulphinpyrazone in the rabbit. *Xenobiotica*, **14**, 815–826.

Strong, H. A., Warner, N. J., Renwick, A. G., and George, C. F. (1985). Sulindac matabolism; the importance of an intact colon. *Clin. Pharmacol. Therap.*, **38**, 387–393.

Strong, H. A., Angus, R., Oates, J., Sembi, J., Howarth, P., Renwick, A. G., and George, C. F. (1986). Effects of ischaemic heart disease, Crohn's disease and antimicrobial therapy on the pharmacokinetics of sulphinpyrazone. *Clin. Pharmacokinet.*, **11**, 402–411.

Strong, H. A., Renwick, A. G., George, C. F., Liu, Y. F., and Hill, M. J. (1987). The reduction of sulphinpyrazone and sulindac by intestinal bacteria. *Xenobiotica*, **17**, 685–696.

Tatsumi, K., Kitamura, S., and Yamada, H. (1982). Involvement of liver aldehyde oxidase in sulfoxide reduction. *Chem. Pharm. Bull.*, **30**, 4585–4588.

Tatsumi, K., Kitamura, S., and Yamada, H. (1983). Sulfoxide reductase activity of liver aldehyde oxidase. *Biochim. Biophys. Acta,* **747**, 86–92.

Traficante, L. J., Siekierski, J., Sakalis, G., and Gershon, S. (1979). Sulfoxidation of chlorpromazine and thioridazine by bovine liver — preferential metabolic pathways. *Biochem. Pharmacol.,* **28**, 621–626.

Vignier, V., Berthou, F., Dreano, Y., and Floch, H. H. (1985). Dibenzothiophene sulphoxidation: a new and fast high performance liquid chromatographic assay of mixed function oxidation. *Xenobiotica,* **15**, 991–999.

Yoshihara, S. and Tatsumi, K. (1985a). Sulfoxide reduction catalysed by guinea pig liver aldehyde oxidase in combination with one electron reducing flavoenzymes. *J. Pharmacobio-Dyn.,* **8**, 996–1005.

Yoshihara, S. and Tatsumi, K. (1985b). Guinea pig liver aldehyde oxidase as a sulfoxide reductase: its purification and characterisation. *Archiv. Biochem. Biophys.,* **242**, 213–224.

Zinder, S. H. and Brock, T. D. (1978). Dimethylsulphoxide reduction by microorganisms. *J. Gen. Microbiol.,* **105**, 335–342.

6

Sulphonium salts

Peter A. Crooks

College of Pharmacy, University of Kentucky, Lexington, KY 40536, USA

SUMMARY

1. The structure, stereochemistry, and configurational and chemical stability of sulphonium salts are discussed. Sulphonium salts are unstable in basic solution, are susceptible to C–S$^+$ bond fission via nucleophilic displacement reactions, and can also undergo β-elimination reactions.

2. Four sulphonium salts, dimethyl-β-propiothetin (DMPT), S-methyl-L-methionine (SMM), S-adenosyl-L-methionine (SAM), and S-adenosyl-(5′)-3-methyl-thiopropylamine (dSAM), occur in nature.

3. The chemistry, biochemistry and the role of each of four natural sulphonium salts in methyltransferase reactions is reviewed. DMPT acts as a methyl donor in the biosynthesis of methionine from homocysteine in mammals and algae; SMM is a cofactor in the biosynthesis of methionine in plants. SAM, a remarkably ubiquitous methyl donor, is the cofactor in a large number of methyltransferase reactions in both plants and animals; dSAM acts as an aminopropyl donor in the biosynthesis of the polyamines spermine and spermidine.

4. The catabolism of the four natural sulphonium salts is discussed. Most catabolic pathways involve cleavage at the C–S$^+$ bond; however, with the adenosyl derivative SAM, cleavage of the N-glycosidic bond and decarboxylation of the amino acid terminus are other routes of catabolism.

5. Very few studies have been carried out on the biotransformation of exogenous sulphonium compounds. There is no evidence of the occurrence of metabolic oxidation at either the S$^+$ or the α-carbon atom. C–S$^+$ bond cleavage appears to be the preferred route of metabolism, to give a thioether that is biotransformed in a predictable manner.

6. The proposed role of 'reactive' sulphonium ion intermediates in the mechanisms of toxicity of vinthionine and 1,2-dihaloethanes is briefly discussed, and a hypothetical involvement of sulphonium ion intermediates in the biotransformation of epoxides is presented.

6.1 INTRODUCTION

Structurally the sulphonium ion (1), like carbon, is tetrahedral in nature. However, the unshared pair of electrons of the sulphur atom, unlike those of nitrogen, can hold configuration at ordinary temperatures. Thus, sulphonium ions with three different substituents around the sulphonium pole can exist in optical isomeric forms (i.e. structures 1 and 2) (Plate 1). Ethylmethylthetin (3) was the first sulphonium salt to be

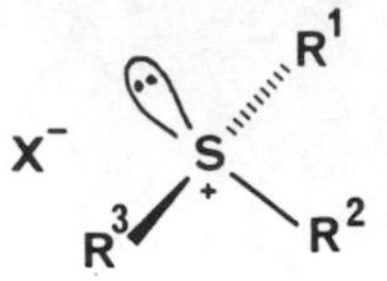

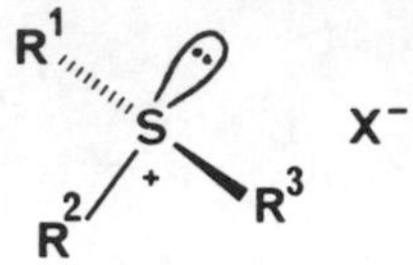

Plate 1

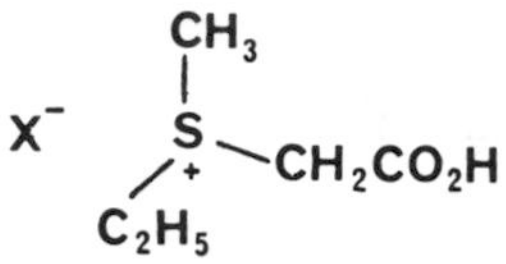

resolved into its enantiomers (Pope and Peachey, 1900). Since then a large number of optically pure sulphonium salts have been prepared, either by resolution or by stereospecific synthesis (Andersen, 1981). However, sulphonium salts are known to undergo configurational change (stereomutation) at sulphur via three major mechanisms (Scheme 1): (a) pyrimidal inversion, (b) reversible dissociation into a carbonium ion and neutral sulphide molecule via an S_N1 mechanism, or (c) S_N2 attack at α-carbon followed by re-formation of the sulphonium salt. 1,2-elimination followed by recombination of the olefin and sulphide has not been detected as a mechanism of configurational inversion.

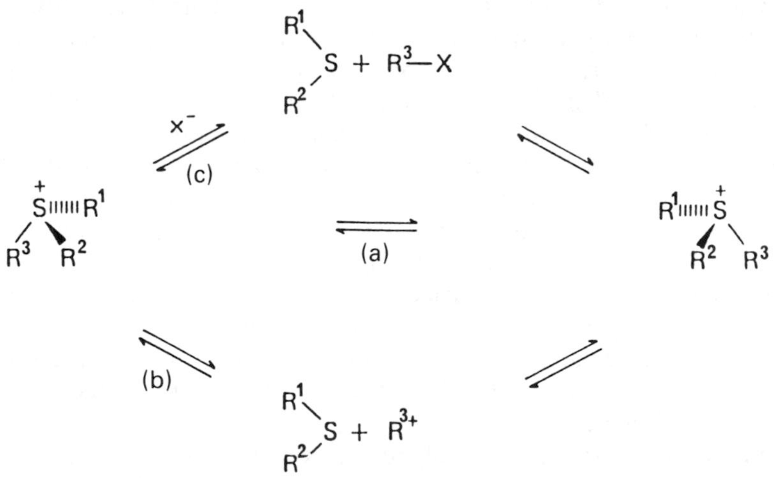

Scheme 1 — Stereomutation of sulplhonium salts. Reproduced with kind permission of John Wiley and Sons, from *The Chemistry of the Sulphonium Group* (eds. C. J. M. Sterling and S. Patai) 1981.

Sulphonium salts are known to undergo several synthetically useful reactions (Scheme 2). They are susceptible to C–S$^+$ bond fission via nucleophilic displacement (Scheme 2, route a), β-elimination reactions to yield the appropriate olefin and thioether (Scheme 2, route b), and α-proton removal to afford the ylide (Scheme 2, route c), which may then undergo molecular rearrangement (Knipe, 1981). As a result of their susceptibility to nucleophilic displacement reactions, sulphonium salts are usually unstable in basic solution.

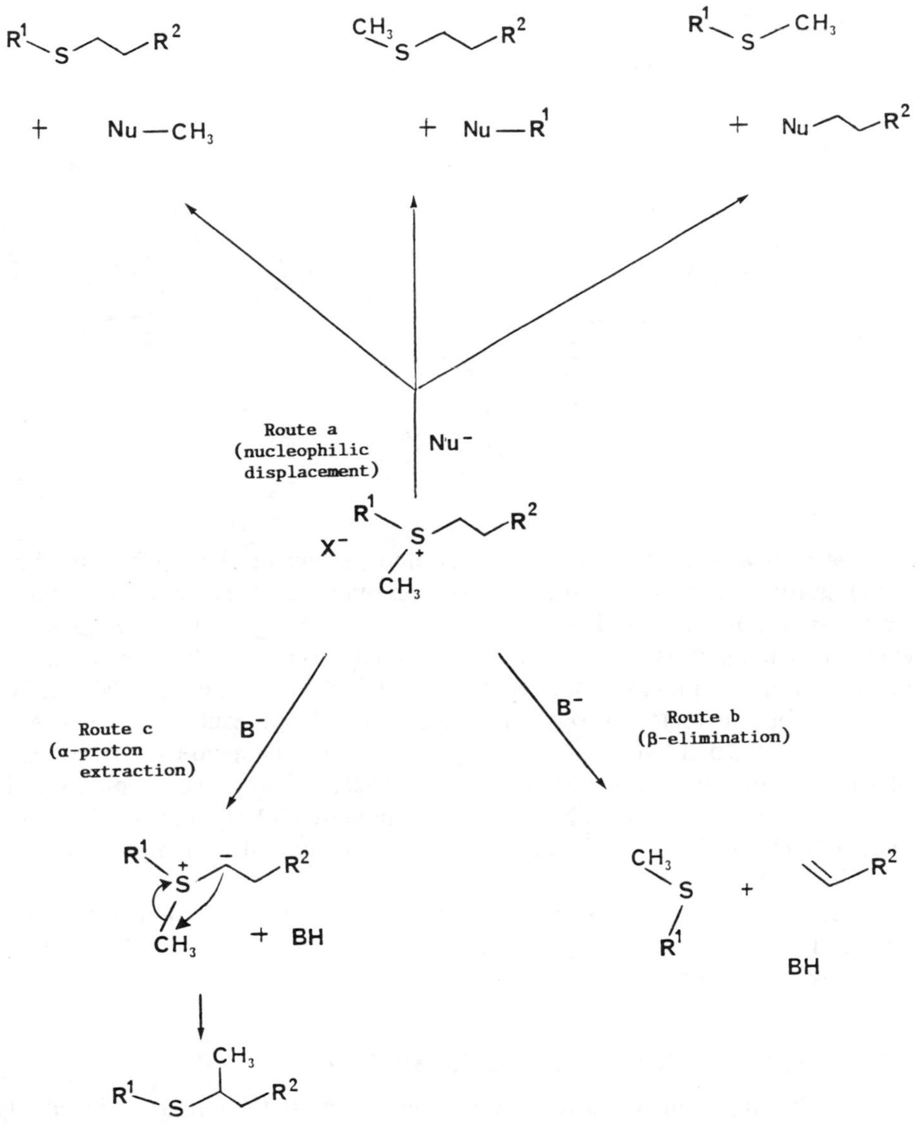

Scheme 2 — Chemical reactions of sulphonium salts.

Only four sulphonium salts have been established as occurring naturally. These
are (Plate 2) 2-carboxyethyldimethylsulphonium (dimethyl-β-priopiothetin,
DMPT) (**4**), methionine methylsulphonium (*S*-methyl-L-methionine, SMM) (**5**),
S-(5′-deoxyadenosyl-(5′)-L-methionine (*S*-adenosyl-L-methionine, SAM) (**6**) and
S-adenosyl-(5′)-3-methylthiopropylamine (decarboxylated SAM, dSAM) (**7**).

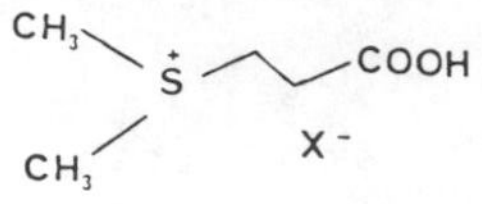

(**4**) 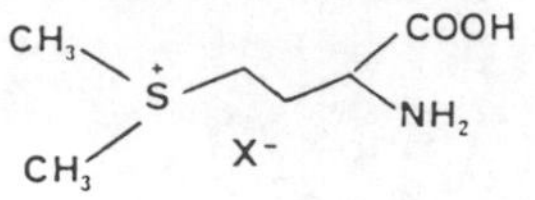   (**5**)

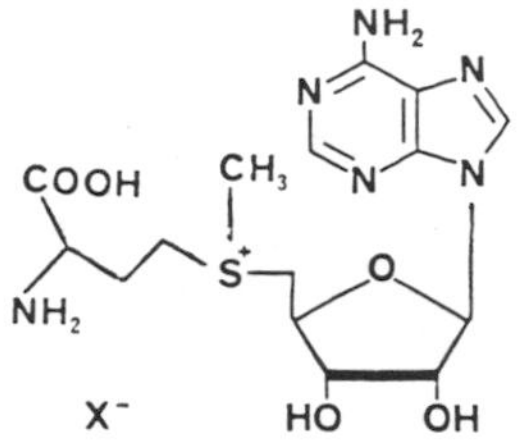

6 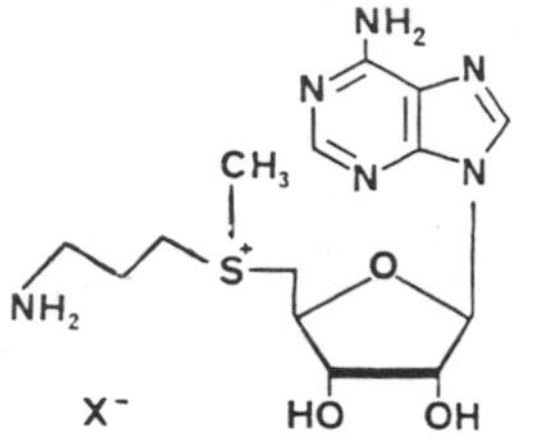   (**7**)

Plate 2

DMPT is widely distributed in marine and freshwater algae (Challenger *et al.*,
1957), marine invertebrates, and many species of plankton (Ackman *et al.*, 1966), as
well as in fish, molluscs and crustacea (Ackman and Hingley, 1968). SMM is found
exclusively in plant tissue, particularly in cabbage, celery roots, kohlrabi and jack
bean, as well as in many other types of vegetables (Kovacheva, 1974). Although
SMM has been detected in cow's milk, this is probably a result of its entry into the
animal from the diet (Keenan and Lindsay, 1968). SAM appears to be a constituent
of virtually all living tissues. It is present in bacteria, fungi, yeast, plants and the
tissues of higher animals. The related compound, dSAM, like SAM, is widely
distributed throughout the plant and animal kingdoms, and is an important intermediate in the biogenesis of polyamines.

This chapter will review the chemistry, role, biosynthesis and catabolism of
endogenous sulphonium compounds, and will deal briefly with the biotransformation of selected exogenous sulphonium salts.

6.2   CHEMICAL AND ANALYTICAL ASPECTS

One of the important features of sulphonium salts is their relative instability in
solution, particularly at pH values above 7. The major decomposition products of
DMPT in alkaline media are dimethylsulphide and acrylic acid (Maw, 1981), formed
presumably via route b shown in Scheme 2.

The stability of a number of salts of SMM at various pH values has been studied by Ramirez *et al.* (1973) in some detail. The major decomposition pathway at pH 7–11 affords dimethylsulphide and α-aminobutyrolactone (**8**) as a result of intramolecular group participation of the carboxyl function via the mechanism shown in Scheme 3; a minor product of this reaction is homoserine. An alternative decomposition route occurs at low pH values, involving nucleophilic attack of dimethylsulphide on the methyl group of SMM, leading to methionine (**9**). Small amounts of homocysteine (**10**) are also formed, probably by a very slow nucleophilic attack of dimethylsulphide on methionine (Scheme 3).

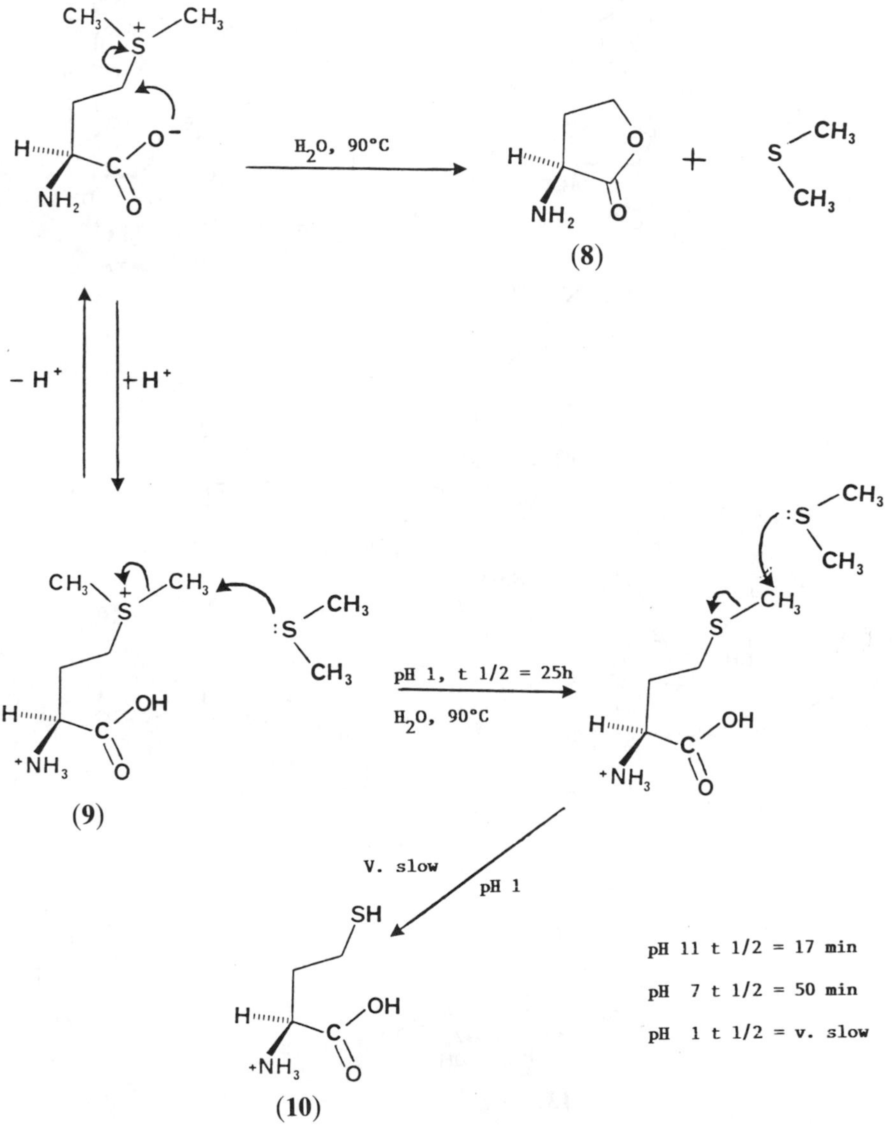

Scheme 3 — Decomposition pathways for *S*-methylmethionine.

SAM is relatively less stable in aqueous solution than SMM (De la Haba *et al.*, 1959), most decompositions giving rise to complex mixtures, some principal components of which have been identified (Cornforth *et al.*, 1977; Borchardt, 1979; Fiecchi, 1979). At slightly acidic pH values SAM degrades via a similar mechanism to SMM, affording 5′-deoxy-5′-methylthioadenosine (MTA) (**11**), α-aminobutyrolactone (**8**) and homoserine via a carboxyl group participation mechanism. At alkaline pH values, e.g. pH 12, the *N*-glycosidic bond is cleaved at room temperature, and adenine (**12**) and a pentosylmethionine (**13**) are formed (see Scheme 4). Halogeno

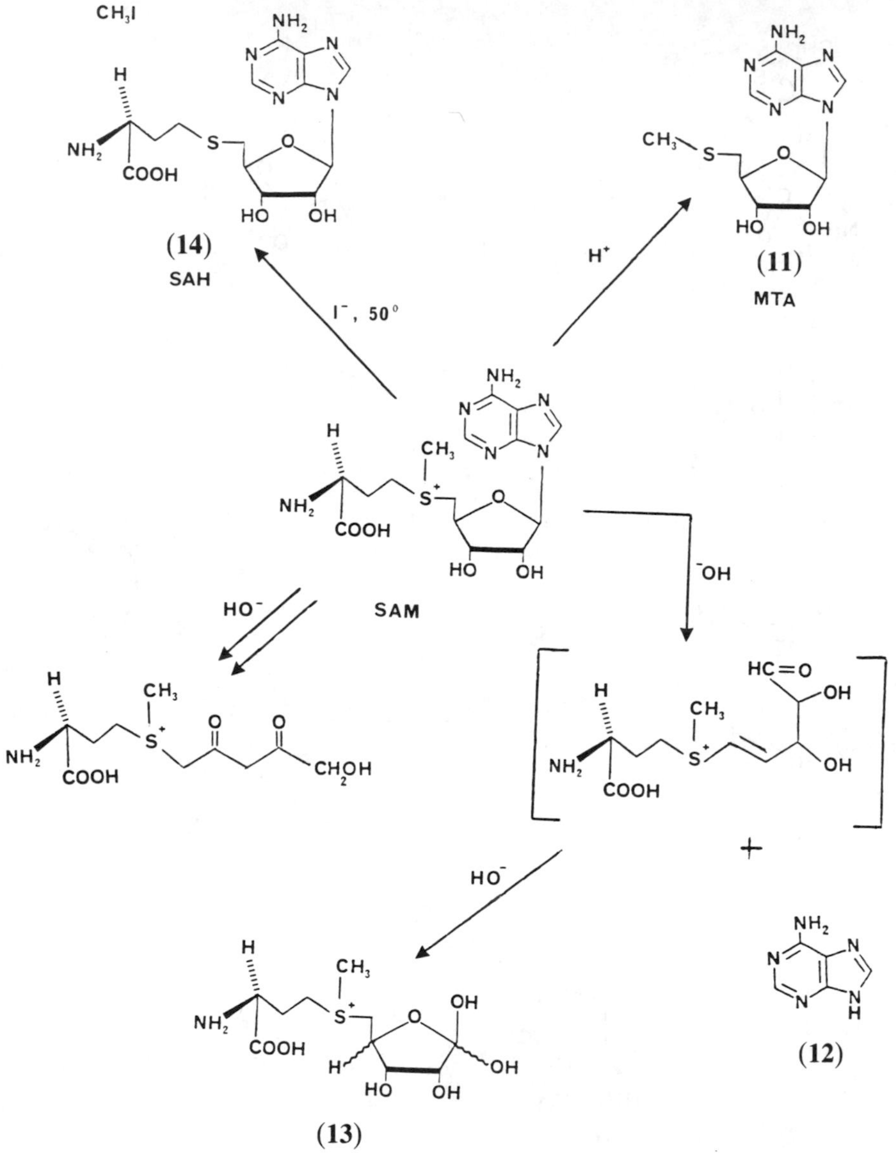

Scheme 4 — Decomposition pathways for *S*-adenosylmethionine.

salts of SAM have been shown to be thermally unstable and chemical S-demethyla-
tion of SAM occurs in fairly good yield by treatment of SAM with iodide ions at 50°C
for 24 h, to give S-adenosyl-L-homocysteine (SAH) (**14**). However, a number of salts
stable enough to allow pharmacological experimentation have been reported, e.g.
SAM. 2RSO$_3$H and SAM . p-CH$_3$–2H$_2$SO$_4$.RSO$_3$H (where R=CH$_3$CH$_2$CH$_2$– or p
CH$_3$–C$_6$H$_4$–) (Fiecchi, 1979).

It should be noted that although SAM is generated enzymically in a chirally pure
form at the sulphonium pole (i.e. in the S configuration), which is thought to be the
enzymatically active form in methyltransferase reactions, Wu *et al.* (1983) have
shown that in solution, at pH 7.5 and 37°C, this form of SAM undergoes simultaneous irreversible conversion to MTA and homoserine with a rate constant of
6×10^{-6} S^{-1}, and a reversible conversion to the enzymatically inactive stereoisomer
with the R configuration at the sulphonium pole with a rate constant at 8×10^{-6} S^{-1}.
Thus it appears that sulphonium salts such as SAM are chirally unstable at the
sulphonium centre when in aqueous solution.

The chemical stability of dSAM has not been examined in any detail. At room
temperature and alkaline pH, cleavage of the glycosidic bond occurs to give adenine,
while, at 100°C, methylthiopropylamine is a major decomposition product (Zappia
et al., 1969b). No significant degradation of the molecule appears to occur at acidic
pH.

The isolation, characterization and quantitation of sulphonium salts are not
without difficulties. Such polar, water-soluble, ionic organicals are less easily
isolated than neutral organic compounds. This, together with their relative instability, presents a formidable challenge to the investigator. Earlier studies utilized
classical chemical methods such as gravimetric determination of insoluble salts, e.g.
perchlorates, tungstosilicates, reineckates and tetraphenyl borates, titrimetric methods, and colorimetric determinations. In addition, polarographic and UV spectroscopic methods have been developed for the analysis of some sulphonium salts
(Ashworth, 1981). More recent techniques for the isolation and quantitation of
sulphonium salts include thin layer chromatography and high-pressure liquid chromatography (Ashworth, 1981; Hassan, 1987; Sheng *et al.*, 1984). The latter technique, in conjunction with fast atom bombardment (FAB) mass spectrometric
analysis, appears to be the most suitable technique for the rapid analysis and
quantitation of sulphonium salts (Sheng *et al.*, 1984).

6.3 BIOCHEMICAL ASPECTS

6.3.1 Biosynthesis of sulphonium salts

DMPT is biosynthesized from methionine in marine algae via the pathway outlined
in Scheme 5 (Greene, 1962). The terminal methylation step that affords the
sulphonium salt is thought to involve $N_{(5)}$-methyltetrahydrofolate as cofactor.

SMM has also been shown to be biosynthesized from methionine in a variety of
plants, but the reaction requires the presence of SAM as the methyl donor (Greene
and Davis, 1960). The products of this reaction are SMM and SAH (Scheme 6).

The biosynthesis of SAM utilizes L-methionine and adenosine triphosphate
(ATP), and involves a nucleophilic transfer of the 5'-deoxyadenosyl moiety of ATP

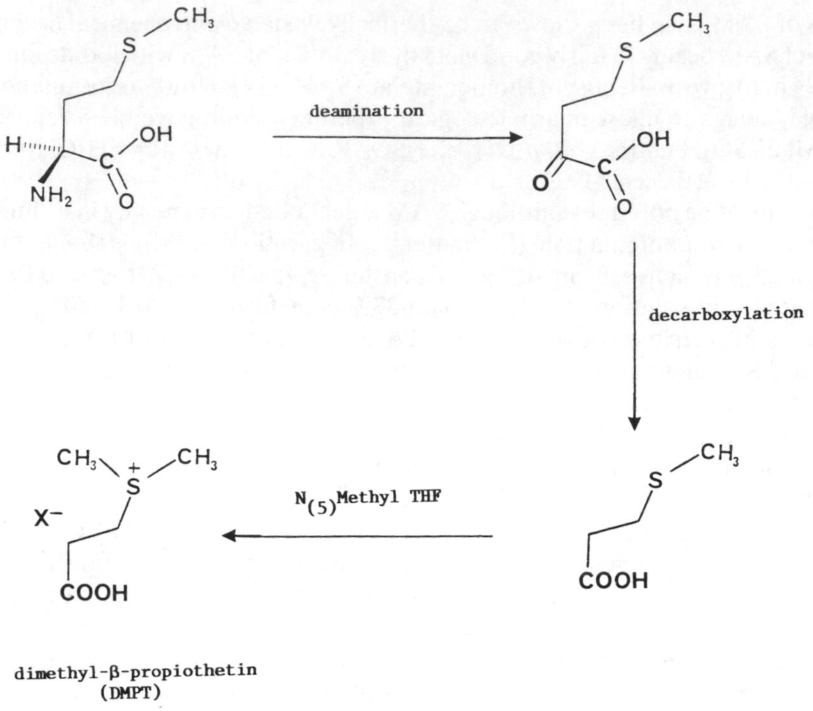

Scheme 5 — Biosynthesis of dimethyl-β-propiothetin.

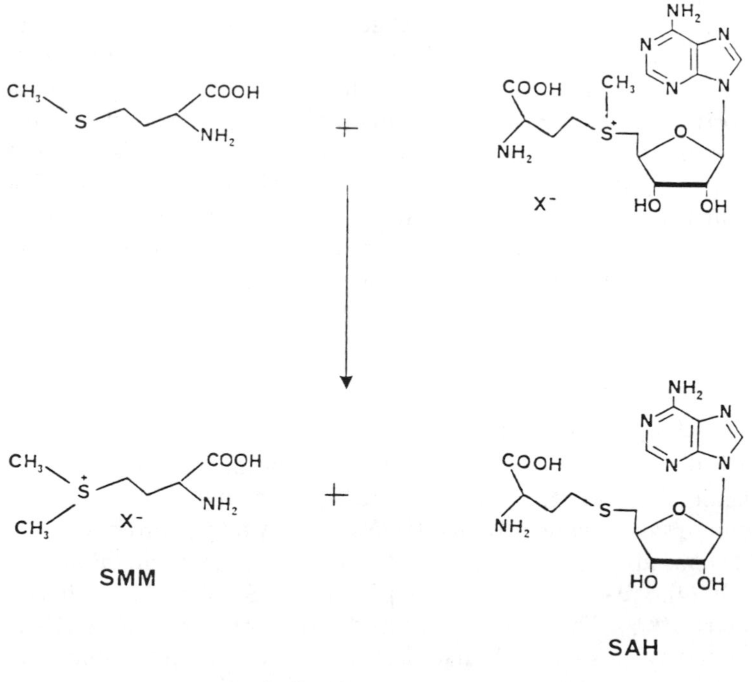

Scheme 6 — Biosynthesis of *S*-methylmethionine in plants.

to the methionine S atom, with release of the terminal phosphate as inorganic phosphate together with the remaining phosphates as pyrophosphate (Mudd, 1965). This reaction has been shown to proceed via a direct nucleophilic attack of the S atom of methionine on the 5'-CH_2 of ATP via a single displacement reaction; no adenosylated enzyme intermediate is formed (Parry and Minta, 1982) (Scheme 7). In addition to the above pathway, at least two other biosynthetic pathways to SAM are known. In several strains of yeast, and in animal tissues, there is evidence that SAM can be formed from SAH via S-methylation with $N_{(5)}$-methyltetrahydrofolate (Finkelstein and Harris, 1973). SAH is obtained from an enzyme-catalysed reaction of adenosine with L-homocysteine. Also, isotopic labelling studies have shown that MTA can act as a precursor of SAM in yeast and bacteria, although the exact mechanism has not been elucidated (Schlenk and Ehninger, 1964). These latter two routes can be regarded as pathways for the recycling of the metabolic breakdown products of SAM, rather than its *de novo* biosynthesis, and may be of importance in conserving sulphur within the organism.

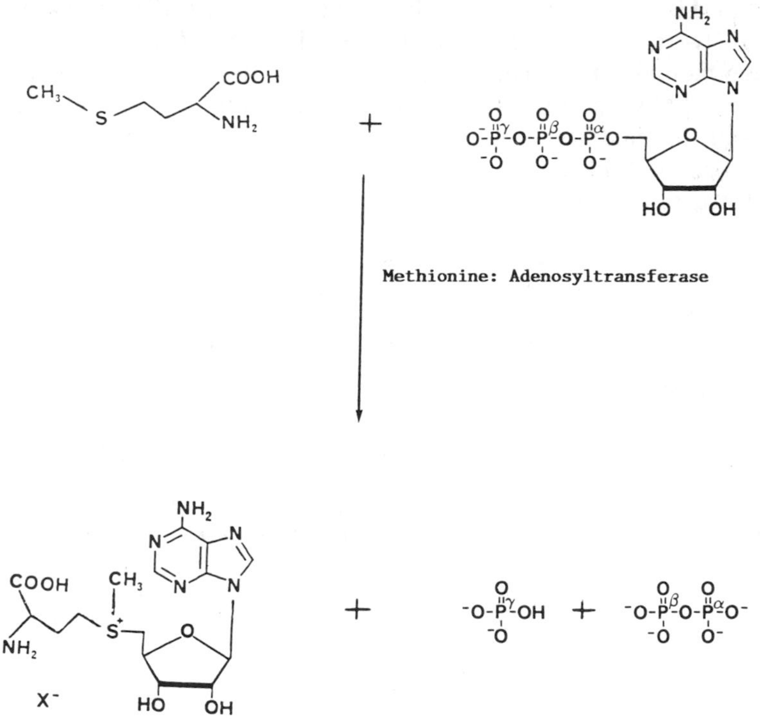

Scheme 7 — Biosynthesis of S-adenosylmethionine.

Decarboxylated SAM is formed from the enzymic decarboxylation of SAM via a pyruvate-dependent decarboxylase, a key enzyme in the regulation of polyamine biosynthesis (Williams-Ashman and Pegg, 1981).

6.3.2 Role of endogenous sulphonium compounds in methyltransferase reactions

Dubnoff and Borsook (1948) established that DMPT acted as a methyl donor in the formation of methionine from homocysteine in mammalian liver and kidney. Subsequent studies showed that the enzyme that catalysed the above reaction, i.e. thetin:L-homocysteine methyltransferase, was present in mammals, including man, and in algae, but was absent in plants and yeasts (Maw, 1981). The enzyme is stereoselective for L-homocysteine, although the D isomer is slightly active as a substrate. Interestingly, the enzyme can utilize other sulphonium compounds as substrates, e.g. SMM and dimethylacetothetin, the latter compound being the most active methyl donor substrate known. This raises the question as to what the function of the enzyme is in mammals. Its presence in the liver in relatively large amounts with a turnover considerably greater than that of other methyltransferases suggests that it may have a role either in methionine biosynthesis or in methyl group transfer. However, it is puzzling that its apparently preferred substrate is neither a constituent of the diet nor a normal metabolite.

Shapiro and Yphantis (1959) and Shapiro (1956) have shown that the methyl group of SMM is transferred specifically to L-homocysteine, the net result being the formation of two molecules of methionine (see Scheme 8). However, the existence of a distinct '*S*-methylmethionine:homocysteine methyltransferase' is still a matter of debate, since SMM is also a highly effective substrate for the related enzyme, *S*-adenosylmethionine:homocysteine methyltransferase (Shapiro *et al.*, 1965). Work by Allamong and Abrahamson (1977) indicates that, at least in plants, there is good evidence for the existence of the two distinct homocysteine methyltransferases.

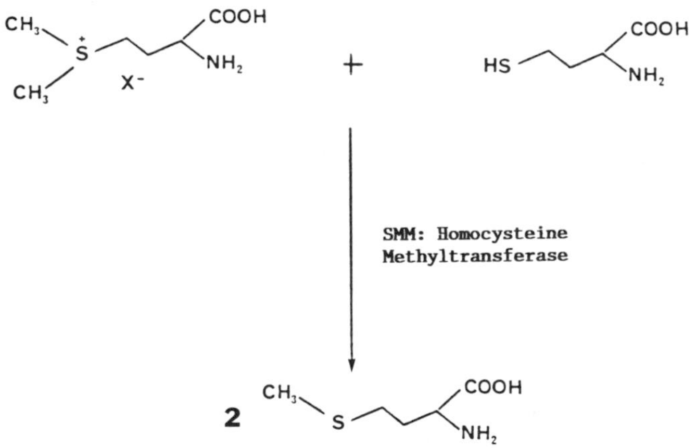

Scheme 8 — The *S*-methylmethionine:homocysteine methyltransferase reaction.

In an elegant study by Kjaer *et al.* (1979), which utilized double-labelled substrates, an interesting stereochemistry was observed in the transfer of methyl group from SMM to homocysteine catalysed by homocysteine methyltransferase from jack bean seeds. It was shown that a stereoselectivity of greater than 90% for the pro-R methyl group is exhibited by the enzyme in the transference of methyl group to L-homocysteine. Since the monochiral SMM contains prochiral, diastereotopic methyl groups, it follows, in principle, that the transfer of these groups to a chiral or achiral acceptor molecule should occur at different rates.

Unlike DMPT and SMM, SAM is a remarkably versatile methyl donor, and it appears to be as ubiquitous as ATP, in that it is implicated in almost every aspect of metabolism. Transfers of methyl group to O, N, S and C atoms are all catalyzed by SAM-dependent methyltransferases. Well over 50 methyltransferases that utilize SAM as cofactor have been officially listed by the International Union of Biochemistry, and there are probably three or four times this number cited in the biochemical literature. Many of these methyltransferases are highly specific with respect to their methyl acceptor substrates. SAM can act as a methyl donor in the methylation of amino acids, catecholamines, purine bases, neurotransmitter amines, sterols and terpenes, carboxylic acids, sugars and many other constituents in both the plant and the animal kingdom. It also serves as the source of methyl group in the methylation of macromolecules such as polysaccharides, proteins, and nucleic acids. The enzyme catechol O-methyltransferase (COMT) may play a significant role in the metabolism of catechol drugs and numerous non-physiological catechols (Borchardt, 1980) and, recently, the identification of non-specific SAM-dependent N-methyltransferases with wide substrate specificities in rabbit liver suggests that SAM-mediated methylations may play an important role in the biotransformation and detoxification of a wide range of xenobiotic amines (Ansher and Jakoby, 1986; Crooks et al., 1988).

In the course of all SAM-dependent methyltransferase reactions, as well as the methylated product, a second product, S-adenosyl-L-homocysteine (SAH) is formed (see Scheme 9). SAH has been shown to be a potent competitive inhibitor of almost all known SAM-dependent methyltransferases and, in many cases, the K_i value for this inhibition is lower than the K_m value for SAM, indicating that the thioether has a greater affinity for the methyltransferase than the donor substrate (Coward and Crooks, 1979). It is now widely accepted that SAH plays a major regulatory role in controlling SAM-mediated cellular methylations. In this respect, the cellular destruction of SAH is controlled by the enzyme SAH hydrolase, which cleaves SAH to adenosine and L-homocysteine (Crooks et al., 1979). This enzyme is widely distributed in animal tissues and in yeasts and is thought to play a complementary role to SAH in regulating cellular methylation.

In recent years, important developments have occurred in the elucidation of the mechanism of methyl group transfer in SAM-dependent enzyme reactions. Initial kinetic studies on the mechanism of the COMT reaction were a matter of some controversy. Investigations by Flohe and Schwabe (1970, 1972) and by Coward et al. (1973) strongly supported a random Bi–Bi mechanism involving a rate-determining methyl transfer directly from SAM to the acceptor substrate. However, inhibition studies with COMT, utilizing tropolone and 8-hydroxyquinoline (Borchardt, 1973), suggested that a ping-pong mechanism was operating, involving initial binding of SAM to the enzyme followed by the formation of a methylated enzyme intermediate which then transferred its methyl group to the acceptor substrate (see Scheme 10). Subsequent studies by Hegazi et al. (1976, 1979) and Rodgers et al. (1982) have

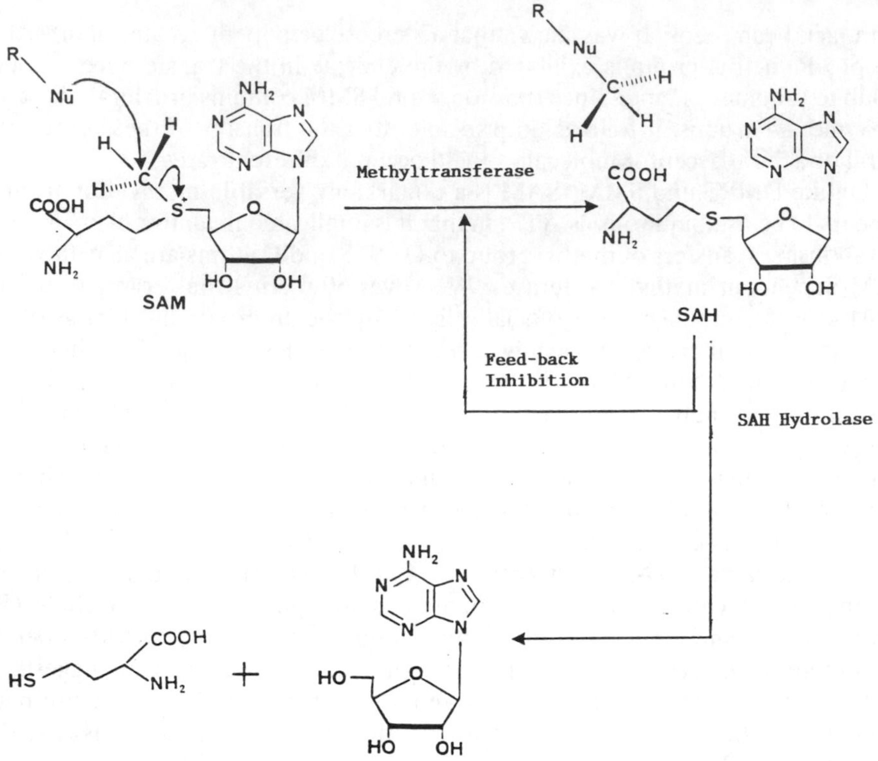

Methyltransferase

SAM

SAH

Feed-back
Inhibition

SAH Hydrolase

HS COOH NH₂ + HO

Scheme 9 — Regulatory role of *S*-adenosylhomocysteine in *S*-adenosylmethionine-dependent methyl-transferase reactions.

$$A + E$$

$$E.A + SAM$$

$$SAM + E \rightleftharpoons E.SAM + A \rightleftharpoons E.SAM.A$$

$$E + SAH + A - CH_3$$

Random Bi–Bi Mechanism

$$SAM + E \rightleftharpoons E.SAM \longrightarrow E - CH_3 + SAH$$

$$A + E - CH_3 \rightleftharpoons A.E - CH_3 \longrightarrow E + A - CH_3$$

Ping Pong Mechanism

Scheme 10 — Random Bi–Bi and ping-pong mechanisms for SAM-dependent methyltransferase reactions.

shown that transfer of the methyl group in the COMT reaction occurs via a tight, in-line symmetrical S_N2 transition state in which the methyl group is located between the leaving group and the nucleophile. The steric course of methyl group transfer catalysed by COMT has also been studied using SAM carrying a methyl group made chiral by labelling with ^{1}H, ^{2}H and ^{3}H in an asymmetrical arrangement (Woodward et al., 1980). The latter study clearly showed that methyl group transfer occurred in an inversion mode, indicating a direct transfer from SAM to the O atom of the catechol substrate via an S_N2 process (Scheme 9). Similar stereochemical results have been obtained in chiral methyl studies with related SAM-dependent reactions (Mascaro et al., 1977; Floss and Tsai, 1979; Asano et al., 1984; Virchaux, 1981), and it can be generally concluded that transfers of sp^3 carbon catalysed by SAM-dependent methyltransferases proceed with inversion of configuration at the migrating carbon, thus involving a direct transfer from donor substrate to acceptor substrate. In the case of the COMT reaction, Olsen et al. (1979) have suggested that catalysis is associated with compression of partial bonds about the transferring methyl group in the transition state. They speculate that compression catalysis may be the way in which COMT catalyses methyl transfer.

In addition to the asymmetric sulphonium centre in SAM, the α-carbon atom of the amino acid moiety is also chiral. Since the configuration of the ribose unit is fixed, this presents four possible diastereomers of SAM. The absolute stereochemistry of naturally occurring SAM has been determined to be 5'-[(3S)-3-amino-3-carboxypropyl)methyl-(S)-sulphonio]-5'-deoxyadenosine by Cornforth et al. (1977) by correlation with the absolute configurations of the diastereomeric S-carboxymethyl-(S)-methionine salts. (+)-S-adenosyl-(S)-methionine (i.e. SR at the sulphonium pole and S at the α-carbon of the amino acid) and (+)-S-adenosyl-(R)-methionine have been prepared, and appear to be identical, with respect to their chemical properties, to natural SAM from yeast (De la Haba et al., 1959). However, (+)-S-methionine exhibited only 50% of the methyltransferase activity of natural SAM. In studies with COMT, a high specificity for (+)-S-adenosyl-(S)-methionine (Borchardt and Cheng, 1978; Borchardt et al., 1976, Borchardt and Wu, 1976) has been observed, and Borchardt and Wu (1976) have shown that the S configuration at the asymmetric sulphonium centre is required for optimal enzymic binding and methyl donation activity. The corresponding R isomer was inactive as a methyl donor but was a potent inhibitor of COMT. Subsequent work (Borchardt et al., 1976) utilizing structural analogues of SAM has clearly shown that the COMT binding site for the methyl donor exhibits strict stereospecificity for the structural features of (S)-S-adenosyl-(S)-methionine. These results are difficult to interpret in a view of the reported chiral instability of SAM in solution (Wu et al., 1983).

dSAM has a central role to play in the biosynthesis of the important cell constituents, the polyamines. Polyamines are widespread in bacteria, plants and animals (Tabor et al., 1961) and consist of putrescine (**15**), spermidine (**16**) and spermine (**17**) (Scheme 11). Bacteria contain high levels of putrescine; however, in animals spermine is the most common polyamine and is present in high concentrations in human seminal fluid and in human prostate. The polyamines are thought to play an important role in cell structure and are normally associated with nucleic acids in animal tissue. They are excreted in abnormally high concentrations from patients with certain forms of cancer, and high levels of spermidine have been detected in the serum of such patients (Russell, 1973; Russell and Russell, 1975).

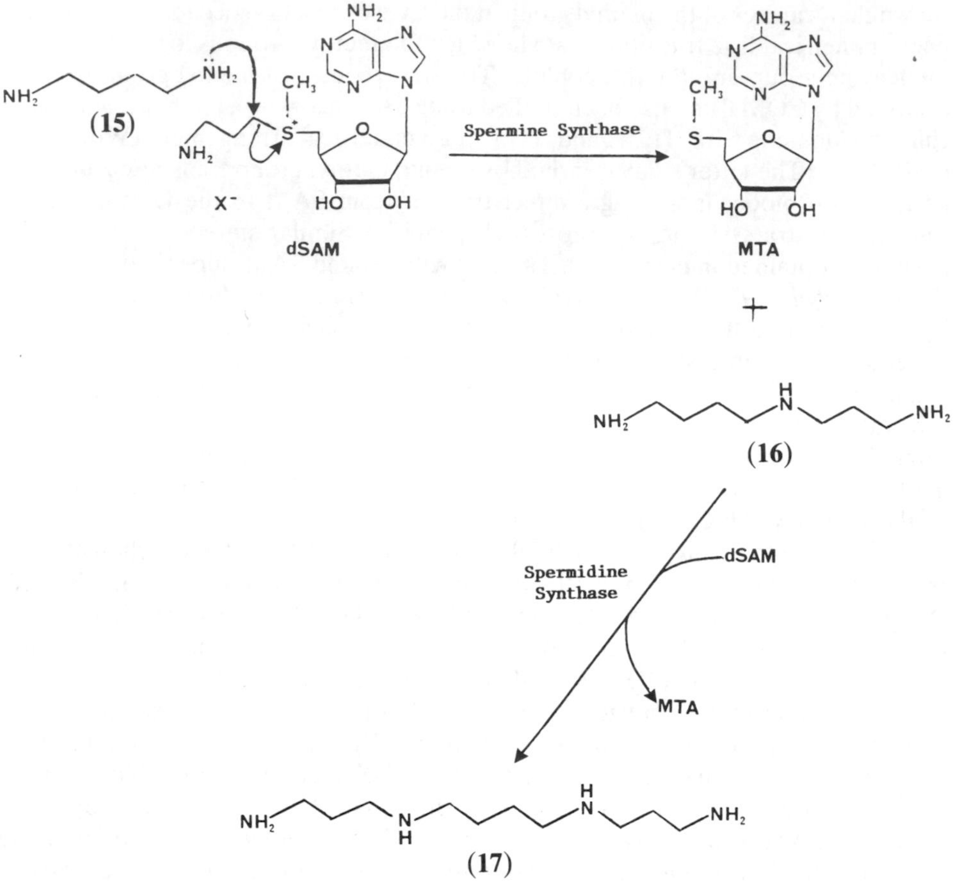

Scheme 11 — Biosynthesis of spermidine and spermine.

The biosynthesis of spermidine is illustrated in Scheme 11. dSAM acts as a aminopropyl group donor, transferring this group to the –NH$_2$ group of the acceptor substrate, putrescine. The reaction is catalysed by the enzyme spermidine synthase and affords MTA as a secondary product. Spermidine can then act as the acceptor substrate in a second aminopropyl group transfer involving dSAM, to give spermine and MTA. This second reaction is catalysed by the enzyme spermine synthase. Tang *et al.* (1981) have shown that, as with some methyltransferase reactions (Benghiat and Crooks, 1983, 1986), transfer of the aminopropyl group to the acceptor substrate can be inhibited by appropriately designed transition state analogues, suggesting that this reaction proceeds via a classical S$_N$2 substitution reaction and does not involve an aminopropylated enzyme intermediate. The secondary product of these reactions, i.e. MTA, appears to act as a feed-back inhibitor of the aminopropyltransferase reactions in much the same way as SAH interacts with SAM-dependent methyltransferases (Pegg and Hibasami, 1979; Schlenk, 1983). Its concentration in tissue is regulated by a phosphate-requiring nucleosidase that is widely distributed in animal tissues and which cleaves MTA to adenine and 5-methylthioribose-1-phosphate; this latter product is eventually recycled into methionine (Schlenk, 1983) via an elaborate sulphur-conserving pathway.

dSAM has been shown to possess some weak methyl donor activity, and it appears to act as a substrate for yeast homocysteine S-methyltransferase (Zappia, 1969a, b).

6.3.3   Catabolism of sulphonium compounds

The catabolism of the endogenous sulphonium compounds invariably involves the enzyme-catalysed cleavage of a C–S$^+$ bond, leading to the formation of a thioether. Additional biotransformations can occur when other moieties are present in the molecule (e.g. an amino acid moiety in the case of SMM and SAM) which may not affect the sulphonium centre. Since the presence of three different groupings around the sulphonium pole is possible, and the cleavage of each of these groupings from sulphur usually requires specific enzymes, it might be expected that each of the natural sulphonium salts is catabolysed via different pathways.

6.3.3.1   *Dimethyl-β-propiothetin*

Catabolism of DMPT occurs via two main pathways: (a) cleavage to dimethyl sulphide and (b) loss of S-methyl group. The former pathway has been demonstrated in algae (Cantoni *et al.*, 1956) and affords acrylic acid as a second product. The second pathway results from the involvement of DMPT in enzyme-catalysed transmethylation reactions together with a second substrate or acceptor molecule. It is unlikely that direct oxidation of the $^+$S–CH$_3$ group of DMPT occurs, since trimethyl-, diethylmethyl- and triethylsulphonium salts are not converted into inorganic sulphate. Although DMPT when administered to rats affords 30% of the $^+$S–CH$_3$ group as respired CO$_2$ (Ferger and Du Vigneaud, 1950) and approximately 60% of the sulphur as urinary inorganic sulphate, it is likely that methyl group transfer is an initial obligatory step prior to subsequent oxidative metabolism.

6.3.3.2   *S-methylmethionine*

As is the case with DMPT, there are two possible catabolic pathways involving C–S fission for the SMM molecule. Loss of the S-methyl group occurs as a consequence of the involvement of SMM as a cofactor in methyltransferase reactions (see earlier), to afford methionine as one of the products (Shapiro, 1956). Removal of the amino acid moiety of SMM by a methioninesulphonium lyase, with the formation of dimethylsulphide, has also been reported in bacteria, yeast and plants. Homoserine is the second product of this reaction; it probably results from ring opening of the intermediate butyrolactone (**8**) (Mazelis *et al.*, 1965; Hattula and Granroth, 1974).

6.3.3.3   *S-adenosylmethionine*

SAM is susceptible to enzymatic breakdown at a number of sites in the molecule, these being represented in Fig. 1. Cleavage of the C–S$^+$ bond at all three sites in the molecule has been observed.

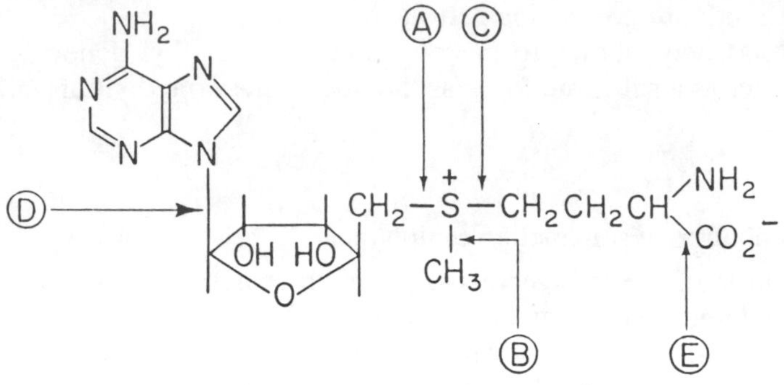

Fig. 1 — . Sites of chemical and enzymatic bond fission in S-adenosylmethionine. Reproduced with kind permission of John Wiley and Sons, from *The Chemistry of the Sulphonium Group* (eds. C. J. M. Sterling and S. Patai) 1981.

(1) Breakage of the ribose C–S bond at A is catalysed by an adenosyltransferase, an enzyme found in extracts of an *E. coli* mutant (Pfeffer and Shapiro, 1962), giving rise to methionine as the sulphur-containing product and resulting in transference of the adenosyl moiety to the acceptor molecule, homocysteine, to afford SAH.

(2) Enzymatic loss of the methyl group at B takes place through the mediation of methyltransferases, provided that a second substrate is present to act as a methyl acceptor (see earlier), and yields SAH as a second product.

(3) The enzyme-catalysed bond fission at C leads to the loss of the α-amino-n-butyryl moiety to yield MTA. This is a characteristic reaction found in many microorganisms. The $C_{(4)}$ group is displaced as α-amino-γ-butyrolactone which is then hydrolysed non-enzymically to homoserine. Nishimura *et al.* (1974) have reported the enzymatic transfer of the α-amino-n-butyryl group of SAM to uridine residues in t-RNA, to afford 3-(α-amino-n-butyryl)uridine.

The enzymatic cleavage of the adenine–ribose linkage at site D has not been firmly established; however, a nucleosidase apparently restricted to procaryotes can cleave this glycosidic bond after removal of the S-methyl group of SAM (Duerre, 1962), to afford adenine and S-ribosylhomocysteine.

The catabolism of SAM subsequent to methyl group transfer is shown in Scheme 12, and this illustrates the effective re-utilization of part of the SAM molecule. SAH, resulting from methyltransferase activity, has three fates; it can be converted into S-adenosyl-γ-thio-α-ketobutyrate by L-amino acid oxidase, into homocysteine and adenosine via the action of SAH hydrolase, or into S-ribosylhomocysteine by a nucleosidase followed by degradation by an appropriate lyase to homocysteine (Crooks *et al.*, 1979). Homocysteine generated in this way may undergo condensation with serine to yield cystathionine which in turn is cleaved to cysteine by cystathionase and further oxidized to inorganic sulphate. Alternatively, homocysteine may be methylated by betaine-, SAM- orr tetrahydrofolate-dependent methyltransferases to give methionine, which is then converted back into SAM via a methionine-activating enzyme.

This re-utilization of SAM from its catabolic products enables it to function

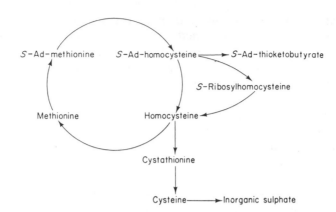

Scheme 12 — Pathways of *S*-adenosylmethionine catabolism. Reproduced with kind permission of John Wiley and Sons, from *The Chemistry of the Sulphonium Group* (eds. C. J. M. Sterling and S. Patai) 1981.

adequately in tissues even though its concentration is very low. In this respect it resembles other co-enzymes such as NAD, FAD and ATP.

Enzymatic decarboxylation of the amino acid moiety at site E is a well-established pathway and results in the formation of dSAM, the cofactor in the biosynthesis of the polyamines. SAM does not appear to act as a substrate for other amino acid-metabolizing enzymes such as the amino acid oxidases.

6.3.3.4 *Decarboxylated SAM*

Unlike SAM, dSAM does not generally act as a cofactor in methyltransferase reactions, and thus examples of catabolism via cleavage of the $^{+}$S–CH$_3$ bond are rare. In fact, dSAM serves as a weak inhibitor of other methyltransferases, and it may play a regulatory role in this capacity. In yeast, however, dSAM does have some activity as a methyl donor for the homocysteine *S*-methyltransferase enzyme.

The major catabolic pathway for dSAM is cleavage of the aminopropyl group. In this capacity dSAM serves as an aminopropyl donor for the biosynthesis of the polyamines spermine and spermidine, each reaction being catalysed by the corresponding synthase enzymes (see earlier). The second product of each of these reactions is MTA; this product can be recycled back to SAM via the salvage pathway illustrated in Scheme 13.

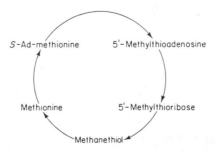

Scheme 13 — Regeneration of *S*-adenosylmethionine in bacteria via methylthioadenosine formation. Reproduced with kind permission of John Wiley and Sons, from *The Chemistry of the Sulphonium Group* (eds. C. J. M. Sterling and S. Patai) 1981.

6.3.4 Metabolism of exogenous sulphonium compounds

Very few studies have been carried out on the biotransformation of exogenous sulphonium compounds. In an early study, trimethylsulphonium chloride, when administered to rats, afforded dimethylsulphide together with the corresponding sulphone (Maw, 1953). This simple sulphonium salt appears to act as a substrate for thetin:homocysteine methyltransferase in rat liver (Maw, 1958), having 17% of the activity of dimethylacetothetin as a methyl donor. Trimethylsulphonium salts are also metabolized to the disulphide by a strain of *Pseudomonas* (Wagner *et al.*, 1966, 1967), a soil bacterium, which uses this compound as its sole source of carbon. The liberated methyl group is utilized for cellular biosynthesis by the organism, being transferred to tetrahydrofolate and then to SAM. Several other sulphonium salts related structurally to either dimethylacetothetin or DMPT, e.g. ethymethylacetoth-·etin chloride, dimethyl-α-propiothetin bromide, ethylmethyl-β-propiothetin bromide and dimethyl-γ-butyrothetin bromide, are also substrates for thetin:homocysteine methyltransferase. Diethylacetothetin chloride, sulphocholine iodide $((CH_3)_2^+SCH_2CH_2OH \cdot I^-)$ and triethylsulphonium iodide are not substrates for the enzyme. There is no evidence of the occurrence of metabolic oxidation at either the sulphur atom or at the α-carbon atom of exogenous sulphonium ions.

Administration of *S*-β-L-alanyltetrahydrothiophenium (as the mesylate salt) to rats afforded 3-hydroxytetrahydrothiophen sulphone as a major urinary metabolite (Roberts and Warwick, 1961). Clearly this sulphonium salt is initially degraded *in vivo* to tetrahydrothiophen, which is further oxidized to the ring hydroxylated sulphone (for structures, see Scheme 5, Chapter 9, this volume). Structurally similar tetrahydrothiophenium salts have been identified as metabolites, e.g. busulfan affords γ-glutamyl-β-(*S*-tetrahydrothiophenium)alanylglycine through glutathione conjugation (Hassan and Ehrsson, 1987a,b). This sulphonium metabolite is also degraded to tetrahydrothiophen, apparently in the gut, and the thioether is reabsorbed and oxidized to afford 3-hydroxytetrahydrothiophen sulphone as a urinary metabolite.

As was observed with DMPT, structural analogues of SAM may also serve as donor substrates in methyltransferase reactions. *S*-adenosyl-L-ethionine (**18**) and *S*-adenosyl-L-priopiothionine (**19**) (Plate 3) can transfer an ethyl group or an *n*-propyl group respectively in a number of methyltransferase-catalysed reactions, although the rate of reaction decreases dramatically with the *n*-propyl sulphonium salt (Maw, 1981; Schlenk and Dainko, 1975).

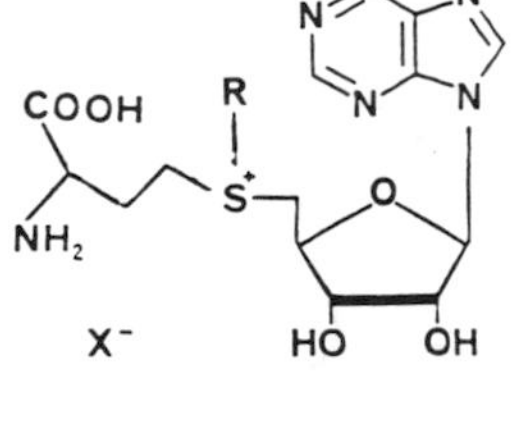

(**18**) R = C₂H₅
(**19**) R = n-C₃H₇

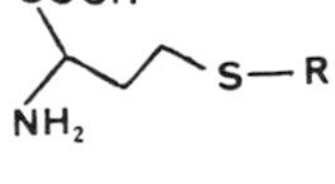

(**20**) R = C₂H₅
(**21**) R = CH=CH₂

Plate 3

6.4 TOXICOLOGICAL ASPECTS

While few toxicological studies have been carried out on sulphonium salts, the involvement of sulphonium intermediates in the mechanisms of toxicity of a number of xenobiotics has been proposed. The abilty of ethionine (**20**) and vinthionine (**21**) to induce hepatic carcinomas in rats has been linked to the bioactivation of these amino acids to their corresponding *S*-adenosylated derivatives (Leopold *et al.*, 1982). In the case of ethionine, its carcinogenicity is thought to be due to the resulting *S*-adenosylethionine having the capability of mimicking SAM and enzymically transferring ethyl groups to macromolecules such as DNA. A proposed mechanism for vinthionine toxicity is its *in vivo* conversion to *S*-adenosylvinthionine (**22**), which would be expected to be much more reactive towards cellular nucleophiles than either vinthionine or *S*-adenosylethionine because of the inductive and electron-sharing stabilization of the carbanion intermediate formed during nucleophilic addition at the β-carbon atom of the vinyl group (Leopold *et al.*, 1982) (see Scheme 14).

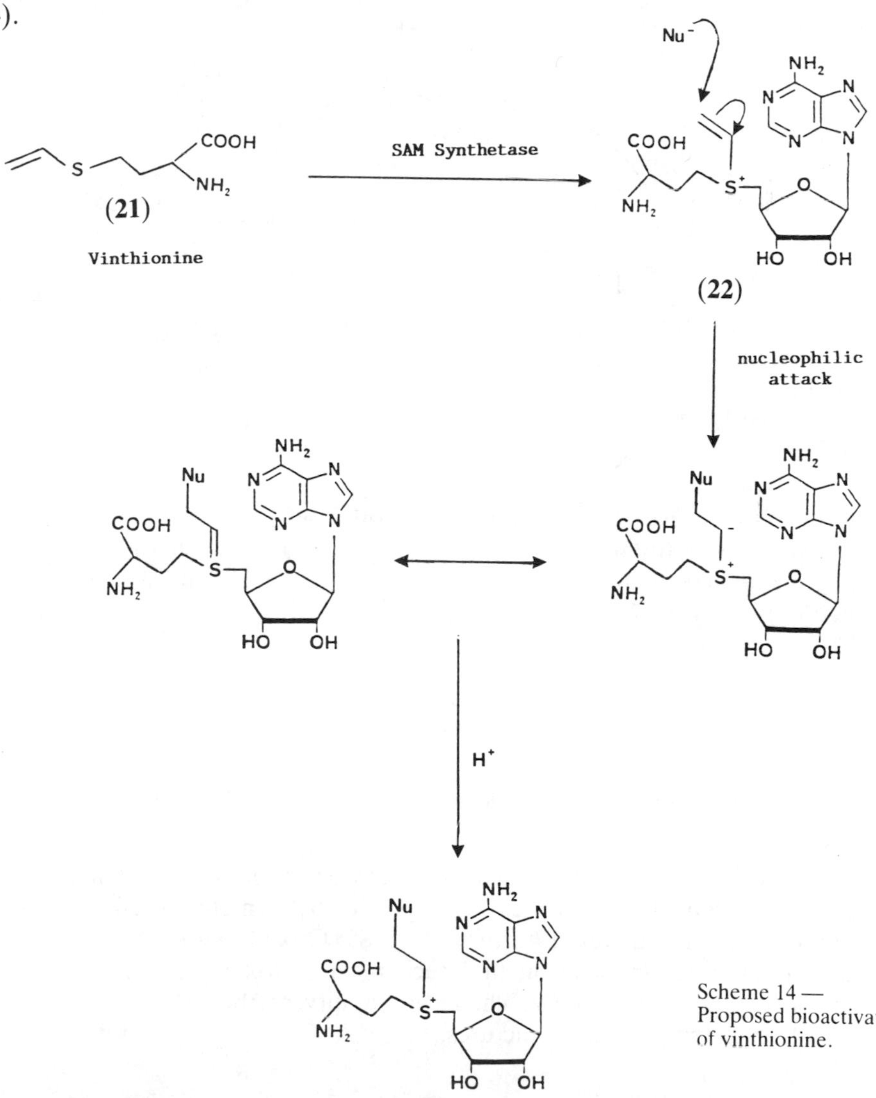

Nu⁻

(**21**)

Vinthionine

SAM Synthetase

(**22**)

nucleophilic attack

H⁺

Scheme 14 — Proposed bioactivation of vinthionine.

The toxicity of 1,2-dihaloethanes has received much attention in the last ten years (Elfarra *et al.*, 1985; Elfarra and Anders, 1985). Such compounds are potent mutagens and nephrotoxins (Rannung *et al.*, 1978) and are biotransformed mainly to the *S*-haloalkyl conjugates **23** and **24** (see Scheme 15). It has been proposed that the formation of these glutathione and cysteine conjugates proceeds via an intermediate episulphonium ion (see structures **25** and **26**, Scheme 15), which is reactive enough to interact with cell macromolecules to form covalent adducts (Webb *et al.*, 1985).

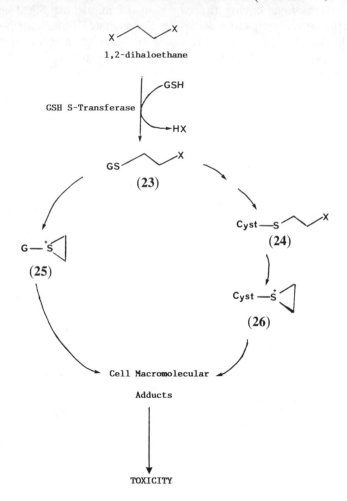

Scheme 15 — Proposed mechanisms of toxicity of 1,2-dihaloethanes.

Although other dihaloalkanes such as 1,3-dibromopropane, 1,4-dihaloalkanes and other 1,4-disubstituted alkanes also appear to be metabolized via similar sulphonium ion intermediates (Elfarra *et al.*, 1985; Onkenhout *et al.*, 1986), these ions are probably not involved in the mechanism of toxicity reported for such compounds, owing to the relative chemical stability of the 4- and 5-membered intermediate cyclic sulphonium structures.

Sheng *et al.* (1984) have proposed a role for sulphonium ion intermediates in the biotransformation of epoxides. Their hypothesis, which is based on the chemical reactivity of thioethers towards epoxides, and the observation that epoxides are usually excreted in the urine as ring-opened methylthio derivatives, is that a methylthio cellular component reacts with the epoxide to form methylsulphonium ions **27** and **28**, and that these ionic intermediates undergo elimination reactions to give methylthio metabolites **29** and **30** respectively (Scheme 16). The identity of the endogenous methylthio compound is speculated to be L-methionine. Application of this hypothesis to the biotransformation of arene oxides is an area worthy of study.

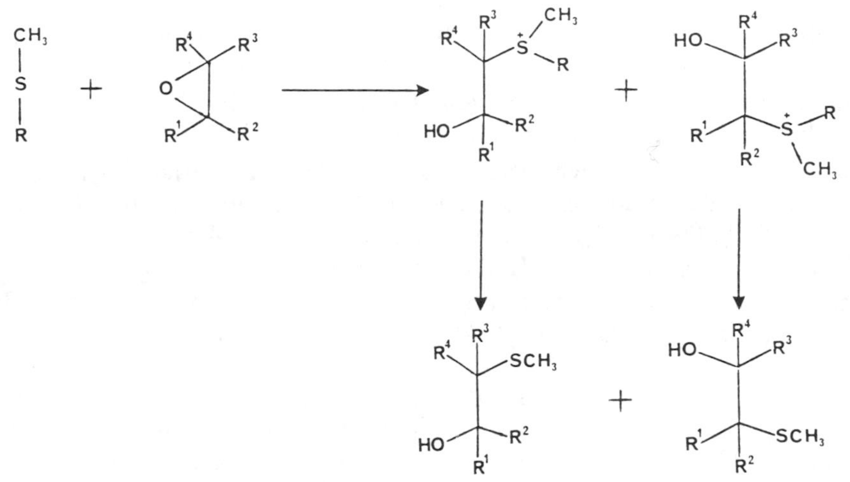

Scheme 16 — Proposed biotransformation of epoxides via methylsulphonium ion intermediates.

REFERENCES

Ackman, R. G. and Hingley, H. J. (1968). The occurrence and retention of dimethyl-β-propiothetin in some filter-feeding organisms. *J. Fish. Res. Bd. Can.*, **25**, 267–284.

Ackman, R. G., Tocher, C. S., and McLachlan, J. (1966). Occurrence of dimethyl-β-propiothetin in marine phytoplankton. *J. Fish. Res. Bd. Can.*, **23**, 357–364.

Allamong, B. D. and Abrahamson, L. (1977). Methyltransferase activity in dry and germinating wheat seedlings. *Bot. Gaz.*, **138**, 46–51.

Andersen, K. K. (1981). Stereochemistry and chiroptical properties of the sulphonium group. In C. J. M. Stirling and S. Patai (eds.), *The Chemistry of the Sulphonium Group*, Vol. 1, Wiley, Chichester, pp. 229–266.

Ansher, S. S. and Jakoby, W. B. (1986). Amine N-methyltransferases from rabbit liver. *J. Biol. Chem.*, **267**, 3996–4001.

Asano, Y., Woodard, R. W., Houck, D. R., and Floss, H. G. (1984). Stereochemical course of the transmethylation catalyzed by histamine N-methyltransferase. *Arch. Biochem. Biophys.*, **23**, 254–256.

Ashworth, M. R. F. (1981). Analysis and determination. In C. J. M. Stirling and S. Patai (eds.), *The Chemistry of the Sulphonium Group*, Vol. 1, Wiley, Chichester, pp. 79–99.

Benghiat, E. and Crooks, P. A. (1983). Multisubstrate adducts as potential inhibitors of S-adenosylmethionine-dependent methylases: inhibition of indole-N-methyltransferase by (5'-deoxyadenosyl-[3(3-indolyl)prop-1-yl]methylsulfonium and (5'-deoxyadenosyl)[4(3-indolyl)but-1-yl]methylsulfonium salts. *J. Med. Chem.*, **26**, 1470–1477.

Benghiat, E. and Crooks, P. A. (1986). Inhibition of Vaccinia mRNA cap-methylating enzyme by compounds designed as multisubstrate adducts. *J. Pharm. Sci.*, **75**, 142–145.

Borchardt, R. T. (1973). Catechol-O-methyltransferase. 1. Kinetics of tropolone inhibition. *J. Med. Chem.*, **16**, 377–382.

Borchardt, R. T. (1979). Mechanism of alkaline hydrolysis of S-adenosyl-L-methionine and related sulfonium nucleosides. *J. Amer. Chem. Soc.*, **101**, 458–463.

Borchardt, R. T. (1980). N- and O-methylation. In W. B. Jakoby (ed.), *Enzymatic Basis of Detoxification*, Vol. II, Academic Press, New York, pp. 43–62.

Borchardt, R. T. and Cheng, C. F. (1978). Purification and characterization of rat heart and brain catechol methyltransferase in rabbit. *Biochem. Biophys. Acta*, **522**, 49–62.

Borchardt, R. T. and Wu, Y. S. (1976). Potential inhibitors of S-adenosylmethionine-dependent methyltransferases. 5. Role of asymmetric sulfonium pole in the enzymatic binding of S-adenosyl-L-methionine. *J. Med. Chem.*, **19**, 1099–1103.

Borchardt, R. T., Wu, Y. S., Huber, J. A., and Wycpalek, A. F. (1976). Potential inhibitors of S-adenosylmethionine-dependent methyltransferases. 6. Structural modifications of S-adenosylmethionine. *J. Med. Chem.*, **19**, 1104–1110.

Cantoni, G. L., Anderson, D. G., and Rosenthal, E. (1956). Enzymatic cleavage of dimethylpropiothetin by *Polysiphonia lanosa*. *J. Biol. Chem.*, **222**, 171–177.

Challenger, F., Bywood, R., Thomas, P., and Hayward, B. J. (1957). Studies on biological methylation. XVII. The natural occurrence and chemical reactions of some thetins. *Arch. Biochem. Biophys.*, **69**, 514–523.

Cornforth, J. W., Reichard, S. A., Talalay, P., Carrell, H. L., and Glusker, J. P. (1977). Determination of the absolute configuration at the sulfonium center of S-adenosylmethionine. Correlation with the absolute configuration of the diastereomeric S-carboxymethyl-(S)-methionine salts. *J. Amer. Chem. Soc.*, **99**, 7292–7300.

Coward, J. K., Slitz, E. P., and Wu, F. Y.-H. (1973). Kinetic studies on catechol-O-methyltransferase. Product inhibition and the nature of the catechol binding site. *Biochemistry*, **12**, 2291–2297.

Coward, J. K. and Crooks, P. A. (1979). *In vitro* and *in vivo* effects of S-tubercidinylhomocysteine: a physiologically stable methylase inhibitor. In E. Usdin, R. T. Borchardt, and R. C. Crevling (eds.), *Transmethylation*, Vol. 5, *Developments in Neuroscience*, North-Holland, Amsterdam, pp. 215–224.

Crooks, P. A., Dryer, R. N., and Coward, J. K. (1979). Metabolism of S-adenosylhomocysteine and S-tubercidinylhomocysteine in neuroblastoma cells. *Biochemistry,* **18**, 2601–2609.

Crooks, P. A., Godin, C. S., Damani, L. A., Ansher, S. S., and Jakoby, W. B. (1988). Formation of quaternary amines by N-methylation of azaheterocycles with homogeneous amine N-methyltransferases. *Biochem. Pharmacol.,* **37**, 1673–1677.

De la Haba, G., Jamieson, G. A., Mudd, S. H., and Richards, H. H. (1959). S-adenosylmethionine: the relation of configuration at the sulfonium center to enzymatic reactivity. *J. Amer. Chem. Soc.,* **81**, 3975–3980.

Dubnoff, J. W. and Borsook, H. (1984). Dimethylation and dimethyl-β-propiothetin in methionine synthesis. *J. Biol. Chem.,* **176**, 789–796.

Duerre, J. A. (1962). A hydrolytic nucleosidase acting on S-adenosylhomocysteine and on 5'-methylthioadenosine. *J. Biol. Chem.,* **237**, 3737–3741.

Elfarra, A. A. and Anders, M. W. (1985). S-(1,2-dichlorovinyl)-L-homocysteine (DCVHC), an analogue of the renal toxin S-(1,2-dichlorovinyl)-L-cysteine (DCVC), is a potent nephrotoxin. *Fed. Proc.,* **44**, 1624.

Elfarra, A. A., Baggs, P. B. and Anders, M. W. (1985). Structure–nephrotoxicity relationships of S-(2-chloroethyl)-DL-cysteine and analogs: role for episulfonium ion. *J. Pharmacol. Exp. Ther.,* **233**, 512–516.

Ferger, M. F. and Du Vigneaud, V. (1950). Oxidation *in vivo* of the methyl groups of choline, betaine, dimethylthetin and dimethyl-β-propiothetin. *J. Biol. Chem.,* **185**, 53–57.

Fiecchi, A. (1979). The chemical properties of sulfonium compounds. In V. Zappia, E. Usdin, and F. Salvatore (eds.), *Biochemical and Pharmacological Roles of Adenosylmethionine in the Central Nervous System*, Pergamon, New York, pp. 17–23.

Finkelstein, J. D. and Harris, B. (1973). Methionine metabolism in mammals: synthesis of S-adenosylhomocysteine in rat tissues. *Arch. Biochem. Biophys.,* **159**, 160–165.

Flohe, L. and Schwabe, K. P. (1970). Kinetics of purified catechol-O-methyltransferase. *Biochim. Biophys. Acta,* **220**, 469–476.

Flohe, L. and Schwabe, K. P. (1972). Catechol-O-methyltransferase, II. *Hoppe-Seyler's Z. Physiol. Chem.,* **353**, 463–475.

Floss, H. G. and Tsai, M. D. (1979). Chiral methyl groups. *Adv. Enzymol.,* **50**, 243–302.

Greene, R. C. (1962). Biosynthesis of dimethyl-β-propiothetin. *J. Biol. Chem.,* **237**, 2251–2253.

Greene, R. C. and Davis, N. B. (1960). Biosynthesis of S-methylmethionine in the jack bean, *Biochim. Biophys. Acta,* **43**, 360–362.

Hassan, S. F. (1987). The synthesis and inhibitory activities of compounds designed as mechanism-based inhibitors of phenylethanolamine N-methyltransferase. *Ph.D. Thesis*, University of Kentucky.

Hassan, M. and Ehrsson, M. (1987a). Metabolism of ^{14}C-busulfan in isolated perfused rat liver. *Eur. J. Drug Metab. Pharmacokinet.,* **12**, 71–76.

Hassan, M. and Ehrsson, H. (1987b). Urinary metabolites of busulfan in the rat. *Drug Metab. Disp.,* **15**, 399–402.

Hattula, T. and Granroth, B. (1974). Formation of dimethyl sulfide from S-methylmethionine in onion seedlings (*Allium cepa*). *J. Sci. Food Agric.*, **25**, 1517–1521.

Hegazi, M. F., Borchardt, R. T., and Schowen, R. L. (1976). S_N2-like transition state for methyl transfer catalyzed by catechol-O-methyltransferase. *J. Amer. Chem. Soc.*, **98**, 3048–3049.

Hegazi, M. F., Borchardt, R. T. and Schowen, R. L. (1979). α-deuterium and carbon-13 isotope effects of methyl transfer catalyzed by catechol O-methyl-transferase. S_N2-like transition state. *J. Amer. Chem. Soc.*, **101**, 4359–4365.

Keenan, T. W. and Lindsay, R. C. (1968). Evidence for a dimethyl sulfide precursor in milk, *J. Dairy Sci.*, **51**, 112–114.

Kjaer, A., Grue-Sorensen, G., Kelstrep, E., and Ogaard Madsen, J. (1979). Stereochemical aspects of transmethylations of potential biological interest. *Pure Appl. Chem.*, **52**, 157–163.

Knipe, A. C. (1981). Reactivity of sulphonium salts. In C. J. M. Stirling and S. Patai (eds.), *The Chemistry of the Sulphonium Group*, Vol. 1, Wiley, Chichester, pp. 313–385.

Kovacheva, E. G. (1974). Method for measuring S-methylmethionine in natural products. *Prikl. Biokhim. Mikrobiol.*, **10**, 129–135.

Leopold, W. R., Miller, J. A., and Miller, E. C. (1982). Comparison of some carcinogenic, mutagenic and biochemical properties of S-vinylhomocysteine and ethionine. *Cancer Res.*, **42**, 4364–4374.

Mascaro, L., Horhammer, R., Eisenstein, S., Sellers, L. K., Mascaro, K., and Floss, H. G. (1977). Synthesis of methionine carrying a chiral methyl group and its use in determining the steric course of the enzymatic C-methylation of indolepyru-vate during indolmycin biosynthesis. *J. Amer. Chem. Soc.*, **99**, 273–274.

Maw, G. A. (1953). The oxidation of dimethylthetin and related compounds to sulphate in the rat. *Biochem. J.*, **55**, 42–46.

Maw, G. A. (1958). Thetin-homocysteine transmethylase. Some further characteristics of the enzyme from rat liver. *Biochem. J.*, **70**, 168–173.

Maw, G. A. (1981). The biochemistry of sulphonium salts. In C. J. M. Stirling and S. Patai (eds.), *The Chemistry of the Sulphonium Group*, Part II, Wiley, Chichester, pp. 703–771.

Mazelis, M., Levin, B., and Mallinson, N. (1965). Decomposition of methyl methionine sulfonium salts by a bacterial enzyme. *Biochem. Biophys. Acta*, **105**, 106–114.

Mudd, S. H. (1965). The mechanism of the enzymatic synthesis of S-adenosylmeth-ionine. In S. K. Shapiro and F. Schlenk (eds.), *Transmethylation and methionine biosynthesis*, University of Chicago Press, Chicago, pp. 330–347.

Nishimura, S., Taya, Y., Kuchino, Y., and Ohashi, Z. (1974). Enzymatic synthesis of 3-(3-amino-3-carboxypropyl)uridine in *Escherichia coli* phenylalanine transfer RNA: transfer of the 3-amino-β-carboxypropyl group from S-adenosyl-methionine. *Biochem. Biophys. Res. Commun.*, **57**, 702–708.

Olsen, J., Wu, Y. S., Borchardt, R. T., and Schowen, R. L. (1979). Transition-state structure and catalytic power in methyl transfer. In E. Usdin, R. T. Borchardt and C. R. Creveling (eds.), *Transmethylation*, North-Holland, Amsterdam, pp. 127–133.

Onkenhout, W., van Loon, W. M. G. M., Buijs, W., Van der Gen, A., and Vermeulen, N. P. E. (1986). Biotransformation and quantitative determination of sulfur-containing metabolites of 1,4-dibromobutane in the rat. *Drug Metab. Disposit.*, **14**, 608–612.

Parry, R. J. and Minta, A. (1982). Studies of enzyme stereochemistry. Elucidation of the stereochemistry of S-adenosylmethionine formation by yeast methionine adenosyltransferase. *J. Amer. Chem. Soc.*, **104**, 871–872.

Pegg, A. E. and Hibasami, H. (1979). The role of S-adenosylmethionine in mammalian polyamine biosynthesis. in E. Usdin, R. T. Borchardt, and C. R. Creveling, C. R. (eds.), *Transmethylation*, North Holland, Amsterdam, pp. 105–116.

Pfeffer, M. and Shapiro, S. K. (1962). Biosynthesis of methionine from S-adenosylmethionine in *Escherichia coli, Biochem. Biophys. Res. Commun.*, **9**, 405–409.

Pope, W. J. and Peachey, S. J. (1900). Asymmetric optically active sulphur compounds. d-methylethylthetin platinichloride. *J. Chem. Soc.*, **77**, 1072–1075.

Ramirez, F., Firan, J. J., and Carlson, M. (1973). *In vitro* decomposition of S-methylmethionine sulfonium salts. *J. Org. Chem.*, **38**, 2597–2603.

Rannung, U., Sundvall, A., and Ramel, C. (1978). The mutagenic effect of 1,2-dichloroethane on *Salmonella typhimurium*. I. Activation through conjugation with glutathione *in vitro. Chem.-Biol. Interact.*, **20**, 1–16.

Roberts, J. J. and Warwick, G. P. (1961). The formation of 3-hydroxytetrahydrothiophen-1,1-dioxide from myleran, S-β-L-alanyltetrahydrothiophenium mesylate, tetrahydrotriophen and tetrahydrothiophen-1,1-dioxide in the rat, rabbit and mouse. *Biochem. Pharmacol.*, **6**, 217–227.

Rodgers, J., Femec, D. A., and Schowen, R. L. (1982). Isotopic mapping of transition-state structural features associated with enzymic catalysis of methyl transfer. *J. Amer. Chem. Soc.*, **104**, 3263–3268.

Russell, D. H. (1973). Polyamines in growth-normal and neoplastic cells. In D. H. Russell (ed.), *Polyamines in Normal and Neoplastic Growth*, Raven Press, New York, pp. 1–13.

Russell, D. H. and Russell, S. D. (1975). Relative usefulness of measuring polyamines in serum, plasma, and urine as biochemical markers of cancer. *Clin Chem.*, **21**, 860–863.

Schlenk, F. (1983). Methylthioadenosine. *Adv. Enzymol.*, **54**, 195–205.

Schlenk, F. and Dainko, J. L. (1975). The S-*n*-propyl analogue of S-adenosylmethionine. *Biochem. Biophys. Acta*, **385**, 312–323.

Schlenk, F. and Ehninger, D. J. (1964). Observations on the metabolism of 5′-methylthioadenosine. *Arch. Biochem. Biophys.*, **106**, 95–100.

Shapiro, S. K. (1956). Biosynthesis of methionine from homocysteine and S-methylmethionine in bacteria. *J. Bacteriol.*, **72**, 730–735.

Shapiro, S. K. and Yphantis, D. A. (1959). Assay of S-methylmethionine and S-adenosylmethionine homocysteine transmethylases. *Biochim. Biophys. Acta*, **36**, 241–244.

Shapiro, S. K., Almenas, A., and Thompson, J. F. (1965). Biosynthesis of methionine in *Saccharomyces cerevisae. J. Biol. Chem.*, **240**, 2512–2518.

Sheng, L.-S., Horning, E. C., and Horning, M. G. (1984). Synthesis and elimination reactions of methylsulfonium ions formed from styrene oxide and methythio compounds related to methionine and cysteine. *Drug Metab. Disposit.*, **12**, 297–303.

Tabor, H., Tabor, C. W., and Rosenthal, S. M. (1961). The biochemistry of the polyamines: spermidine and spermine. *Ann. Rev. Biochem.*, **30**, 579–604.

Tang, K. C., Marriuzza, R., and Coward, J. K. (1981). Synthesis and evaluation of some stable multisubstrate adducts as specific inhibitors of spermidine synthase. *J. Med. Chem.*, **24**, 1277–1284.

Virchaux, P. N. H. (1981). Undersuchung des Stereochemischen Varlaufs Biologischer Methylierugen mit Hilfe Chiralor Methylgruppen. *Ph.D. Thesis*, University of Zurich.

Wagner, C., Lusty, S. M., Kung, H.-F., and Rogers, N. L. (1966). Trimethylsulfonium-tetrahydrofolate methyltransferase. A novel enzyme in the utilization of 1 carbon units. *J. Biol. Chem.*, **241**, 1923–1924.

Wagner, C., Lusty, S. M., Kung, H.-F., and Rogers, N. L. (1967). Preparation and properties of trimethylsulfonium–tetrahydrofolate methyltransferase. *J. Biol. Chem.*, **242**, 1287–1293.

Webb, W., Elfarra, A., Thom, R., and Anders, M. W. (1985). S-(2-chloroethyl)-DL-cysteine (CEC)-induced cytotoxicity: a role for the episulfonium ion. *Pharmacologist*, **27**, 228.

Williams-Ashman, H. G. and Pegg, A. E. (1981). Aminopropyl group transfers in polyamine biosynthesis. In D. R. Morris and L. J. Marton (eds.), *Polyamines in Biology and Medicine*, Dekker, New York, pp. 43–73.

Woodard, R. W., Tsai, M.-D., Floss, H. G., Crooks, P. A., and Coward, J. K. (1980). Stereochemical course of the transmethylation catalyzed by catechol O-methyltransferase. *J. Biol. Chem.*, **255**, 9124–9127.

Wu, S. E., Huskey, W. P., Borchardt, R. T., and Schowen, R. L. (1983). Chiral instability at sulfur of S-adenosylmethionine. *Biochemistry*, **22**, 2828–2831.

Zappia, V., Zydeck-Cwick, C. R., and Schlenk, F. (1969a). The specificity of S-adenosyl-L-methionine sulfonium stereoisomers in some enzyme systems. *Biochim. Biophys. Acta*, **178**, 185–187.

Zappia, V., Zydeck-Cwick, C. R., and Schlenk, F. (1969b). The specificity of S-adenosylmethionine derivatives in methyl transfer reactions. *J. Biol. Chem.*, **244**, 4499–4509.

7

Sulphonamides

Peter A. Crooks
College of Pharmacy, University of Kentucky, Lexington, KY 40536, USA

SUMMARY

1. The structure and chemistry of the sulphonamido group is briefly discussed. The sulphonamido group is a relatively stable chemical entity.
2. The potential pathways for metabolism of the sulphonamido group are outlined. The most important pathway is N-conjugation, with N-glucuronidation being the major route of metabolism.
3. Metabolic cleavage of the sulphonamido S–N bond has not been reported. However, N-dealkylation, N-dearylation and N-hydroxylation reactions have been observed.
4. In some thiazole- and imidazole-2-sulphonamides, displacement of the 2-sulphonamido group by glutathione has been reported.

7.1 INTRODUCTION

Structurally, the sulphonamido group can be represented as a mix of the canonical forms **1** and **2** (Plate 1). The presence of the adjacent S–O grouping renders the N atom non-basic (*cf.* the structure of the amide group). In fact, sulphonamides that have an N–H grouping are weakly acidic and dissolve readily in aqueous alkali solution, whereas those that contain a fully substituted N atom are insoluble in basic solution. The sulphonyl ($R–SO_2$) group, like the acyl (R–CO) group, will deactivate an attached nitrogen, and amines are often protected by conversion to the corresponding sulphonamide with sulphonyl chlorides. However, sulphonamides are much more difficult to hydrolyse back to the amine than are carboxamides.

Sulphonamides have medicinal value as antibacterial agents, and this group of compounds represents the largest group of xenobiotics of this type that man is exposed to. Other compounds in this class are the hypoglycaemic sulphonyl ureas, the heterocyclic sulphonamide diuretics, and synthetic sweetening agents such as saccharin.

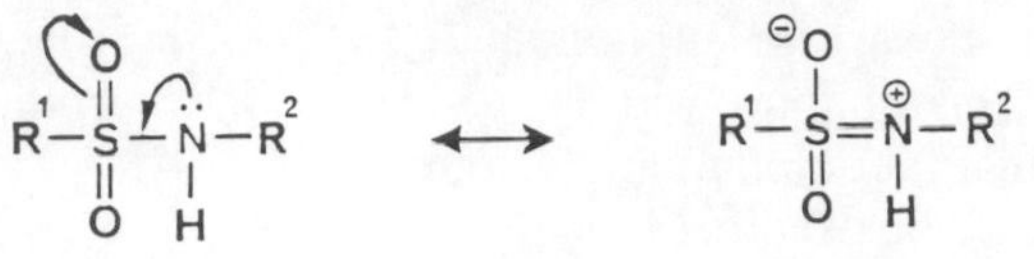

$$R^1-\underset{\underset{H}{|}}{\overset{\overset{O}{||}}{S}}-\underset{\underset{H}{|}}{N}-R^2 \quad \longleftrightarrow \quad R^1-\underset{\underset{O}{|}}{\overset{\overset{\ominus}{O}}{S}}=\underset{\underset{H}{|}}{\overset{\oplus}{N}}-R^2$$

(1)   (2)

Plate 1

Research on the antibacterial sulphonamides was triggered by the discovery that Prontosil (**3**), (Plate 2), an azo product of the dye industry of I.G. Fabenindustrie, had a curative effect when injected into *Streptococci*-infected mice (Domagk, 1935). This discovery was honored in 1939 by the award of the Nobel prize in medicine. Work by Tréfouel *et al.* (1935) at the Pasteur Institute subsequently showed that Prontosil was biodegraded in tissues to *p*-aminobenzene sulphonamide, **4** (sulphanilamide), and Fourneau *et al.* (1936) found sulphanilamide to be as effective as Prontosil as a curative agent. These discoveries heralded the beginnings of the development of the orally active sulphonamides of today.

As early as 1937, pioneering studies on the metabolism of sulphonamides had begun. In that year, Marshall *et al.* (1937) showed that orally administered sulphanilamide was partly excreted in a conjugated form that required hydrolysis to make the N^4-amine function titratable by diazo coupling. This metabolite was later identified as N^4-acetylsulphanilamide. James (1940) observed that excretion of organic sulphate and glucuronic acid increased after sulphanilamide treatment followed by acid hydrolysis of the urine, suggesting the presence of both sulphate and glucuronide conjugates. These early studies indicated that sulphonamides were biotransformed mainly into polar, conjugated metabolites.

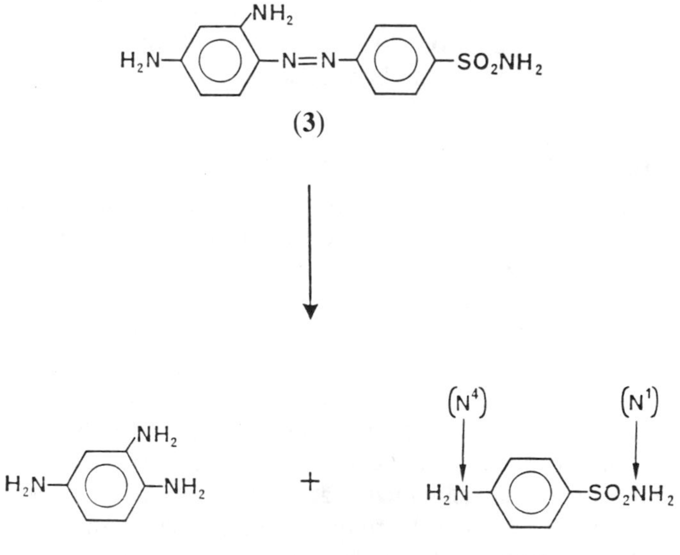

$$H_2N-\text{⟨⟩}(NH_2)-N=N-\text{⟨⟩}-SO_2NH_2$$

(3)

$$H_2N-\text{⟨⟩}(NH_2)-NH_2 \quad + \quad H_2N-\text{⟨⟩}-SO_2NH_2$$

(N^4) (N^1)

(4)

Plate 2

The potential pathways for sulphonamide metabolism are summarized in Scheme 1. The scope of this chapter will be to review the current knowledge on the biotransformation of the sulphonamide group in those groups of drugs and xenobiotics that man is most exposed to.

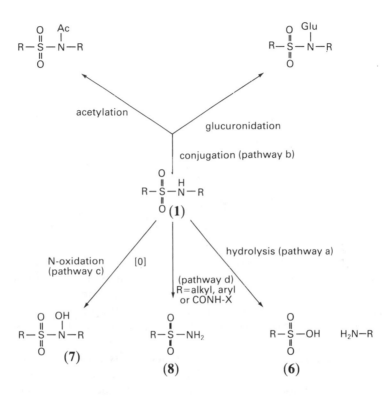

Scheme 1 — Potential pathways for sulphonamide metabolism.

7.2 CHEMICAL AND ANALYTICAL ASPECTS

Chemically, the sulphonamido group is a relatively stable entity and can only be cleaved under harsh conditions such as strong mineral acid treatment; it is resistant to S–N bond fission by base unless the N moiety is a very good leaving group, or under extreme conditions (i.e. fusion with NaOH). Oxidative cleavage of the S–N bond has not been reported. This stability is reflected in the metabolic profile of sulphonamides, where, as yet, there have been no examples cited of metabolic cleavage of the S–N bond. Thus pathway a (Scheme 1) appears not to operate in the metabolism of the sulphonamido group.

The weak acidity of the sulphonamides is attributable to the ability of the S–O bond to delocalize the resulting negative charge on the N atom following proton loss (See **5** (Plate 3)). The nucleophilicity of the anion **5** and its concentration at physiological pH are probably important factors to be considered in the mechanism of formation of N-conjugated products (pathway b, Scheme 1).

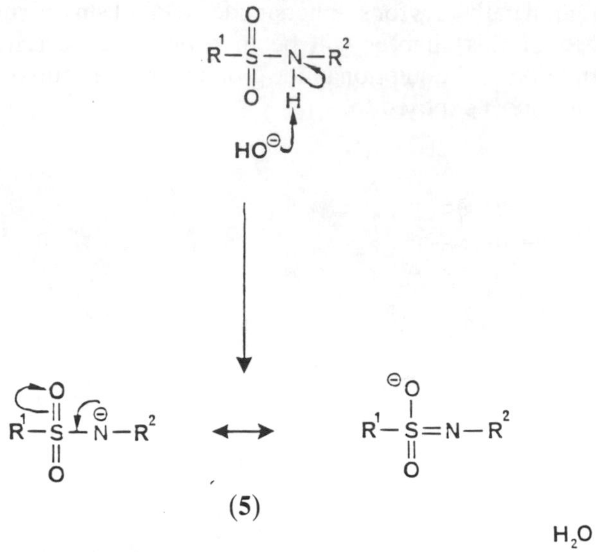

$$R^1 - \overset{\overset{\displaystyle O}{\|}}{\underset{\underset{\displaystyle O}{\|}}{S}} - \overset{\overset{\displaystyle}{|}}{\underset{\underset{\displaystyle H}{|}}{N}} - R^2$$

$$HO^{\ominus}$$

$$R^1 - \overset{\overset{\displaystyle O}{\|}}{\underset{\underset{\displaystyle O}{\|}}{S}} - \overset{\ominus}{N} - R^2 \quad \longleftrightarrow \quad R^1 - \overset{\overset{\displaystyle \ominus O}{|}}{\underset{\underset{\displaystyle O}{\|}}{S}} = N - R^2$$

(5)

$$H_2O$$

Plate 3

Since sulphonamides are structurally related to carboxamides, a potential oxidative metabolite of the former group of compounds is an *N*-hydroxy derivative (pathway c, Scheme 1); this is analogous to the formation of hydroxamic acids from amides. Sulphonamides that contain an *N*-alkyl group may also be susceptible to oxidative *N*-dealkylation, liberating the parent sulphonamide (pathway d, Scheme 1).

Early investigations on the *in vivo* metabolism of sulphonamides uitilized analytical methodology that was relatively unsophisticated (Hirtz, 1971), usually involving the extraction of metabolites from large volumes of urine. Identification and quantification was carried out utilizing classical methods of organic analysis, paper and thin-layer chromatography or conventional column chromatography. Particular difficulty was experienced with the isolation and characterization of the glucuronide conjugates owing to their polarity, water solubility and relative instability (Bridges, *et al.*, 1965). The advent of more recent analytical techniques such as high-pressure liquid chromatography, gas–liquid chromatography and gas–liquid chromatography–mass spectrometry has resulted in the development of more rapid, sensitive and unambiguous methodologies for the identification and quantitication of sulphonamides and their metabolites (Sharma *et al.*, 1976; Meffin and Miners, 1980; Kaye, 1980; Paulson *et al.*, 1981).

7.3 BIOCHEMICAL ASPECTS

N-conjugation of the sulphonamide group (pathway b) is the most important *in vivo* biotransformation pathway for this functional group. This route of metabolism has been studied extensively within the antibacterial sulphonamide group of drugs (Shepherd, 1970). Since these compounds offer two possible sites of conjugation (i.e. at N^1 and N^4), it is possible to obtain both N^1- and N^4-conjugation products. *In vivo* studies in man have clearly shown that, in this series of compounds, N^1-glucuronidation is the primary route of biotransformation of the $-SO_2NH-$ moiety and, in some cases, constitutes the major urinary metabolite of the drug (see Table 1). It should be noted that both N^4- and N^1-glucuronides of sulphonamide drugs are detectable in urine. The N^4-glucuronides are minor urinary components (1–4%) and are extremely labile in nature. They have been shown to be formed spontaneously by the reaction of glucuronic acid and the free aromatic amine compound in solution (Bridges, 1963). In addition, when the amine is added to drug-free urine, the N^4-glucuronide can be detected. Thus it is widely accepted that these aromatic amine glucuronides are not true metabolites but are artifacts formed during the collection and storage of biological samples (Caldwell, 1982). Conversely, the N^4-glucuronides are relatively stable compounds and are not considered to be artifacts (Adamson *et al.*, 1970).

Table 1 — Extent of urinary N^1-glucuronide formation in a series of sulphonamide drugs

Sulphonamide	Percent of N^1-glucuronide in urine, based on original dose	Reference
4-sulphanilamido-2,6-dimethoxypyrimidine (**9**) (sulfadimethoxine)	80–85	Uno and Sekine, 1966
5-sulphanilamido-1-phenylpyrazole (**10**) (sulfaphenazole)	70–95	Shepherd, 1966, 1970
3-sulphanilamido-6-methoxypyridazine (**11**) (sulphamethoxypyridazine)	15–20	Shepherd, 1966
2-sulphanilamidothiazole (**12**) (sulphathiazole)	4	Zbinden, 1964
5-sulphanilamido-3,4-dimethoxyisoxazole (**13**) (sulphonisoxazole)	5	Zbinden, 1964
5,6-dimethoxy-4-sulphanilamidopyrimidine (**14**)	2	Bridges *et al.*, 1966

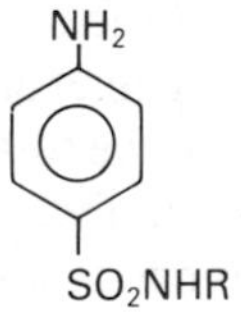

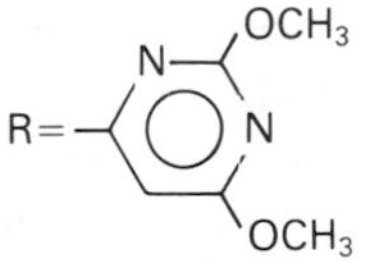

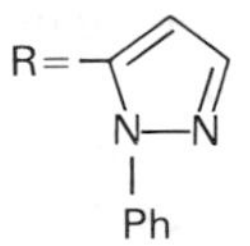

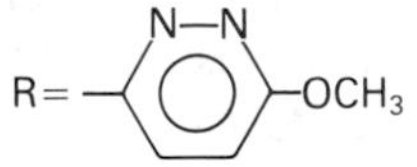

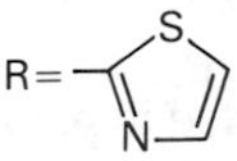

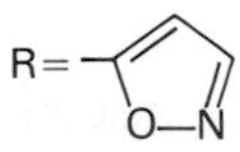

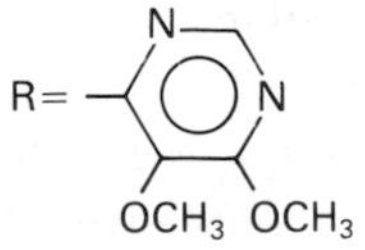

Although N^4-acetylation appears to be the major route of metabolism for many of the bacteriostatic sulphonamides, N^1-acetylation of the sulphonamido moiety has also been observed for some compounds (Parke, 1968). Sulphanilamide (**15**) has long been known to form both N^4 and N^1-acetyl derivatives (**16** and **17**), as well as the N^1,N^4-diacetyl derivative (**18**), *in vivo* (see Scheme 2) (Boyer *et al.*, 1956).

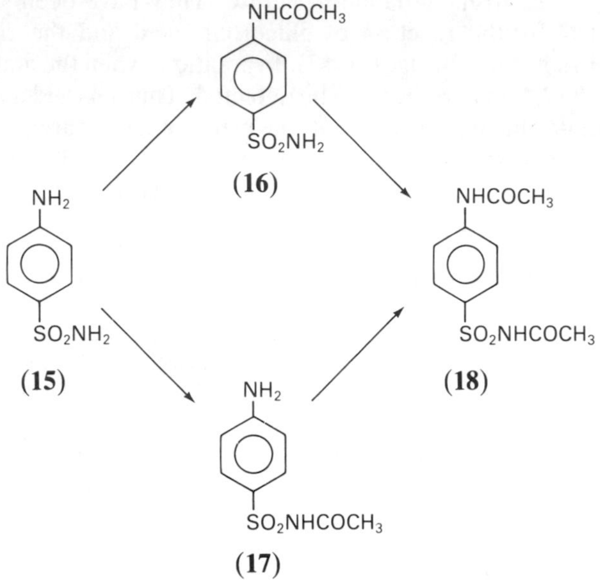

Scheme 2 — *N*-acetylation products of sulphanilamide metabolism.

The formation of *N*-conjugates of sulphonamido drugs appears to be species-dependent. In man and other primate species, including Old and New World monkeys and lemurs, at least 45% of the total 24 h excretion dose of sulphadimethoxine (**9**) is in the form of the N^1-glucuronide. In non-primates, this metabolite generally accounts for less than 5% of the total excretion, except in the cat, where the value is 19% (Adamson *et al.*, 1970). The *N*-acetylation of sulphonamido drugs is also species-dependent. Unlike most animal species, the dog does not excrete significant amounts of aromatic amines as acetylated derivatives. This may be due to an inhibitor of arylamine acetyl transferase that is present in dog liver and kidney (Leibman and Anaclerio, 1961) or to the high aromatic deacylase activity of dog liver, which leads to rapid deacetylation of administered *N*-acetyl compounds. However, dogs readily acetylate aliphatic amines and the levels of aliphatic deacylase are correspondingly low in this species. Both dogs and foxes form N^1-acetylated derivatives of sulphonamides; in these species, sulphanilamide is excreted as N^1-acetylsulphanilamide; none of the N^4-conjugate is detectable (Bridges and Williams, 1963).

N^1-acetylation of sulphanilamide has also been detected in plants (Parke, 1968); the broad bean (*Vicia fava*) *N*-acetylates both the aromatic amino and sulphamoyl groups of sulphanilamide to give all three possible metabolites (see Scheme 2).

Several studies describing the N-dealkylation of the sulphonamido group have been reported. Wiseman *et al.* (1962), studied the *in vivo* metabolism of the benzthiadiazine diuretic **19** (Plate 4) and showed it to be smoothly and quantitatively N-demethylated in the dog and rat to the parent drug, polythiazide **20**, which is a potent diuretic. Similarly, **21** also underwent *in vivo* N-demethylation at the sulphamoyl group that was *ortho* to the chloro substituent although interestingly, in the same study, N-demethylation of the SO_2–$NHCH_3$ group adjacent to the aromatic amino function in **22** was not observed (Wiseman *et al.*, 1962). These metabolically removed methyl groups were expired from the animal as carbon dioxide. The metabolic cleavage of polythiazide **20** (Wiseman *et al.*, 1962) to the ring-opened product **22** may also be considered as an N-dealkylation of the cyclic sulphonamide function.

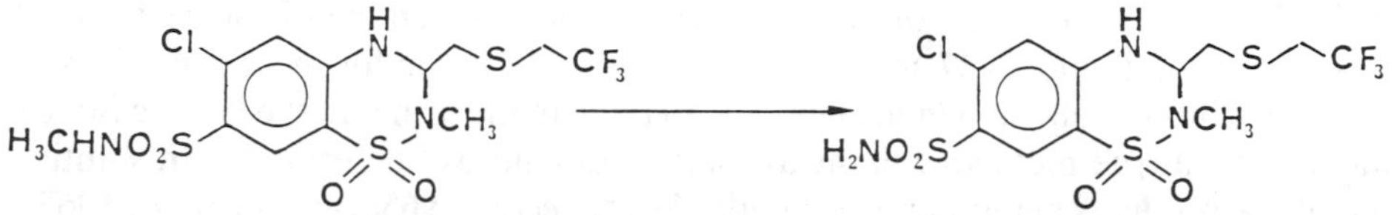

(**19**) $\longrightarrow$ (**20**)

(**21**) $\longrightarrow$ (**22**)

(**22**) $\xrightarrow{\quad\times\quad}$ [product]

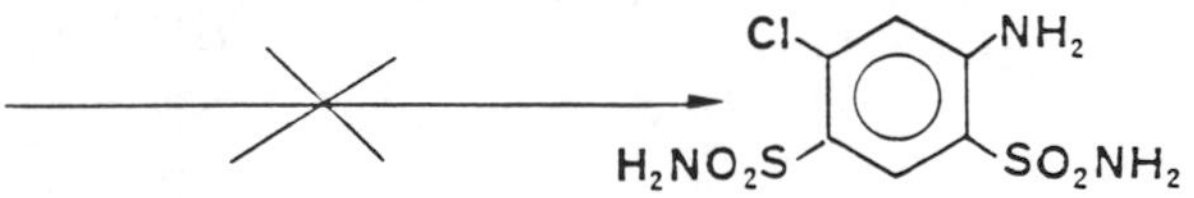

Plate 4

From a study in mice by Smith *et al.* (1965), the anticonvulsant activities of a series of N-alkyl sulphonamides (see structure **23**) were found to be directly related to their extent of metabolism to 4-bromobenzenesulphonamide. The degree of *in vivo* dealkylation varied inversely with the bulk of the alkyl substituent. The observed metabolism of **24** (Plate 5) to the parent sulphonamide **25** indicates that a free hydrogen atom on the sulphonamido nitrogen is not a prerequisite for *in vivo* dealkylation, as has been postulated by Wiseman *et al.* (1962).

Br   Br

(structure)   (structure)

SO$_2$N<R,R   SO$_2$NHR
                                     (23)

(24) R = CH$_3$ R = Me, Et, CH$_2$CH=CH$_2$,

(25) R = H   i-Pr, n-Pr, n-Bu.

Plate 5

Studies by Uno *et al.* (1963) and Uno and Sekine, (1966) and by Bruck *et al.* (1960) have shown that loss of the N^1-heteroaromatic group of both sulfadiazine 26 (Plate 6) and sulfamoxol 27 are minor routes of metabolism for these drugs in man, small amounts of sulphanilamide being detectable in the urine. In the latter case, this may not be a true metabolic pathway, since sulphamoxol deteriorates in solution, one of the products being sulphanilamide (Seydel *et al.*, 1965; Buttner *et al.*, 1965).

NH$_2$

(structure)

SO$_2$NHR

(26) R =   (structure)

(27) R =   (structure) CH$_3$
                         CH$_3$

Plate 6

Tolbutamide 28 (Plate 7), a sulphonylurea with hypoglycaemic properties, is metabolized in man, rat and rabbit by oxidation of the *para*-methyl group (Thomas and Ikeda, 1966); no biotransformation of the sulphonylurea group occurs in these species. However, in dogs, tolbutamide is metabolized into *para*-toluenesulphony-lurea 29 and *para*-toluenesulphonamide 30 (Scholtz and Häussler, 1964).

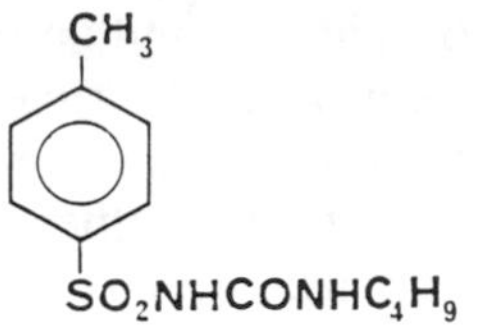

(28)

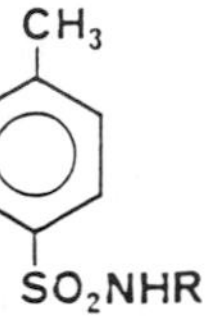

(29) R = CONH$_2$

(30) R = H

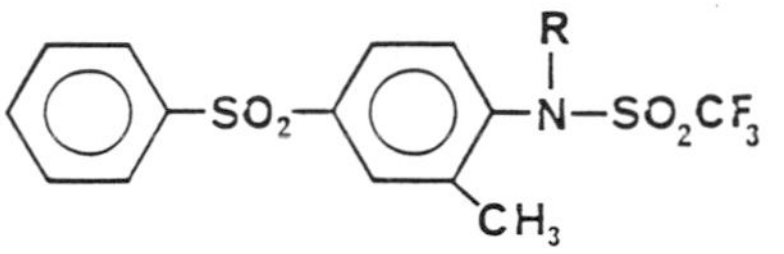

(31) R = OH

(32) R = H

Plate 7

A report by Paulson *et al.* (1977) has demonstrated the *N*-hydroxylation of a sulphonamide group. The *N*-hydroxy compound **31** was found to be an important biotransformation product of perfluidone **32** in the rat. *N*-hydroxylation is a well established route of metabolism for amines and amides. Considering the structural similarities between sulphonamides and amides, it is surprising that **31** is the only documented report of this type of biotransformation product.

(33)

(34)

(35)

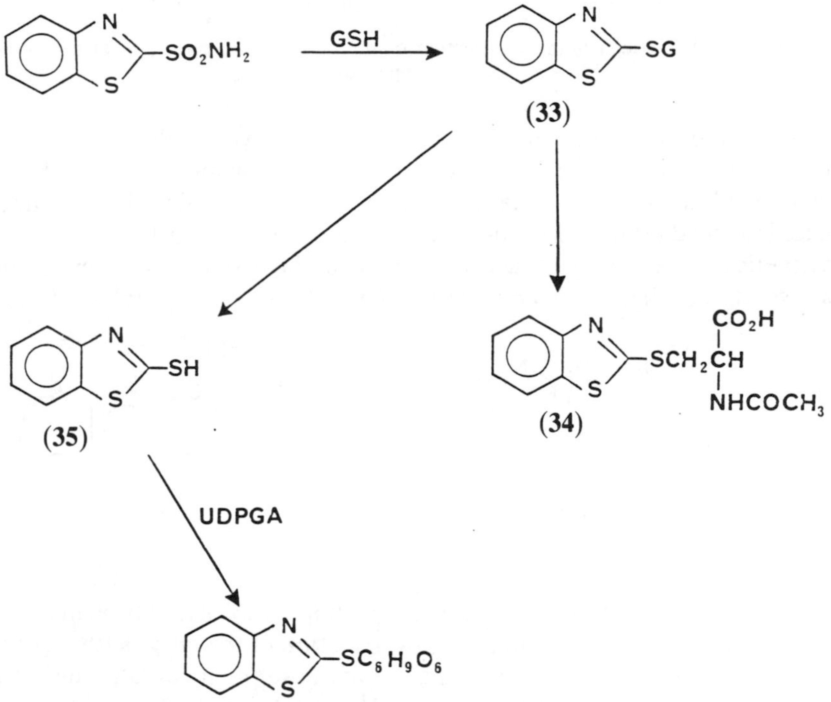

Scheme 3 — Biotransformation of benzothiazole-2-sulphonamide.

An interesting and unusual transformation of thiazole- and imidazole-2-sulpho-namides has been reported in which there is a cleavage of the ring carbon–sulphur bond (see Scheme 3) (Clapp, 1956; Colucci and Buyske, 1965; Conroy *et al.*, 1984) by glutathione or cysteine. The resulting conjugate **33** is then converted to a mercapturic acid **34**, or via C–S lyases to a mercaptan **35**, which can then undergo *S*-glucuronidation. This type of metabolic pathway appears to be restricted to thiazoles and imidazoles bearing a 2-sulphonamido group. The displacement of this group is not unexpected, in view of the reported chemistry of these heteroaromatic systems (Stirling, 1974), and probably occurs via the mechanism shown in Scheme 4.

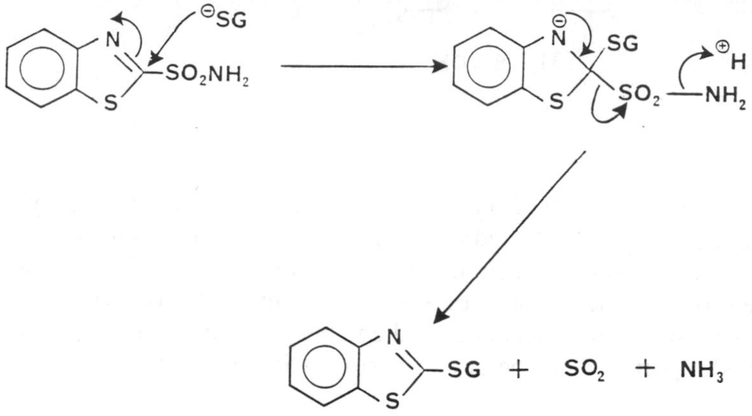

Scheme 4 — Mechanism of formation of the gluthathione conjugate of benzothiazole-2-sulphonamide.

It is important to note that, because of the relative stability of the sulphonamido functionality, there are several examples of sulphonamide drugs that are excreted unchanged in the urine. Notable examples are some cyclic sulphonamides, i.e. the benzthiazine diuretics, chlorthiazide **36** (Plate 8) and hydrochlorthiazide **39**, and the synthetic sweetening agent saccharin **38**. All of these xenobiotics have been shown to be excreted unchanged in dogs, rats and man (Parke, 1968; Ball *et al.*, 1977).

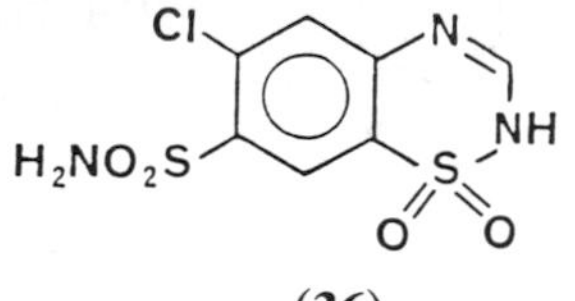

(36)

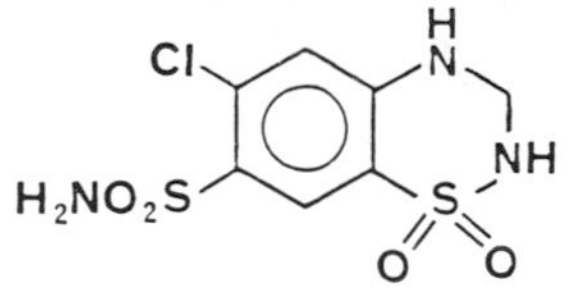

(37)

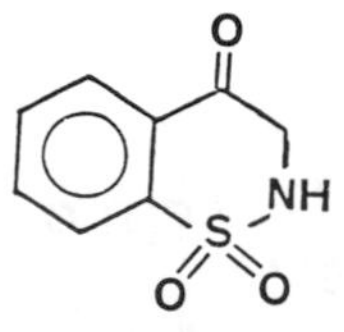

(38)

Plate 8

Only a few studies on the *in vitro* metabolism of sulphonamido compounds have been reported. Suspensions of liver microsomes do not appear to metabolize the $-SO_2NH_2$ group in sulphanilamide, although the aromatic hydroxyamino derivative is formed (Thauer *et al.*, 1965). Studies have been carried out on the *in vitro* acetylation of a number of bacteriostatic sulphonamides (Krebs *et al.*, 1947; Govier, 1965; Notter and Roland, 1978; Morland and Olsen, 1977; Suolinna, 1980). According to Govier (1965) the *N*-acetyltransferase in rabbit appears to be associated with the reticuloendothelial cells, and this was supported by the studies of Notter and Roland (1978). Later studies (Morland and Olsen, 1977; Suolinna, 1980) which utilized suspensions of pure rat and rabbit liver parenchymal and non-parenchymal cells, indicated that the *N*-acetylation of sulphonamides was carried out exclusively by the parenchymal cells. It is not known whether acetylations at N^1 and N^4 in sulphanilamide are carried out by the same *N*-acetyltransferase. No *in vitro* studies appear to have been carried out on the *N*-glucuronidation of sulphonamido compounds.

7.4 TOXICOLOGICAL ASPECTS

Since *N*-acetylation is remarkable among the major conjugation reactions, in that it does not increase the polar nature or the hydrophobicity of the substrate (Caldwell, 1980) but rather reduces the basicity of the amine, this has been the major cause of toxicity in the bacterial sulphonamides. It was realized many years ago that certain sulphonamides were found to produce renal toxicity owing to their *N*-acetyl conjugates' crystallizing within the kidney tubules. In this respect, the low toxicity of sulphadimethoxine **9** may be related to the fact that this drug is excreted in man mainly as the N^1-glucuronide, a relatively polar, water-soluble metabolite. Variation of N^1-substituents markedly influences the metabolic fate of the sulphonamides in man (Anand, 1979; however, no clear-cut structure–activity relationships have been found in this regard.

REFERENCES

Adamson, R. H., Bridges, J. W., Kibby, M. R., Walker, S. R., and Williams, R. T. (1970). The fate of sulphadimethoxine in primates compared with other species. *Biochem. J.*, **118**, 41–46.

Anand, N. (1979). Sulfonamides and sulfones. In M. E. Wolff (ed.), *Burgers Medicinal Chemistry*, Part II. Wiley-Interscience, New York, pp. 32–34.

Ball, L. M., Renwick, A. G., and Williams, R. T. (1977). The fate of [^{14}C]saccharin in man, rat and rabbit and of 2-sulphamoyl[^{14}C]benzoic acid in the rat. *Xenobiotica*, **1**, 189–203.

Boyer, F., Saviard, M., and Dechavassine, M. (1956). Metabolism of sulphanilamide. *Ann. Inst. Pasteur.*, **90**, 339–346.

Bridges, J. W. (1963). Aspects of the metabolism of aromatic amines, particularly sulphonamide drugs. *Ph.D. Thesis*, University of London.

Bridges, J. W. and Williams, R. T. (1963). Species differences in the acetylation of sulphanilamide. *Biochem. J.*, **87**, 19P–20P.

Bridges, J. W., Kibby, M. R., and Williams, R. T. (1965). The structure of the glucuronide of sulphadimethoxine formed in man. *Biochem. J.*, **96**, 829–836.

Bridges, J. W., Kibby, M. R., and Williams, R. T. (1966). Species differences in the metabolism of some methoxy-6-sulphonamidopyrimidines. *Biochem. J.*, **98**, 14P.

Bruck, C. G. V., Delfs, F. M., Serick, E., and Wolf, V. (1960). Verhalten des Sulfadimethyloxazols im Menschlichen Organismus. *Arzneimitt. Forsch.*, **10**, 621–626.

Buttner, H., Portwich, F., and Seydel, J. (1965). Stabilitats- Und Stoffwechselunter-suchungen an 2-Sulfanilamido-4,5-dimethyloxazol. *Chemotherapia*, **10**, 1–11.

Caldwell, J. (1980). The conjugation reactions and their significance in biochemical pharmacology and toxicology. In P. Jenner and B. Testa (ed.), *Concepts in Drug Metabolism*, Dekker, New York, Part A, pp. 211–250.

Caldwell, J. (1982). Conjugation reactions of nitrogen centres. In W. B. Jakoby, J. R. Bend and J. Caldwell (eds.), *Metabolic Basis of Detoxification–Metabolism of Functional Groups*, Academic Press, New York, pp. 294–301.

Clapp, J. W. (1956). A new metabolic pathway for a sulphonamide group. *J. Biol. Chem.*, **233**, 207–214.

Colucci, D. F. and Buyske, D. A. (1965). The biotransformation of a sulphonamide to a mercaptan and to a mercapturic acid and glucuronide conjugates. *Biochem. Pharmacol.*, **14**, 457–466.

Conroy, C. W., Schwamm, H., and Maren, T. H. (1984). The nonenzymic displace-ment of the sulfamoyl group from different classes of aromatic compounds by glutathione and cysteine. *Drug Metab. Disposit.*, **12**, 614–618.

Domagk, G. (1935). Chemotherapy of bacterial infections. *Deut. Med. Wochschr.*, **61**, 250–253.

Fourneau, E., Tréfouel, J., Nitti, F., and Bovet, D. (1936). Chemotherapy of Streptocci infections by derivatives of p-aminophenylsulphonamide. *Compt. Rend. Soc. Biol.*, **122**, 258–259.

Govier, W. C. (1965). Reticuloendothelial cells as the site of sulfanilamide acetyla-tion in the rabbit. *J. Pharmacol. Exp. Ther.*, **150**, 305–308.

Hirtz, J. L. (1971). Sulfonamides. In E. R. Garret (ed.), *Analytical Metabolic Chemistry of Drugs*, Vol. 4, *Medicinal Research: a Series of Monographs*, New York, pp. 159–180.

James, G. V. (1940). The isolation of some oxidation products of sulphanilamide from urine. *Biochem. J.*, **34**, 640–647.

Kaye, C. M. (1980). The gas chromatographic analysis of drugs in biological fluids. In J. W. Bridges and L. F. Chasseaud (eds.), *Progress in Drug Metabolism*, Vol. 4, Wiley, New York, pp. 165–259.

Krebs, H. A., Sykes, W. O., and Bartley, W. C. (1947). Acetylation and deacetyla-tion of the p-amino group of sulphonamide drugs in animal tissues. *Biochem. J.*, **41**, 622–630.

Leibman, K. C. and Anaclerio, A. M. (1961). Comparative studies on sulphanila-mide acetylation; an inhibitor in dog liver. *Proc. 1st Int. Pharmacological Meet.*, Vol. 6, pp. 91–96.

Marshall Jr., E. K., Emerson Jr., K., Cutting, W. C., and Babbit, B. (1937). Para-aminobenzenesulfonamide. Adsorption and excretion: method of determi-nation in urine and blood. *J. Am. Med. Assoc.*, **108**, 953–957.

Meffin, P. J. and Miners, J. P. (1980). The high performance liquid chromatographic measurement of drugs in biological fluids. In J. W. Bridges and L. F. Chasseaud, (eds.), *Progress in Drug Metabolism*, Vol. 4, Wiley, New York, pp. 261–307.

Morland, J. and Olsen, H. (1977). Metabolism of sulfadimidine, p-aminobenzoic acid, and isoniazid in suspensions of parenchymal and nonparenchymal rat liver cells. *Drug Metab. Disposit.*, **5**, 511–517.

Notter, D. and Roland, E. (1978). Localization of N-acetyltransferases in sinusoidal liver cells. Effect of zymosan on the acetylation of sulfamethazine in the rat and in isolated perfused liver. *C. R. Sci. Soc. Biol. Ses. Fil.*, **172**(3), 531–533.

Parke, D. (1968). *The Biochemistry of Foreign Compounds*, Vol. 5, *International Series of Monographs in Pure and Appied Biology*, Pergamon, Oxford, pp. 180–184.

Paulson, G. D., Jacobsen, A. M., and Zaylskie, R. G. (1977). Herbicide perfluinone (1,1,1-trifluoro-N-[2-methyl-4-(phenyl)]methylsulfonamide: its metabolism in rats and chickens. *Pestic. Biochem. Physiol.*, **7**, 62–72.

Paulson, G. D., Giddings, J. M., Lamoureaux, C. H., Mansager, E. R., and Struble, C. B. (1981). The isolation and identification of ^{14}C-sulfamethazine {4-amino-N-(4,6-dimethyl-2-pyrimidinyl)[^{14}C]benzenesulfonamide} metabolites in the tissues and excreta of swine. *Drug Metab. Disposit.*, **9**, 142–146.

Scholtz, J. and Häussler, A. (1964). Metabolic fate of several sulphonylureas in different species. *Proc. European Society for the Study of Drug Toxicity*, Vol. 4, *Some Factors Affecting Drug Toxicity*, Excerpta Medica, Amsterdam, pp. 23–28.

Seydel, J., Buttner, H., and Portwich, F. (1965). Zur Stabilitat des 2-Sulfanilamido-4,5-dimethyloxazol. *Klin. Wschr.*, **43**, 1060–1061.

Sharma, J. P., Perkins, E. G., and Bevill, R. F. (1976). High-pressure liquid chromatographic separation, identification and determination of sulfa drugs and their metabolites in urine. *J. Pharm. Sci.*, **65**, 1606–1608.

Shepherd, R. G. (1966). Synthetic antibacterial agents. *Ann. Rept. Med. Chem. 1965*, 118–128.

Shepherd, R. G. (1970). Sulfonamides and other p-aminobenzoic acid antagonists. In (ed.), *Medicinal Chemistry*, A. Burger, (Ed.) 3rd edn., Part I, Wiley-Interscience, New York, pp. 297–304.

Smith, D. L., Keasling, H. H., and Forist, A. A. (1965). The metabolism of N-alkyl-4-bromobenzenesulfonamides in the mouse. Correlation with anticonvulsant activity. *J. Med. Chem.*, **8**, 520–524.

Stirling, C. J. M. (1974). The sulfinic acids and their derivatives. *Int. J. Sulfur Chem.*, **6**, 277–316.

Suolinna, E.-M. (1980). Metabolism of sulfanilamide and related drugs in isolated rat and rabbit liver cells. *Drug Metab. Disposit.*, **8**, 205–207.

Thauer, R. K., Stoffler, G., and Uehleke, H. (1965). N-hydroxylierung von Sulfanilamid zu p-Hydroylaminobenzolsulfonamid durch Lebermikrosomen. *Naunyn Schmiedebergs Arch. Exp. Pathol. Pharmacol.*, **252**, 32–41.,

Thomas, R. C. and Ikeda, G. J. (1966). The metabolic fate of tolbutamide in man and in the rat. *J. Med. Pharm. Chem.*, **9**, 507–510.

Tréfouel, J., Tréfouel, J., Nitti, F., and Bovet, D. (1935). Action of p-aminophenyl-sulphamide in experimental *streptococcus* infections of mice and rabbits. *Compt. Rend. Soc. Biol.*, **120**, 756–758.

Uno, T. and Sekine, Y. (1966). Studies on the metabolism of sulfadiazine. Quantitative separation of excrements in the human urine after administration of sulfadiazine. *Chem. Pharm. Bull.*, **14**, 687–691.

Uno, T., Yasudo, H., and Sekine, Y. (1963). Metabolism of sulphadiazine. I. Separation and identification of products in human urine after oral administration. *Chem. Pharm. Bull.*, **11**, 872–875.

Wiseman, E. H., Schrieber, E. C., and Pinson Jr., R. (1962). Studies of N-dealkylations of some aromatic sulfonamides. *Biochem. Pharmacol.*, **11**, 881–886.

Zbinden, G. (1964). Molecular modification in the development of newer antiinfective agents. The sulfa drugs. In *Molecular Modification in Drug Design*, Vol. 45, *Advances in Chemistry Series*, American Chemical Society, Washington, DC, pp. 25–38.

8

Sulphamates, sulphonates and sulphate esters

A. G. Renwick
Clinical Pharmacology, University of Southampton, Medical and Biological
Sciences Building, Bassett Crescent East, Southampton SO9 3TU, UK

SUMMARY

1. Sulphamic acids, sulphonic acids and sulphuric acid esters are strongly acidic functional groups. Their presence in a molecule produces a centre of high polarity, which increases excretion and decreases metabolism.
2. Most information on the metabolism of sulphamates has come from the intense sweetener cyclamate (cyclohexylsulphamate). This compound undergoes negligible metabolism by the tissues but is hydrolysed by a microbial enzyme which may be induced in the intestinal microflora by a period of chronic administration.
3. The sulphonic acid moiety is not a site for metabolism by mammalian tissues and therefore the metabolism of sulphonates usually involves reactions at sites distal to this polar group. Some microorganisms can metabolize the sulphonate moiety directly, but the relevance of this to metabolism by the intestinal flora *in vivo* is unclear.
4. The sulphate esters of simple phenols appear to be metabolically stable. In contrast, steroid sulphates are labile and may represent a circulatory form of the steroid.

8.1 INTRODUCTION

The compounds reviewed in this chapter are organic acids of the general structures in Fig. 1. The presence of three strongly electronegative oxygen atoms results in the sulphur atom tending to acquire an electropositive character. This results in an electron displacement towards the sulphur and thus a ready loss of the proton. Thus this group of compounds are much stronger acids than the corresponding carbamic and carboxylic acids. Their pK_a values are therefore comparable with those of inorganic acids and they are highly ionized at physiological pH. It is this high polarity which dominates their fate in the body.

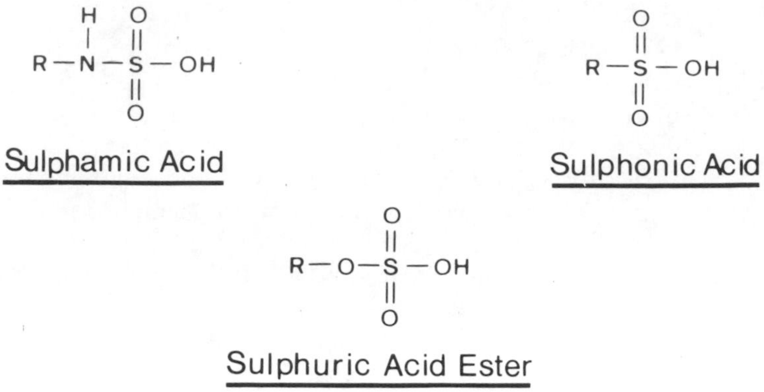

$$R - N - S - OH$$

with H and O substituents

Sulphamic Acid

$$R - S - OH$$

with O substituents

Sulphonic Acid

$$R - O - S - OH$$

with O substituents

Sulphuric Acid Ester

Fig. 1 — The general structures of the acids considered in this chapter.

8.2   N-SULPHAMATES

Sulphamates (Fig. 1) are strong organic acids which may be found as phase-2 sulphate conjugates of aromatic and aliphatic amines (Iwasaki *et al.*, 1986). Sulphamate formation may provide a major route for the detoxication of carcinogenic aromatic amines such as 2-amino-3-methylimidazol[4,5-f]quinoline (Turesky *et al.*, 1986).

The sulphamate group is present also in acidic heteropolysaccharides and specific enzymes exist for their synthesis and degradation. For example, the lysosomal enzyme has been purified from human liver which hydrolyses the sulphamate bond of 2-sulphaminoglucosamine (Freeman and Hopwood, 1986). However, the natural substrates for the enzyme are complex heteropolysaccharides, such as heparin and heparan sulphate, since these are hydrolysed up to 370000 times faster than the simple monosaccharides. The sulphamatase is an exoenzyme which is strongly inhibited by sulphate. The potential for this enzyme to hydrolyse simple xenobiotic sulphamates is not known, but the stability of such compounds *in vivo* indicates that they are not substrates.

The sulphamate moiety, like the sulphate ester group (see section 8.4), is highly polar and xenobiotic sulphamates show slow and/or poor absorption from the gut, negligible metabolism by the tissues and rapid elimination from the circulation, principally in urine. Most of our knowledge concerning the fate of sulphamates has developed from studies on the sulphamate produced from cyclohexylamine, i.e. cyclohexylsulphamate or cyclamate. This chemical is about 30 times sweeter than sucrose and achieved widespread popularity in the 1960s as a sugar substitute. It was banned from use as a food additive in the USA and UK in 1969, but has continued to be available in many other coutries around the world including Germany, Switzerland and Australia. The Joint Expert Committee on Food Additives (JECFA) of the World Health Organisation (WHO) has established an acceptable daily intake of 0–11 mg/kg/day and thus there is clearly no consistent international consensus concerning this compound.

Cyclamate was first discovered accidentally by researchers in the USA (Audrieth and Sveda, 1944). Its pleasant sweet taste led to its undergoing toxicological testing, which included metabolism studies, prior to its use as a component of the human diet (Bopp *et al.*, 1986). The early studies were consistent with what would be expected for a highly polar chemical, i.e. it was incompletely absorbed from the gut, and the fraction which was absorbed was excreted from the body without undergoing significant metabolism. The absence of metabolism was demonstrated in animals using ^{35}S and ^{14}C-labelled cyclamate (Taylor *et al.*, 1951; Miller *et al.*, 1966). Early studies in man (Richards *et al.*, 1951; Schoenberger *et al.*, 1953) demonstrated the urinary excretion of unchanged cyclamate but were not adequate by modern criteria (Renwick, 1983a) to conclude that the compound was not hydrolysed *in vivo*. Prior to 1966 the most important reported toxicological property of cyclamate was the production of diarrhoea at very high doses, due to the unabsorbed compound (Hwang, 1966). However, in 1966, Kojima and Ichibagase reported the presence of low levels of cyclohexylamine in the urines from dogs and man given cyclamate, indicating the possibility of hydrolysis (Fig. 2). Although cyclohexylamine accounted for less than 1% of the administered dose, the greater toxicity and pharmacological activity of the metabolite (see Bopp *et al.*, 1986) was of considerable interest and concern.

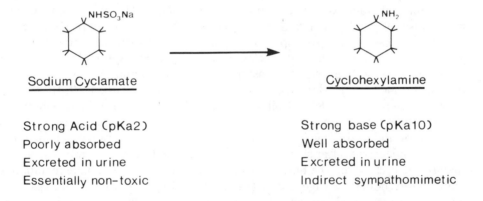

Sodium Cyclamate

Strong Acid (pKa2)

Poorly absorbed

Excreted in urine

Essentially non-toxic

Cyclohexylamine

Strong base (pKa10)

Well absorbed

Excreted in urine

Indirect sympathomimetic

Fig. 2 — The hydrolysis of cyclamate to cyclohexylamine.

In 1969, cyclamate was banned following the discovery of an increased incidence of bladder tumours in male rats fed a 10:1 mixture of cyclamate and saccharin (Price *et al.*, 1970). With the discovery of cyclamate metabolism during this study, one-half of the animals had their diets supplemented with cyclohexylamine. Clearly if the study had been negative then the safety of the sweetener mixture plus the metabolite would have been assured. However, the finding of bladder tumours raised safety questions which are unanswerable from the data available, i.e. was saccharin, cyclamate or cyclohexylamine responsible? The regulatory agencies in the USA and UK responded by banning cyclamate, because this was a more recent innovation than saccharin, and because increased, although non-significant, incidences of bladder tumours had been found in animals fed cyclohexylamine (1 out of 8 rats given 15 mg/kg/day; Price *et al.*, 1970) or cyclamate alone (3 out of 23 rats; Friedman *et al.*, 1972). Studies conducted since 1969 have failed to support the conclusion that cyclamate or its metabolite was involved in the development of bladder tumours. Recent reviews in the USA (by the Cancer Assessment Committee of the Center for Food Safety and Applied Nutrition), in the UK (by the Committee on Toxicity) and by the WHO (JECFA) have concluded that cyclamate is not a carcinogen (see Bopp *et al.*, 1986). Thus approval for the use of cyclamate is determined by the toxicity of the sweetener and its metabolite. Cyclohexylamine produces testicular toxicity in rats when fed at doses greater than 200 mg/kg/day which is equivalent to about 400 mg/kg/day of cyclamate assuming 100% metabolism (but see later). This occurs at a much lower dose than cyclamate-induced diarrhoea and thus the safe level of use for cyclamate must be set by the formation of its metabolite. Thus information on the extent, site and factors affecting cyclamate metabolism is essential to any discussion of its animal toxicity data and its safe regulation as a food additive.

8.2.1 The influence of prior treatment with cyclamate on its metabolism

The discrepancy between early studies which failed to detect metabolism, (Taylor *et al.*, 1951; Miller *et al.*, 1966) and subsequent studies was the influence of prior treatment with the sweetener. Renwick and Williams (1972b) found negligible metabolism of [^{14}C]cyclamate in normal rats, rabbits and guinea pigs but variable and in some cases extensive (up to 40%) metabolism if the radiolabelled dose was given to animals pretreated with cyclamate in the drinking water. Wide interindividual variability has been found in most studies on the metabolism of cyclamate in rats during chronic administration (Oser *et al.*, 1968; Sonders *et al.*, 1969; Bickel *et al.*, 1974; Renwick, 1976, 1986a). Collings (1971) failed to induce cyclamate metabolizing activity in the rat colony at Unilever Research Laboratories, Colworth House, during a period of 6 months of cyclamate ingestion, but the same animals rapidly acquired the activity after sharing cages with imported 'converter rats' which could metabolize cyclamate. The time taken to acquire significant cyclamate metabolizing ability varies both from cage to cage (Bickel *et al.*, 1974; Renwick, 1976, 1986a) and also within individuals sharing the same cage (Renwick, 1976, 1986a). Chronic administration for up to one year may be necessary before some groups of animals display significant metabolism (Bickel *et al.*, 1974; Renwick 1976, 1986a). The maximum reported level in rats is about 40–50% (Sonders *et al.*, 1969; Renwick and Williams, 1972b; Bickel *et al.*, 1974) or 200 mg/kg/day (Renwick, 1986a). Thus although the extent of metabolism may be highly variable it is potentially of considerable quantitative importance.

The studies in rats are predictive for the data obtained in man. However, in man, some metabolism has been found following a single oral dose. This has usually been in the second to fourth day after dosing (Asahina *et al.*, 1971; Renwick and Williams, 1972b) and at a much lower level than following regular intake. During regular intake the daily excretion of cyclohexylamine in the urine increased until a 'plateau' was reached after about 5–10 days (see Renwick, 1983b, for review). Although wide day-to-day fluctuations occur (of up to 50–60% about the mean value) these intrasubject variations are low in comparison with the intersubject differences, which can range from no detectable metabolism to up to 60% conversion to cyclohexylamine. The highest values were found in studies conducted in the late 1960s when cyclamate was freely available as a food additive. Under such circumstances the actual intake may have exceeded the dose given and the calculated percentage metabolism may have considerably overestimated the true percentage metabolism. For example, Collings (1971) reported percentage metabolism data greater than 60% in some individuals in one study, but in a separate study showed that certain individuals still ingested cyclamate and excreted cyclohexylamine even when asked to avoid all sources of artificially sweetened food and drink. In contrast, other workers (Litchfield and Swan, 1971) found only low levels of metabolism in their population.

In order to assess the significance of cyclamate metabolism in man, information on the incidence and extent of metabolism is necessary. Although some studies have reported detecting cyclohexylamine in the urine following a single dose, such data are likely to underestimate the true incidence since detection of metabolism would probably relate more to previous intake than to the test dose. Reviews of the literature indicate that approximately 25% of the population (307 out of 1223) are potential metabolizers of cyclamate (Renwick, 1983b; Bopp et al., 1986). More reliable information can be obtained from studies in which the urine was collected following a period of chronic intake. Although nearly all published studies used different doses and durations of intake, it is possible to draw broad conclusions. When the data from all published studies are pooled, approximately 26% of the population showed some ability to metabolize cyclamate, but only 3–4% metabolized a major proportion of the dose (>20%) (Table 1). This distribution is similar to that obtained from a study in which 24 h urines were collected from Japanese subjects, when cyclamate was widely available as a sweetener (Asahina et al., 1971). Although there is a tendency for a higher apparent metabolism during the dietary intake (Table 1), calculation of these data assumes that the individuals have a regular daily intake and are at steady state. Since most urinary cyclamate is excreted within

Table 1 — The incidence of cyclamate hydrolysis in man

% dose converted	As fixed doses [a]		As a sweetener[b]	
	N	%	N	%
0–0.1	163	73.8	7	14
0.1–1.0	26	11.8	10	20
1–20	24	10.9	27	54
20–60	6	2.7	4	8
60+	2	0.9	2	4

[a]Adapted from the review by Renwick (1986a) for subjects given 0.5–5g/day for a period of at least 3 days.
[b]Adapted from Asahina et al. (1971) assuming that 40% of the unmetabolized cyclamate was excreted in the urine, i.e. the total daily intake of cyclamate=urinary cyclamate/0.4+2×urinary cyclohexylamine.

the first 24 h while most cyclohexylamine formation and excretion occurs between 24 and 72 h, a delay of 48 h between intake and urine collection would severely bias the percentage metabolism figures. It is of interest that the maximum amount of cyclohexylamine excreted was 129 mg/day which is equivalent to about 2 mg/kg/day of cyclohexylamine or 4 mg/kg/day of cyclamate — a value well below the maximum acceptable daily intake recommended by the WHO. This latter analysis, which makes no assumptions concerning the total dose of cyclamate or the temporal relationship between intake and urine collection is a more valid method of interpreting these findings.

8.2.2 The site of cyclamate hydrolysis during chronic intake

A number of factors have been shown to affect cyclamate hydrolysis *in vivo* (Table 2). These are all consistent with the intestinal flora being the principal and probably only site of extensive hydrolysis. Additional data which support this conclusion are

Table 2 — Factors affecting the hydrolysis of cyclamate *in vivo* during chronic intake

Factor	Effect	Species	Reference[a]
Route of administration	Greatly reduced hydrolysis of parenteral doses	Rat Pig Guinea pig	1,2,3 4 5
Oral antibiotics	Suppression of metabolism	Rat Pig Guinea pig Man	1,3 4 5 4
Cessation of intake	Rapid loss of acquired metabolising capacity	Rat Man	2,3 2
Neonatal treatment	Greater metabolism in offspring than parents (?)	Rat	6
Constipation	Increased excretion of cyclohexylamine	Man	7

References: [a]1, Sonders *et al.*, 1969; 2, Renwick and Williams, 1972b; 3, Bickel *et al.*, 1974; 4, Collings, 1971; 5, Asahina *et al.*, 1972b; 6, Renwick, 1976; 7, Davis *et al.*, 1969.

obtained from the *in vitro* hydrolysis of cyclamate using mixed cultures and isolated strains of intestinal bacteria (Table 3). The tissues of animals shown to hydrolyse cyclamate *in vivo* were unable to hydrolyse cyclamate *in vitro* (Prosky and O'Dell, 1971; Asahina *et al.*, 1972b; Drasar *et al.*, 1972). The reports from Kojima, Ichibagase and their colleagues of the formation of trace levels of cyclohexylamine on *in vitro* incubation of liver samples from animals pretreated with cyclamate are questionable owing to the absence of appropriate controls and the possibility that cyclohexylamine was present due to the pretreatment (Kojima and Ichibagase, 1968; Ichibagase *et al.*, 1972).

Table 3 — The hydrolysis of cyclamate by intestinal flora of chronically treated
animals

Species	Source of organisms or isolated strain	Reference[a]
Rat	Intestinal contents–faeces; *Clostridium* sp.	1,2,3,4,5,6,7
Guinea pig	Intestinal contents–faeces; *Pseudomonas*; Corynebacterium; (*Clostridium*+*Campylobacter*+*Propionibacterium* sp. in mixed culture).	2,8,9,10
Rabbit	Intestinal contents–faeces; *Enterobacteria: Clostridia; Streptococcus faecalis; E. coli; Baccillus.*	2,11,12
Dog	Large intestine contents; *Clostridium* sp.	13
Pig	Colon contents	14
Man	Faeces; *Enterococci*	2

References: [a]1, Dalderup *et al.*, 1970; 2, Drasar *et al.*, 1972; 3, Bickel *et al.*, 1974; 4, Tesoriero and Roxon, 1975; 5, Renwick, 1976; 6, McGlinchey *et al.*, 1982; 7, Mallett *et al.*, 1985; 8, Asahina, *et al.*, 1972a; 9, Asahina *et al.*, 1972b; 10, Matsui *et al.*, 1981; 11, Tsuchiya, 1981; 12, Tokeida *et al.*, 1979; 13, Goldberg *et al.*, 1969; 14, Collings, 1971.

8.2.3   The microbial hydrolysis of sulphamates

The microbial hydrolysis of cyclamate *in vitro* is affected by various factors including the substrate and cysteine (Table 4). The influence of cysteine suggests that the gut flora may hydrolyse cyclamate to provide a source of sulphur.

The stoichiometry of cyclamate hydrolysis suggests that the enzyme splits the sulphamate into equimolar amounts of the amine and sulphate (Niimura *et al.*, 1974; Tsuchiya, 1981). However, both moieties may undergo further microbial metabolism especially with *in vitro* cultures. The sulphate group is reduced in whole cell

Table 4 — Factors affecting the hydrolysis of cyclamate by intestinal organisms *in vitro*

Factor	Effect	Source of organisms	Reference[a]
Glucose	Suppression of induced metabolism	Rat (faeces?)	1
Nutrient medium	Failure to induce metabolism *in vitro*	Rat faeces	2
Cyclamate	Loss of induced activity if cyclamate is absent from medium	Rat faeces	3
Cysteine	Suppression of induced metabolism	Rat faeces	4,5
Cyclohexylamine	Suppression of induced metabolism	Rat faeces	3
Cyclamate	Loss of cyclamate assimilating activity if cyclamate is not sole source of C and N	*Pseudomonas* from guinea pig faeces	6
Cyclamate	Loss of induced metabolism if present in medium	Rabbit faeces	3
Cyclamate	Loss of induced metabolism if present in medium	Human faeces	3

References: [a]1, Dalderup *et al.*, 1970; 2, Mallett *et al.*, 1985; 3, Drasar *et al.*, 1972; 4, Tesoriero and Roxon, 1975; 5, Renwick, 1976; 6, Asahina *et al.*, 1972a.

suspensions under anaerobic conditions to produce volatile S compounds (probably H_2S) and may also be incorporated into protein (Tesoriero and Roxon, 1975). The cyclohexylamine moiety may undergo oxidative deamination to cyclohexanone (Asahina et al., 1972a; Tsuchiya, 1981) which may then act as a sole source of carbon during in vitro culture (Asahina et al., 1972a). However, the further metabolism of cyclohexylamine is not likely to occur in the reductive, nutrient-containing environment of the intestinal lumen.

The pattern of metabolites of ^{14}C-cyclamate in vivo in cyclamate pretreated rats is consistent with initial hydrolysis to cyclohexylamine in the gut lumen followed by oxidative metabolism of the amine in the tissues (Renwick and Williams, 1972a, b).

The substrate specificity of the enzyme splitting cyclamate is given in Table 5.

Table 5 — In vitro substrate specificities of microbial sulphamatase

	Pseudomonas[a]	Rat faeces[b]	Rat faecal extract[c]
N-alkylsulphamates			
n-propyl-	1.7	—	—
n-amyl-	1.8	—	—
Cyclopentyl-	—	0.3	0.5
n-hexyl-	1.8	—	—
3-methylcyclopentyl-	—	0.6	—
Cyclohexyl-	1.0	1.0	1.0
Cycloheptyl-	—	1.3	0.5
Cyclohexylmethyl-	—	0.7	—
Benzyl-	—	0.2	—
n-octyl-	4.3	—	6.4
Cyclooctyl-	—	—	0.0
3-phenylpropyl-	1.3	—	—
n-hexadecyl-	w	—	—
n-octadecyl-	w	—	—
N-arylsulphamates			
Phenyl	w	1.6	5.3
Tetrahydronaphthyl-	0.6	—	—
N,N-dialkylsulphamates			
Di-n-propyl	w	—	—
Methyl-, cyclohexyl-	w	—	—
Other compounds			
Sulphamate ion	0	—	—
Cyclohexanolsulphate	0	—	—
Sulphanilamide	0	—	—
Saccharin	—	0	—

—, not determined; w, weak or uncertain activity.
[a]From Niimura et al. (1974) and Tsuchiya (1981).
[b]From Renwick (1976) and Ball et al. (1977).
[c]From McGlinchey et al. (1982).

Despite the different substrates employed, there are clear differences between the pure enzyme isolated from a strain of *Pseudomonas* from guinea pigs and the activity in crude rat faecal preparations. Both enzymes showed greatest activity against *n*-octyl sulphamate; however, only the rat faecal culture, not the *Pseudomonas* enzyme, was able to hydrolyse the simple arylsulphamate, phenylsulphamate.

Although a number of pure strains of bacteria with sulphamate hydrolysing activity have been isolated from different animals it is probable that some form of symbiosis is involved *in vivo*. Mixed cultures of isolated active strains have been reported to show a greater activity than was present in each individual strain (Matsui *et al.*, 1981; Tsuchiya, 1981; McGlinchey *et al.*, 1982).

8.2.4   The *in vivo* hydrolysis of sulphamates other than cyclamate

The substrate specificity of sulphamate hydrolysis *in vivo* has received little attention. The studies of McGlinchey *et al.* (1982) detected low levels of hydrolysis after oral administration of a number of alkyl sulphamates (Table 6). However, these levels (generally <0.1%) were similar to those reported by Kojima and colleagues following oral administration of cyclamate to non-induced animals (Kojima and Ichibagase, 1968; Ichibagase *et al.*, 1972). Thus these data probably refer to the uninduced enzyme activity. The site of this activity has not been defined and could be a tissue sulphamatase, but it is of negligible significance compared with the extensive hydrolysis possible following pretreatment (see Renwick, 1988, for further details). The only report on the *in vivo* metabolism of a second substrate (3-methylcyclopentylsulpahamate) in cyclamate-induced animals (Renwick, 1976) showed an excellent correlation in the extent of hydrolysis of the two sulphamates (Table 6). This finding

Table 6 — *In vivo* hydrolysis of alkylsulphamates

Substrate	% metabolism
In uninduced rats[a]	
Cyclooctyl-	0.297
Cyclopentyl-	0.083
Cycloheptyl-	0.064
3-methylcyclohexyl-	0.024
Cyclohexyl-	0.015
Cyclopentylmethyl-	0.014
4-methylcyclohexyl-	0.009
2-methylcyclohexyl-	<0.001
2-cyclohexenyl-	<0.001
In cyclamate induced rats[b]	
Cyclohexyl-	0.4, <0.1, 4.0, 0.4, 0.3
3-methylcyclopentyl-	1.3, 0.1, 6.9, 0.4, 1.7

[a]Adapted from data compiled in McGlinchey *et al.* (1982); mean data.
[b]Adapted from Renwick (1976); paired data for five individual animals pretreated with cyclamate.

suggests that the *in vitro* data given in Table 5 will be applicable to the *in vivo* metabolism of these sulphamates in cyclamate induced rats. The ability of other sulphamates to induce the microbial enzyme has not been studied. The recent report by Mallett *et al.* (1985) of the induction of microbial cyclamate hydrolysis using an *in vitro* system offers a novel way of determining the substrate specificity of the induction process.

8.2.5 The toxicological consequences of sulphamate hydrolysis

The toxicological consequences of sulphamate hydrolysis centre around the conversion of a highly polar organic acid into an amine. This transformation opens up the possibility of a neurotransmitter activity, e.g. the indirect sympathomimetic effects of aliphatic amines such as cyclohexylamine, or other toxicity such as methaemoglobinaemia or carcinogenicity from an aromatic amine. Again the greatest attention has focused on cyclamate owing to its use as an artificial sweetener.

The potential for cardiovascular toxicity in individuals consuming cyclamate was investigated by Litchfield and Swan (1971), but their study was of limited value since they failed to find very high converters (i.e. 20% or more). Thus the potential for cardiovascular changes has to be assessed from the potency of cyclohexylamine itself and the percentage conversion of cyclamate to cyclohexylamine. Single oral doses of 5 mg/kg or 10 mg/kg of cyclohexylamine caused a significant increase in the mean arterial blood pressure in human volunteers (Eichelbaum *et al.*, 1974), but a lower dose of 2.5 mg/kg was without detectable effect. This 'no effect' dose is equivalent to 5 mg/kg of cyclamate assuming 100% conversion to cyclohexylamine. However, this simple analysis is not valid since it ignores the biological variables implicit in cyclamate metabolism. Firstly, approximately 40% of cyclamate is absorbed from the gut and therefore is not available for microbial metabolism. Thus 60% can be taken as a realistic maximum average conversion and this is supported by data from controlled metabolism studies (see Bopp *et al.*, 1986). Secondly, and of greater importance, is the fact that cyclamate is not metabolized until it reaches the hind gut flora. Large and rapid changes in blood levels of cyclohexylamine are unlikely. The formation of [14C]cyclohexylamine from a single dose of [14C]cyclamate given during chronic cyclamate intake occurred over a prolonged period of 2–3 days (Renwick and Williams, 1972b). Thus a single oral dose of cyclohexylamine itself would give much higher peak levels of cyclohexylamine than would occur with cyclamate metabolism, and it is the peak levels which caused the cardiovascular effects (Eichelbaum *et al.*, 1974). A more realistic approach would be by comparison of the minimum concentration necessary to produce an effect (0.7–0.8 μg/ml; Eichelbaum *et al.*, 1974) with the plasma levels during the metabolism of cyclamate over a period of 24 h. Using the WHO-recommended acceptable daily intake (ADI) of 11 mg/kg/day with 60% metabolism, this would give a dose of cyclohexylamine of 3.31 mg/kg/day. From this intake and the plasma clearance of cyclohexylamine (810 ml/min at 2.5 mg/kg; Eichelbaum *et al.*, 1974) assuming a mean body weight of 78 kg it is possible to calculate the average steady state plasma concentration. This value (0.2 μg/ml) clearly demonstrates that cardiovascular effects would not occur, even at the maximum rate of conversion in an individual consuming cyclamate at the ADI. The third variable which also supports the conclusion that cyclamate is safe is the well-recognized tolerance which develops to indirectly acting sympathomimetics. Since cyclamate metabolism increases over a period of 7–10 days it is likely that tolerance to any pharmacological effects of cyclohexylamine would occur during the induction of the microbial sulphamatase enzyme.

The other form of cyclohexylamine toxicity which has been the focus of attention for regulators considering the safety of cyclamate is that of testicular toxicity in rats given regular doses of 200 mg/kg/day for 90 days. Applying the normal 100-fold safety factor (used to establish the safe level of intake for a food additive) to the no-effect level of 100 mg/kg/day gives an acceptable daily intake for cyclohexylamine of 1 mg/kg/day which is equivalent to 2 mg/kg/day of cyclamate. The 100-fold safety factor is designed to allow both for interspecies differences between rat and man in pharmacokinetics and pharmacodynamics (sensitivity) and for differences between human individuals in these variables. The WHO took a pragmatic approach (see Renwick, 1983b) of applying a 30% conversion figure to the 60% unabsorbed to derive a percentage conversion value of 18%. Application of this to the 1 mg/kg/day of cyclohexylamine gave the ADI of 11 mg/kg/day of cyclamate. An alternative way of viewing the same data base proposed by Renwick (1983b) was to use the maximum steady state conversion value of 60% but then to apply a safety factor which allowed for the fact that differences between human individuals in the formation of the active metabolite cyclohexylamine had been excluded by the use of a worst case figure. The 100-fold safety factor is usually regarded as being 10-fold for animals to man and 10-fold for interindividual differences in humans. Each 10-fold factor could therefore be divided into two 3.16-fold factors for differences in kinetics and dynamics. Therefore logically if 60% conversion is used then a safety factor of 10×3.16 should be adequate. This would convert a no-effect level for cyclohexylamine of 100 mg/kg/day in rats into an ADI for cyclamate of 10.5 mg/kg/day in man assuming 60% conversion. However, again knowledge of the biological processes involved in cyclamate metabolism suggests that this represents an overconservative approach. The ADI assumes that the same individuals show high levels of cyclamate metabolism throughout their lives. The limited data from long-term studies in man suggest that a high converter does not necessarily remain a high converter despite continuous intake (Wills *et al.*, 1981), and this is consistent with the animal data. Also, since cessation of intake results in a rapid loss of the acquired sulphamatase activity (Table 2), it is unlikely that very large amounts of cyclohexylamine would be formed following the use of cyclamate as a food additive. The data of Asahina *et al.* (1971) on the urinary excretion of cyclohexylamine in 50 Japanese individuals following 'uncontrolled' intake of cyclamate found a maximum excretion of cyclohexylamine in one subject of 129 mg/day. This is equivalent to about 3 mg/kg/day which is still 50-fold less than the no-effect level in rats. An alternative approach has become available recently from data on the plasma concentrations of cyclohexylamine in rats fed toxic doses (Roberts and Renwick, 1987). Animals fed cyclohexylamine to provide intakes equivalent to just less than the minimally effective dose of 200 mg/(kg day) had peak plasma concentrations of approximately 4 μg/m. Comparison of this value with that calculated above (0.2 μg/ml) for humans converting 60% of the ADI of 11 mg/kg/day provides a safety margin of 20-fold. In this comparison the major interspecies differences, i.e. in kinetics, have been allowed for as well as interindividual differences in metabolism in man — the maximum conversion has been assumed. The presence of a large residual safety factor confirms that cyclamate would be safe at the WHO-recommended ADI.

Clearly, from the detailed discussion given above, knowledge of the incidence, inducibility and activity of the microbial sulphamatase enzyme is central to establishing safe levels of use of this still available artificial sweetener.

8.3 SULPHONIC ACIDS

Williams (1959) in his classic text *Detoxication Mechanisms* concluded that the sulphonic acid group is not a site for metabolism and that simple sulphonic acids are excreted unchanged. A review of the current literature reveals that this remains true although other sites within more complex sulphonic acid derivatives may be subject to biotransformation.

The high polarity of the sulphonic acid group results in slow absorption following oral administration. Thus this functional group is not found in many therapeutic drugs but is common in colours and detergents. Sulphonation of azo dyes results in water-soluble colours which can have a wide range of uses. Their use as food colourings requires that the compound and the products of azo reduction are of low toxicity. Although the metabolism of such dyes may be complex, the sulphonic acid moiety is not an important site of metabolism; rather, it helps to provide rapid elimination of that part of the molecule (Walker, 1970). The property of rapid urinary elimination of sulphonic acids has led to the development of the drug Mesna (sodium 2-mercaptoethanesulphonate). This compound is a readily excretable thiol which is therefore able to protect the urinary bladder from the cytotoxicity of oxazaphosphorine alkylating agents such as cyclophosphamide (ABPI, 1986).

Simple alkyl sulphonates such as dodecyl- and hexadecyl-sulphonate are almost completely absorbed from the gut and undergo extensive ω- and β-oxidation *in vivo* to produce short-chain metabolites, e.g. butyric acid-4-sulphonate (Taylor *et al.*, 1978). No evidence of desulphonation was found in the rat using [35]-labelled compounds.

Similarly, simple aromatic sulphonates such as the 1- and 6-sulphonates of 2-naphthylamine are eliminated in the urine largely unchanged, with negligible (<0.1%) desulphonation (Batten, 1979).

The metabolism of ^{35}S-labelled mixed linear alkyl benzene sulphonates (containing 10–14 carbon side chains) and branched chain alkyl benzene sulphonates (containing largely a 12 carbon side chain) in the rat showed no detectable desulphonation to ^{35}S-sulphate. Metabolism was by ω- and β-oxidation to yield sulphophenylbutanoic and sulphophenylpentanoic acids in the case of the linear alkyl compounds and sulphophenyl acids with longer side chains in the case of the branched alkyl compounds (Michael, 1968).

Microbial desulphonation of alkyl- and aryl-sulphonates is important in their biodegradation in the environment. For example, *n*-alkane sulphonates can be used by green algae such as *Chlorella fusca* as a sole source of sulphur (Biedlingmaier and Schmidt, 1983). Indeed, this organism can grow on many of the classes of sulphur compounds discussed in this book (Krauss and Schmidt, 1987). Simple *n*-alkane-1-sulphonates are catabolized by *Pseudomonas* via oxidation on carbon 1 to produce a 1-hydroxy-*n*-alkane sulphonate. This compound (Fig. 3) is equivalent to an aldehyde

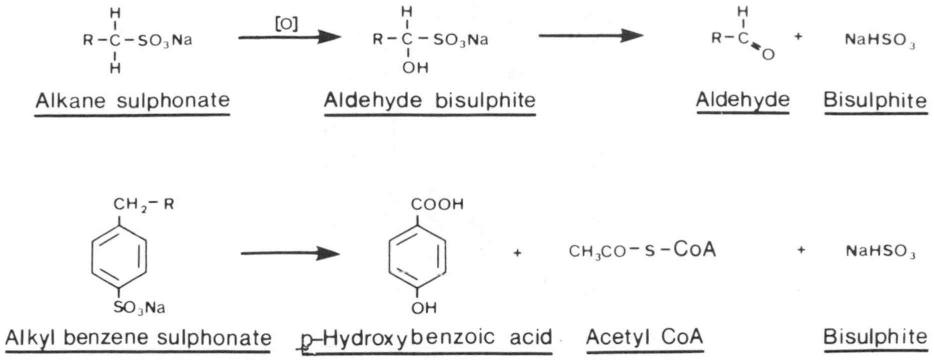

$$R-\overset{\overset{\textstyle H}{|}}{\underset{\underset{\textstyle H}{|}}{C}}-SO_3Na \quad \xrightarrow{[O]} \quad R-\overset{\overset{\textstyle H}{|}}{\underset{\underset{\textstyle OH}{|}}{C}}-SO_3Na \quad \longrightarrow \quad R-\overset{\overset{\textstyle H}{|}}{C}\diagup_O \quad + \quad NaHSO_3$$

Alkane sulphonate   Aldehyde bisulphite   Aldehyde   Bisulphite

Alkyl benzene sulphonate (CH_2-R / SO_3Na) $\longrightarrow$ p-Hydroxybenzoic acid (COOH / OH) $+$ $CH_3CO-S-CoA$ (Acetyl CoA) $+$ $NaHSO_3$ (Bisulphite)

Fig. 3 — The hydrolysis of alkyl and aryl sulphonates.

bisulphite complex which therefore readily decomposes to the corresponding aldehyde plus bisulphite (Thysse and Wanders, 1974). In contrast, the mechanism of microbial desulphonation of aromatic sulphonates either as simple aromatic compounds (Ripin *et al.*, 1971) or as alkyl benzene sulphonates (Willetts and Cain, 1972) is unclear but is due to action of an inducible enzyme system. The desulphonation is apparently an early step which precedes aromatic ring cleavage (Fig. 3). The sulphur moiety which accumulates may be either sulphite or sulphate, depending on the presence of sulphite oxidase (Ripin *et al.*, 1971). Many of these studies employed the sulphonate as the sole source of carbon and sulphur and the organisms studied were selected on the basis of their ability to grow under such conditions. Thus these observations on specific organisms cannot be assumed to be applicable to the intestinal microflora *in vivo* under anaerobic conditions.

8.4   SULPHATE ESTERS

Sulphate esters represent important phase-2 metabolites for a wide range of alcohols and phenols, as well as occurring naturally in intermediary metabolism, e.g. heteropolysaccharides.

Sulphate conjugates are major urinary metabolites of steroids which also represent important transport forms within the body by acting as circulating substrates for steroid synthesis (Briggs and Brotherton, 1970). Although metabolism of the steroid nucleus of steroid sulphates can occur without hydrolysis of the sulphate ester bond (Roberts and Lieberman, 1970), many important reactions occur subsequent to the action of a tissue sulphatase. Early studies showed that the steroid sulphatase enzyme is a distinct form of arylsulphatase and is present in high concentrations in liver, adrenals, ovaries, testes and placenta (Roy, 1970), The lability of steroid sulphates *in vivo* in humans was shown in 1951, when the ^{35}S label of ^{35}S-oestrone sulphate was recovered in the urine largely as inorganic sulphate (see Diczfalusy and Levitz, 1970).

Mammalian hepatocytes contain a number of aryl sulphatases which are either microsomal or lysosomal in origin (Farooqui and Mandel, 1977) and which show a

Table 7 — Aryl sulphatase enzymes in human tissues

Type	A	B	C
Location	Lysosomal	Lysosomal	Microsomal
Nature	Glycoprotein	Glycoprotein	Lipoprotein
Molecular weight	102 000 413 000	60 000	105 000
Effect of SO_4^{2-}	Competitive inhibition	Non-competitive inhibition	No effect
Effect of Ag^+	Inhibition	Stimulation	Inhibition
Typical substrates	4-nitrocatechol sulphate Methylumbeliferone sulphate Ascorbic acid-2-sulphate[a] Cerebroside-3-sulphate[a] Seminolipid[a]	4-nitrocatechol sulphate Methylumbeliferone sulphate UDP N-acetyl galactosamine-4-sulphate[a] Glucosamine-4,6-disulphate	Nitrophenol sulphate Oestrone sulphate[a] Dehydroepiandrosterone sulphate[a]

Adapted from Farooqui and Mandel (1977) and Farooqui (1980).
[a]Postulated natural substrates.

range of substrate specificities (Table 7). Their physiological importance is illustrated by the diseases which can result from a deficiency of these enzymes (Farooqui and Mandel, 1977; Farooqui, 1980). The most important enzyme with respect to xenobiotics is the microsomal sulphatase C which primarily acts on the sulphate esters of steroids. This enzyme is probably identical to oestrone sulphatase, whilst dehydroepiandrosterone sulphatase represents a second microsomal enzyme (Farooqui, 1980). The arylsulphatase C present in rat liver has been purified following solubilization and shown to be a glycoprotein of molecular weight 72 000. The native enzyme probably exists as a tetramer (Moriyasu et al., 1982) with its subunits positioned across the microsomal membrane. Thus it is exposed to both inner and outer surfaces of the membrane with the catalytic site exposed to the cytoplasm (Moriyasu and Ito, 1982). The in vivo substrate specificity of the enzyme may therefore be limited to some extent by the ability of the sulphate ester to cross the cell membrane. Therefore, large steroid esters may possess sufficient lipophilicity to enter the hepatocyte and undergo microsomal hydrolysis, while small molecules may either be excluded or undergo cytoplasmic reactions. This may explain why simple phase-2 metabolites, such as phenylsulphate, are metabolically stable and excreted without undergoing significant hydrolysis (see Renwick, 1983a).

The regulation of the activity of the microsomal steroid sulphatases is complex and may involve both hormonal activation and the presence of intracellular soluble protein enzyme inhibitors (Moutaourakkil and Adessi, 1986). The circulating levels of steroid sulphates represent a balance between synthesis and hydrolysis. The microsomal enzymes oestrone sulphatase (steroid sulphate sulphohydrolase EC 3.1.6.1) and dehydroeandrosterone sulphatase (steroid sulphate sulphohydrolase EC 3.1.6.2) have been purified and isolated following solubilization from the livers of normal individuals and patients with cirrhosis (Prost et al., 1984). The enzyme activity corresponded to fractions with molecular weights of about 330 000. Lower levels of the hydrolytic enzymes were found in the patients with cirrhosis suggesting that the decreased levels of oestrone sulphate and increased levels of oestrone, which may be linked to the feminization associated with cirrhosis, do not arise from an excess of the hepatic sulphatase enzyme but rather a deficiency in the conjugation reaction.

Hydrolysis of steroid sulphates may also be mediated by the intestinal microflora as part of an enterohepatic circulation (see Renwick, 1986b). The intestinal flora are also able to hydrolyse the 3-sulphate esters of bile salts (Huijghebaert et al., 1984).

The metabolism of simple alkyl sulphate esters involves primarily ω- and β-oxidation at sites distant from the sulphate moiety. Even-numbered carbon chain primary alkyl sulphate esters are degraded principally to butanoate-4-sulphate and in man and dogs some acetate-2-sulphate (Denner et al., 1969; Merits, 1975), while odd-numbered chains are degraded to propionate-3-sulphate (Burke et al., 1976). The metabolite, butyric acid 4-sulphate, undergoes a pH dependent (i.e. at pH>5) non-enzymatic hydrolysis to sulphate and γ-butyrolactone (Fig. 4) (Ottery et al., 1970). The extent of this reaction in vivo would be limited by the rapid excretion of the alkyl sulphate. Toxicity due to γ-butyrolactone was not produced following the administration of butyric acid 4-sulphate to rats (Ottery et al., 1970). The metabolism of secondary alkyl sulphate esters is more complex (Maggs et al., 1982; Bains et al., 1987) but the sulphate ester bond is largely metabolically stable. Although there

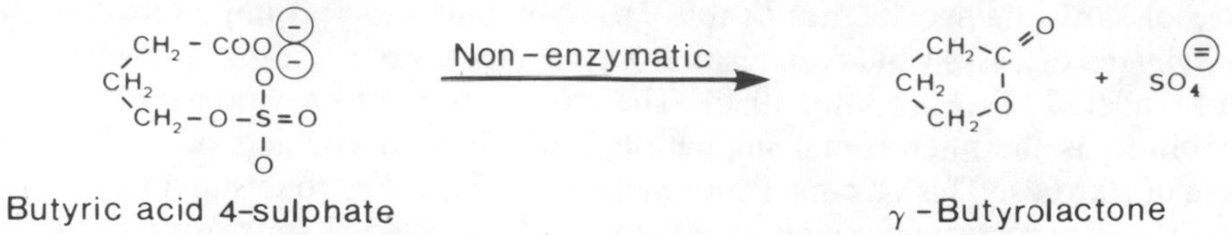

Fig. 4 — Non-enzymatic hydrolysis of butyric acid-4-sulphate.

was elimination of ^{35}S inorganic sulphate following administration of [^{35}S]nonan-5-sulphate this was of limited extent (<3%) and was similar after both oral and intravenous administration (Bains *et al.*, 1987). The limited ability of this ester to cross membranes was shown by the slow absorption from the gut and limited tissue distribution, such that no tissue contained greater concentrations of radioactivity than were present in blood (Bains *et al.*, 1987).

The pharmacological and toxicological implications of the hydrolysis of sulphate bonds are extremely substrate dependent. Clearly in the case of steroid sulphates these represent an important circulating form of the steroid nucleus and also may exhibit activity in their own right. For the majority of xenobiotics sulphate conjugation is an important and largely irreversible detoxication mechanism. In such cases sulphatase activity would represent a potential intoxication or re-toxication reaction. However, for some chemicals such as fluorenylacetamide (Smith *et al.*, 1987), and metabolites of polycyclic aromatic hydrocarbons (Watabe *et al.*, 1985) and allylbenzenes via the 1'-hydroxy metabolite (Randerath *et al.*, 1984), the sulphate ester is an active metabolite and thus its hydrolysis would represent a detoxication reaction.

REFERENCES

ABPI (1986). *ABPI Data Sheet Compendium 1986–87*, Datapharm Publications, pp. 238–239.

Asahina, M., Yamaha, T., Watanabe, K., and Sarrazin, G. (1971). Excretion of cyclohexylamine, a metabolite of cyclamate in human urine. *Chem. Pharm. Bull.*, **19**, 628–632.

Asahina, M., Niimura, T., Yamaha, T. and Takahashi, T. (1972a). Formation of cyclohexylamine and cyclohexanone from cyclamate by microorganisms isolated from the feces of guinea pigs. *Agr. Biol. Chem.*, **36**, 711–718.

Asahina, M., Yamaha, T., Sarrazin, G., and Watanabe, K. (1972b). Conversion of cyclamate to cyclohexylamine in guinea pigs. *Chem. Pharm. Bull.*, **20**, 102–108.

Audrieth, L. F. and Sveda, M. (1944). Preparation and properties of some N-substituted sulfamic acids. *J. Org. Chem.*, **9**, 89–101.

Bains, S. K., Olavesen, A. H., Black, J. G., Howes, D., Curtis, C. G., and Powell, G. M. (1987). Metabolism in the rat of potassium nonan-5-sulphate, a symmetrical anionic surfactant. *Xenobiotica*, **17**, 709–723.

Ball, L. M., Renwick, A. G., and Williams, R. T. (1977). The fate of [^{14}C]saccharin in man, rat and rabbit and of 2-sulphamoyl[^{14}C]benzoic acid in the rat. *Xenobiotica*, **7**, 189–203.

Batten, P. L. (1979). Metabolism of 2-naphthylamine sulphonic acids. *Toxicol. Appl. Pharmacol.*, **48**, A171.

Bickel, M. H., Burkard, B., Meier-Strasser, E., and Van Den Broek-Boot, M. (1974). Enterobacterial formation of cyclohexylamine in rats ingesting cyclamate. *Xenobiotica*, **4**, 425–439.

Biedlingmaier, S. and Schmidt, A. (1983). Alkylsulfonic acids and some sulfur-containing detergents as sulfur sources for growth of *Chlorella fusca*. *Arch. Microbiol.*, **136**, 124–130.

Bopp, B. A., Sonders, R. C., and Kesterson, J. W. (1986). Toxicological aspects of cyclamate and cyclohexylamine. *Critical Rev. Toxicol.*, **16**, 213–306.

Briggs, M. H. and Brotherton, J. (1970). In *Steroid Biochemistry and Pharmacology*, Academic Press, London, pp. 80–81.

Burke, B., Olavesen, A. H., Curtis, C. G., and Powell, G. M. (1976). The biodegradation of the surfactant undecyl sulphate. *Xenobiotica*, **6**, 667–678.

Collings, A. J. (1971). The metabolism of sodium cyclamate. In G. C. Birch, L. F. Green, and C. B. Coulson (eds.), *Sweetness and Sweeteners*, Applied Science, London, pp. 51–68.

Dalderup, L. M., Keller, G. H. M., and Schoeten, F. (1970). Cyclamate and cyclohexylamine. *Lancet*, **1**, 845.

Davis, T. R. A., Adler, N., and Opsahl, J. C. (1969). Excretion of cyclohexylamine in subjects ingesting sodium cyclamate. *Toxicol. Appl. Pharmacol.*, **15**, 106–116.

Denner, W. H. B., Olavesen, A. H., Powell, G. M., and Dodgson, K. S. (1969). The metabolism of potassium dodecyl [^{35}S]sulphate in the rat. *Biochem. J.*, **111**, 43–51.

Diczfalusy, E. and Levitz, M. (1970). Formation, metabolism and transport of estrogen conjugates. In S. Bernstein and S. Solomon (eds.), *Chemical and Biological Aspects of Steroid Conjugation*, Springer, Berlin, pp. 291–320.

Drasar, B. S., Renwick, A. G. and Williams, R. T. (1972). The role of the gut flora in the metabolism of cyclamate. *Biochem. J.*, **129**, 881–890.

Eichelbaum, M., Hengstmann, J. H., Rost, H. D., Brecht, T., and Dengler, H. J. (1974). Pharmacokinetics, cardiovascular and metabolic actions of cyclohexylamine in man. *Arch. Toxikol.*, **31**, 243–263.

Farooqui, A. A. (1980). Sulfatases, sulfate esters and their metabolic disorders. *Clin. Chim. Acta*, **100**, 285–299.

Farooqui, A. A. and Mandel, P. (1977). On the properties and role of arylsulphatases A, B and C in mammals. *Int. J. Biochem.*, **8**, 685–691.

Freeman, C. and Hopwood, J. J. (1986). Human liver sulphamate sulphohydrolase. Determinations of native protein and subunit M_r values and influence of substrate aglycone structure on catalytic properties. *Biochem. J.*, **234**, 83–92.

Friedman, L., Richardson, H. L., Richardson, M. E., Lethco, E. J., Wallace, W. C., and Sauro, M. (1972). Toxic response of rat to cyclamates in chow and semisynthetic diets. *J. Natl. Cancer. Inst.*, **49**, 751–764.

Goldberg, L., Parekh, C., Patti, A., and Soike, K. (1969). Cyclamate degradation in mammals and *in vitro*. *Toxicol. Appl. Pharmacol.*, **14**, 654.

Huighebaert, S., Parmentier, G., and Eyssen, H. (1984). Specificity of bile salt sulfatase activity in man, mouse and rat intestinal microflora. *J. Steroid Biochem.*, **20**, 907–912.

Hwang, K. (1966). Mechanism of the laxative effect of sodium sulphate, sodium cyclamate and calcium cyclamate. *Arch. Int. Pharmacodyn.*, **163**, 302–340.

Ichibagase, H., Kojima, S., Suenaga, A., and Inoue, K. (1972). Studies on synthetic sweetening agents. XVI. Metabolism of sodium cyclamate (5). The metabolism of sodium cyclamate in rabbits and rats after prolonged administration of sodium cyclamate. *Chem. Pharm. Bull.*, **20**, 1093–1101.

Iwasaki, K., Shiraga, T., Tada, K., Noda, K., and Noguchi, H. (1986). Age and sex-related changes in amine sulphoconjugations in Sprague-Dawley strain rats. Comparison with phenol and alcohol sulphoconjugations. *Xenobiotica*, **16**, 717–723.

Kojima, S. and Ichibagase, H. (1966). Studies on synthetic sweetening agents. VIII. Cyclohexylamine, a metabolite of sodium cyclamate. *Chem. Pharm. Bull.*, **14**, 971–974.

Kojima, S. and Ichibagase, H. (1968). Studies on synthetic sweetening agents. XIII. Metabolism of sodium cyclamate (2). Detection of metabolites of sodium cyclamate in rabbit and rat by gas–liquid chromatography. *Chem. Pharm. Bull.*, **16**, 1851–1854.

Krauss, F. and Schmidt, A. (1987). Sulfur sources for growth of *Chlorella fusca* and their influence on key enzymes of sulfur metabolism. *J. Gen. Microbiol.*, **133**, 1209–1220.

Litchfield, M. H. and Swan, A. A. B. (1971). Cyclohexylamine production and physiological measurements in subjects ingesting sodium cyclamate. *Toxicol. Appl. Pharmacol.*, **18**, 535–541.

Maggs, J. L., Powell, G. M., Dodgson, K. S., Howes, D., Black, J. G., and Olavesen, A. H. (1982). Metabolism in the rat of potassium DL-octan-2-sulphate, a secondary alkylsulphate. *Xenobiotica*, **12**, 101–109.

Mallett, A. K., Rowland, I. R., Bearne, C. A., Purchase, R., and Gangolli, S. D. (1985). Metabolic adaptation of rat faecal microflora to cyclamate *in vitro*. *Fd. Chem. Toxicol.*, **23**, 1029–1034.

Matsui, M., Tanimura, A., and Kurata, H. (1981). Identification of cyclamate-converting bacteria. (Studies on the metabolism of food additives by micro-organisms inhabiting the gastrointestinal tract. VI). *J. Fd. Hyg. Soc., Jpn.*, **22**, 215–222.

McGlinchey, G., Coakley, C. B., Gestaustus-Tansey, V., Gault, J., and Spillane, W. J. (1982). *In vivo* and *in vitro* studies with sulfamate sweeteners. *J. Pharm. Sci.*, **71**, 661–665.

Merits, I. (1975). The metabolism of labelled hexadecyl sulphate salts in rat, dog and human. *Biochem. J.*, **148**, 219–228.

Michael, W. R. (1968). Metabolism of linear alkylate sulfonate and alkyl benzene sulfonate in albino rats. *Toxicol. Appl. Pharmacol.*, **12**, 473–485.

Miller, J. P., Crawford, L. E. M., Sonders, R. C., and Cardinal, E. V. (1966). Distribution and excretion of ^{14}C-cyclamate sodium in animals. *Biochem. Biophys. Res. Comm.*, **25**, 153–157.

Moriyasu, M. and Ito, A. (1982). Transmembranous disposition of arylsulphatase C in microsomal membranes of rat liver. *J. Biochem.*, **92**, 1197–1204.

Moriyasu, M., Ito, A., and Omura, T. (1982). Purification and properties of arylsulfatase C from rat liver microsomes. *J. Biochem.*, **92**., 1189–1195.

Moutaourakkil, M. and Adessi, G. L. (1986). A dehydroepiandrosterone sulfatase inhibiting activity in soluble proteins of guinea pig livers. *Steroids*, **47**, 401–411.

Niimura, T., Tokeida, T., and Yamaha, T. (1974). Partial purification and some properties of cyclamate sulfamatase. *J. Biochem.*, **75**, 407–417.

Oser, B. L., Carson, S., Vogin, E. E., and Sonders, R. C. (1968). Conversion of cyclamate to cyclohexylamine in rats. *Nature*, **220**, 178–179.

Ottery, J., Olavesen, A. H., and Dodgson, K. S. (1970). Metabolism of dodecyl sulphate in the rat: non-enzymatic liberation of sulphate and γ-butyrolactone from the major metabolite, butyric acid 4-sulphate. *Life Sci.*, **9** (11), 1335–1340.

Price, J. M., Biava, C. G., Oser, B. L., Vogin, E. E., Steinfeld, J., and Ley, H. L. (1970). Bladder tumors in rats fed cyclohexylamine or high doses of a mixture of cyclamate and saccharin. *Science*, **167**, 1131–1132.

Prosky, L. and O'Dell, R. G. (1971). *In vivo* conversion of ^{14}C-labeled cyclamate to cyclohexylamine. *J. Pharm. Sci.*, **60**, 1341–1343.

Prost, O., Ottignon, Y., Remy-Martin, A., Vintton, D., Miguet, P., and Adessi, G. L. (1984). Steroid sulfatase activities in normal and cirrhotic livers and plasma levels of estrone sulphate, estrone and estradiol-17β in man. *Steroids*, **43**, 189–199.

Randerath, K., Haglund, R. E., Phillips, D. H., and Reddy, M. V. (1984). Phosphrus 32 post labelling analysis of DNA adducts formed in the livers of animals treated with safrole, estragole and other naturally-occurring alkenyl-benzenes: 1. Adult female CD-1 mice. *Carcinogenesis*, **5**, 1613–1622.

Renwick, A. G. (1976). Microbial metabolism of drugs. In D. V. Parke and R. L. Smith (eds.), *Drug Metabolism from Microbe to Man*, Taylor and Francis, London, pp. 169–189.

Renwick, A. G. (1983a). Unmetabolised compounds. In J. Caldwell and W. B. Jakoby (eds.), *Biological Basis of Detoxication*, Academic Press, New York, pp. 151–179.

Renwick, A. G. (1983b). The fate of non-nutritive sweeteners in the body. In T. H. Grenby, K. J. Parker, and M. G. Lindley (eds.), *Developments in Sweeteners 2*, Elsevier London, pp. 179–224.

Renwick, A. G. (1986a). The metabolism of intense sweeteners. *Xenobiotica*, **16**, 1057–1071.

Renwick, A. G. (1986b). Gut bacteria and the enterohepatic circulation of foreign compounds. In M. J. Hill (ed.), *Microbial Metabolism in the Digestive Tract*, CRC Press, pp. 135–153.

Renwick, A. G. (1988). Intense sweeteners and the gut microflora. In I. R. Rowland (ed.), *Role of the Gut Flora in Toxicity and Cancer*, Academic Press, London, pp. 175–206.

Renwick, A. G. and Williams, R. T. (1972a). The metabolites of cyclohexylamine in man and certain animals. *Biochem. J.*, **129**, 857–867.

Renwick, A. G. and Williams, R. T. (1972b). The fate of cyclamate in man and other species. *Biochem. J.*, **129**, 869–879.

Richards, R. K., Taylor, J. D., O'Brien, J. L., and Duescher, H. O. (1951). Studies on cyclamate sodium (sucaryl sodium), a new noncaloric sweetening agent. *J. Amer. Pharmac. Assoc.*, **40**, 1–6.

Ripin, M. J., Noon, K. F., and Cook, T. M. (1971). Bacterial metabolism of arylsulfonates 1. Benzene sulfonate as growth substrate for *Pseudomonas testosteroni* H-8. *Appl. Microbiol.*, **21**, 495–499.

Roberts, K. D. and Lieberman, S. (1970). The biochemistry of the 3β-hydroxy-Δ^5-steroid sulphates. In S. Bernstein and S. Solomon (eds.), *Chemical and Biological Aspects of Steroid Conjugation*, Springer, Berlin, pp. 219–290.

Roberts, A. and Renwick, A. G. (1987). The fate of cyclohexylamine in rat and mouse in relation to testicular toxicity. *Human Toxicol.*, in press.

Roy, A. B. (1970) Enzymological aspects of steroid conjugation. In S. Bernstein and S. Solomon (eds.), *Chemical and Biological Aspects of Steroid Conjugation*, Springer, Berlin, pp. 74–130.

Schoenberger, J. A., Rix, D. M., Sakamoto, A., Taylor, J. P. and kark, R. M. (1953). Metabolic effects, toxicity and excretion of calcium N-cyclohexylsulfamate (sucaryl) in man. *Amer. J. Med. Sci.*, **225**, 551–559.

Smith, B. A., Springfield, J. R. and Gutmann, H. R. (1987). Solvolysis and metabolic degradation, by rat liver, of the ultimate carcinogen, N-sulfonoxy-2-acetylaminofluorene. *Mol. Pharmacol.*, **31**, 438–445.

Sonders, R. C., Netwal, J. C., and Wiegand, R. G. (1969). Site of conversion of cyclamate to cyclohexylamine. *Pharmacologist*, **11**, 241.

Taylor, A. J., Olavesen, A. H., Black, J. G., and Howes, D. (1978). The metabolism of the surfactants dodecylsulfonate and hexadecylsulfonate in the rat. *Toxicol. Appl. Pharmacol.*, **45**, 105–117.

Taylor, J. D., Richards, R. K., and Davin, J. C. (1951). Excretion and distribution of radioactive S^{35}-cyclamate sodium (sucaryl sodium) in animals. *Proc. Soc. Exp. Med.*, **78**, 530–533.

Tesoriero, A. A. and Roxon, J. J. (1975). [^{35}S]Cyclamate metabolism: incorporation of ^{35}S into protein of intestinal bacteria *in vitro* and production of volatile ^{35}S-containing compounds. *Xenobiotica*, **5**, 25–31.

Thysse, G. J. E. and Wanders, T. H. (1974). Initial steps in the degradation of *n*-alkane-1-sulphonates by *Pseudomonas*. *Antonie van Leeuwenhoek*, **40**, 25–37.

Tokeida, T., Niimura, T., Yamaha, T., Hasegawa, T., and Suzuki, T. (1979). Anaerobic deamination of cyclohexylamine by intestinal microorganisms in rabbits. *Agric. Biol. Chem.*, **43**, 25–32.

Tsuchiya, T. (1981). Studies on the metabolism of sodium cyclamate by intestinal bacteria. *Memoirs Tokyo Univ. Agric.*, **23**, 1–55.

Turesky, R. J., Skipper, P. L., Tannenbaum, S. R., Coles, B., and Ketterer, B. (1986). Sulfamate formation is a major route for detoxification of 2-amino-3-methylimidazo[4,5-f]quinoline in the rat. *Carcinogenesis*, **7**, 1483–1486.

Walker, R. (1970). The metabolism of azo compounds: a review of the literature. *Fd. Cosmet. Toxicol.*, **8**, 659–676.

Watabe, T., Fujieda, T., Hiratsuka, A., Ishizuke, T., Hakamata, Y., and Ogura, K. (1985). The carcinogen, 7-hydroxymethyl-12-methylbenz[a]anthracene, is activated and covalently binds to DNA via a sulphate ester. *Biochem. Pharmacol.*, **34**, 3002–3005.

Willetts, A. J. and Cain, R. B. (1972). Microbial metabolism of alkylbenzene sulphonates. Bacterial metabolism of undecylbenzene-p-sulphonate and dode-cylbenzene-p-sulphonate. *Biochem. J., ***129**, 389–402.

Williams, R. T. (1959). *Detoxication Mechanisms. The Metabolism and Detoxication of Drugs, Toxic Substances and other Organic Compounds,* Chapman and Hall, London, pp. 497–500.

Wills, J. H., Serrone, D. M. and Coulston, F. (1981). A 7-month study of ingestion of sodium cyclamate by human volunteers. *Regulatory Toxicol. Pharmacol., ***1**, 163–176.

9

Sulphur heterocycles

D. J. Rance
Department of Drug Metabolism, Pfizer Central Research, Sandwich, UK

SUMMARY

1. The metabolism of heterocyclic sulphur compounds is classified in terms of ring size, the degree of saturation, the presence of other heteroatoms, and fusion to carbocyclic or other heterocyclic rings.
2. Biotransformation options open to sulphur heterocycles consist of attack at sulphur, at another heteroatom such as nitrogen, or at an annular carbon atom. Ring scission may then follow as a secondary reaction.
3. These reactions are discussed with reference to chemical reactivity of the ring systems, which reveals that the enzyme systems responsible for metabolism are generally electrophilic in character.
4. Some general conclusions are offered as to the basis of predictive rules, but these should be applied cautiously in view of the limited data base from which they are derived.
5. Areas particularly worthy of future study are the relative importance of cytochrome P-450 and the FAD containing monooxygenase in *S*-oxygenation pathways, and stereochemical aspects of the formation and disposition of sulphoxide metabolites.

9.1 INTRODUCTION

Sulphur is a common element in medicinal agents and other xenobiotic compounds. In many of these compounds, the sulphur atom forms part of a heterocyclic ring, often in combination with other heteroatoms such as nitrogen. The biotransformation of these sulphur ring systems is the subject of this review. Although the metabolism of sulphur heterocycles has been briefly discussed by Damani and Case (1984) and Damani (1987), the present review constitutes the first systematic treatment of the subject.

Special emphasis is paid here to a consideration of how the chemistry of sulphur heterocycles underlies their routes of metabolism. A wide variety of sulphur functionalities is represented in this class of compounds, and four general aspects have been employed in their classification, i.e. ring size, the degree of saturation, the presence of other heteroatoms, and fusion to carbocyclic or other heterocyclic rings. In each section, a discussion of the metabolism of sulphur heterocycles is prefaced by a brief overview of the relevant chemistry; although not intended to tax the average reader, these chemical subsections may be omitted without serious loss of continuity.

The detailed metabolic fate of substituent groups is beyond the scope of this chapter, and discussion is restricted to examples where the heterocyclic ring is a major determinant of substituent metabolism.

9.2 THREE- AND FOUR-MEMBERED HETEROCYCLES CONTAINING SULPHUR

9.2.1 Chemical considerations

The parent three- and four-membered sulphur heterocycles are known as thiirens (1) and thiets (2) respectively; the corresponding reduced derivatives are thiiranes (3) and thietanes (4). The chemistry of these systems (Walshe, 1979; Dittmer, 1984; Block, 1984) is dominated by the effects of ring strain which accounts for their reactivity; the unsubstituted thiiren (1) is unknown. The S-oxidized derivatives (sulphoxides and sulphones) are normally more stable than the parent compounds but still degrade thermally with liberation of sulphur monoxide and sulphur dioxide. Attack on the sulphur atom of thiiranes by electrophiles can yield cyclic sulphonium salts (5) which in the presence of a nucleophile undergo ring opening reactions.

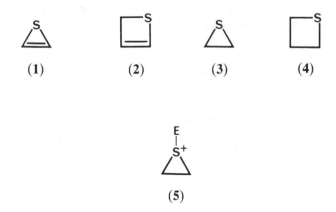

9.2.2 Metabolism of compounds containing a three- or four-membered sulphur ring

These ring systems do not generally feature in xenobiotic molecules and consequently the paucity of data on their metabolism is not surprising. However, the involvement of an episulphonium ion intermediate (7) in a route of metabolism can be inferred from the recent identification of a DNA-derived adduct S-[2-(N[7]-guanyl)ethyl]glutathione (8) in livers of rats treated with 1,2-dibromoethane (6);

Koga *et al.*, 1986); the suggested pathway for the production of this adduct thought to be responsible for the carcinogenicity of **(6)** is shown in Scheme 1. (Also, see Chapter 6 on sulphonium salts in this volume.)

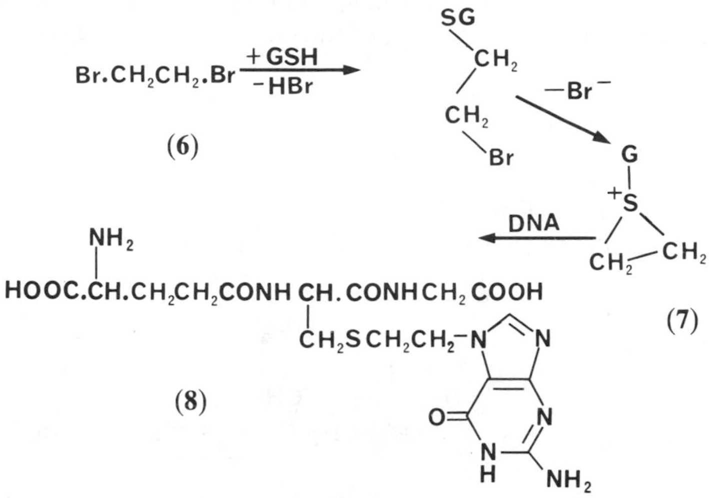

Scheme 1

A three-membered thioxirane ring has been suggested as an intermediate in the metabolic desulphuration reactions of thiones (Damani, 1987). The initial reaction is thought to be *S*-oxygenation of the thione **(9)** to produce a sulphene **(10)** which then cyclizes to a thioxirane **(11)**. This may then decompose to give the oxo derivative **(12)** and highly reactive singlet atomic sulphur which is able to form hydrosulphides with endogenous thiol groups (Scheme 2). This sequence may be responsible for the tissue necrosis shown by various thiones (e.g. thioacetamides and thiocarbamides). A similar mechanism via an intermediate phosphoxythiirane has been proposed for the desulphuration reaction of parathion to paraoxon (Kamataki *et al.*, 1976).

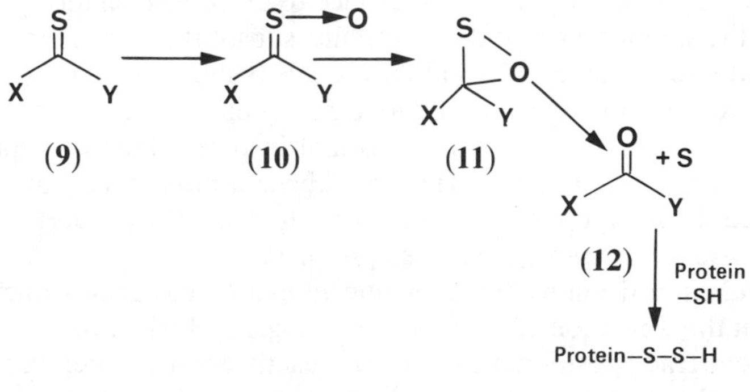

Scheme 2

Notwithstanding these proposed reactive intermediates, the only reported example of metabolism of a 'small-ring' sulphur heterocycle is that of the artificial sweetner alitame (13). This compound contains a four-membered thietane ring, and sulphoxide and sulphone metabolites (14) have been isolated as major metabolites from urine after its administration to man and during safety evaluation in animals (Figdor and Caldwell, 1987); no cleavage of the thietane ring was observed.

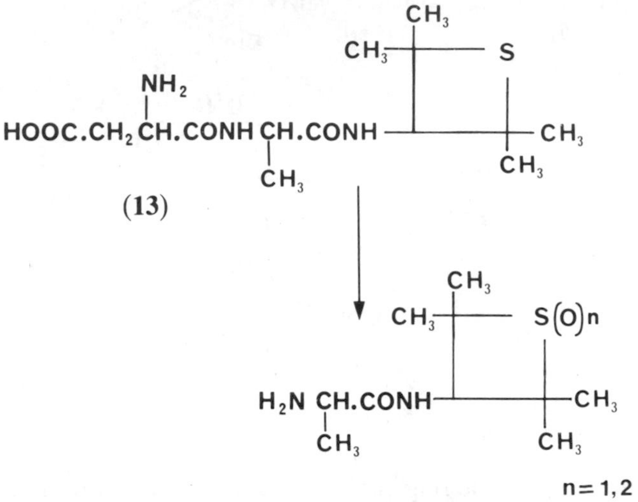

(13)

(14)

$$n = 1, 2$$

9.3 FIVE-MEMBERED HETEROCYCLES CONTAINING SULPHUR AS THE ONLY HETEROATOM

9.3.1 Chemical considerations

The chemistry of thiophen (15) and its fused analogues has been much studied and the interested reader is referred to more detailed reviews (Rajappa, 1984; Meth-Cohn, 1979). Reactions of principal relevance to biotransformation studies are the oxidation of the sulphur atom and electrophilic substitution of the ring.

The 1-oxide (sulphoxide; (16)) and 1,1-dioxide (sulphone; (17)) of thiophen itself are highly reactive, although the sulphone has been isolated. The corresponding derivatives of some substituted thiophens and in particular the sulphones of the fused ring systems benzothiophen (18) and dibenzothiophen (19) are more stable. Moreover, tetrahydrothiophen (THTP; (20)), which has the properties of an aliphatic sulphide, also forms stable S-oxidized products.

Electrophilic substitution of the aromatic thiophen ring occurs principally at the 2 position, or at the 5 position of derivatives bearing a 2-alkyl or 2-acyl substituent. In benzothiophen, the positional order of reactivity has been determined as 3>2>6>5>4>7, all positions being more reactive than in benzene but with substitution in the hetero ring being favoured over that in the carbocyclic ring.

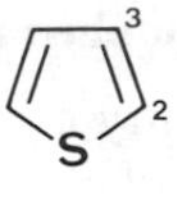

(15) 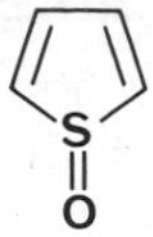　　　　　　　　　(16) 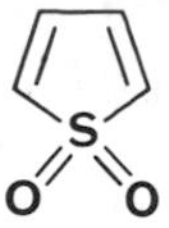　　　　　　　　　(17)

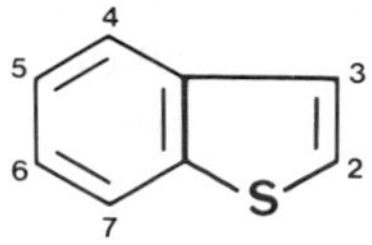

(18) 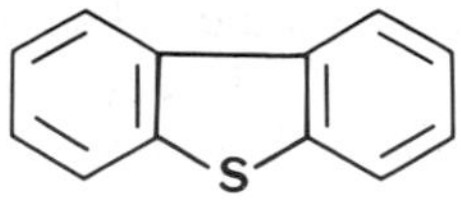　　　　　　　　　(19) 　　　　　　　　　(20)

Direct hydroxylation at the 2- or 3-position of the thiophen ring is not normally possible by chemical means; synthetic routes to these potential metabolites of thiophen compounds have been described (Grondowitz and Hörnfeldt, 1986). 2-Hydroxy-thiophens (**21a**) and their 3-hydroxy isomers (**22a**) normally decompose on standing at room temperature, a reaction which may be facilitated by keto–enol tautomerization. Thus, for 2-hydroxythiophens, the non-aromatic γ-thiolactone (**21c**) is generally preferred, although the position of equilibrium between the three tautomeric forms (**21a-c**) depends on the substitution pattern. With the 3-hydroxythiophens only two tautomers are feasible and, in general, the hydroxy form (**22a**) is more favoured than in the 2-series.

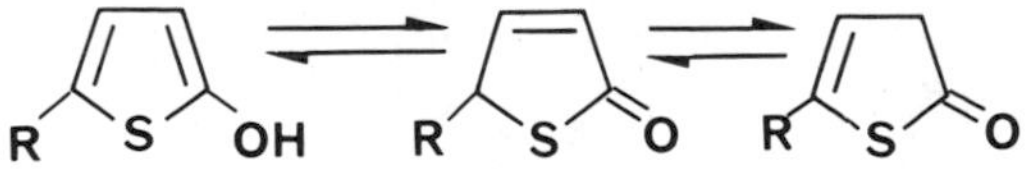

(21a)　　　　　　　　(21b)　　　　　　　　(21c)

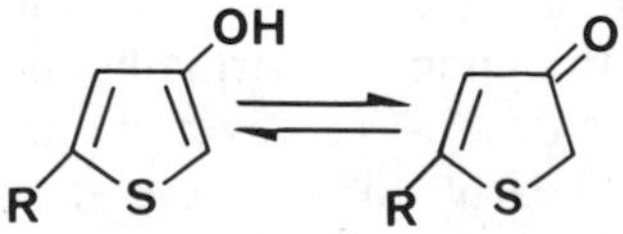

(22a)　　　　　　　　(22b)

Ring systems containing two sulphur atoms are the isomeric 1,2- and 1,3-dithioles (**23**) and (**24**). The parent 1,2-dithiole is unknown but the 1,3-dithioles are generally chemically more stable. The chemistry of these ring systems has been reviewed by Walshe (1979).

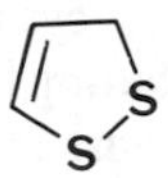

(23) 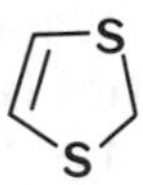　　　　　　　　(24)

9.3.2   Metablosim of compounds containing an unfused thiophen ring

The metabolism of thiophen itself was first studied a century ago (Heffter, 1886). Oral administration to dogs resulted in an increased output of ethereal sulphates, an observation consistent with hydroxylation of the thiophen ring to produce metabolites which are then conjugated with inorganic sulphate; more direct evidence for this pathway is still awaited. Similar administration of thiophen or its 2-bromo analogue to rabbits did not result in increased ethereal sulphate production, but analysis of urine suggested the presence of dihydrodiol and mercapturic acid metabilites. Subsequent studies by Bray *et al.* (1971) confirmed these results in the case of thiophen and identified the metabolites, which were also produced by rats, as 2-thienylmercapturic acid (25) and a premercapturic acid, 3-hydroxy-2,3-dihydro-2-thienylmercapturic acid (26), the latter predominating.

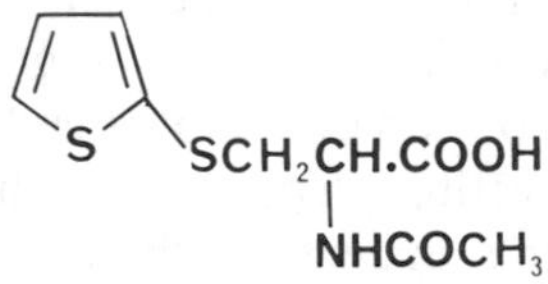

(25) 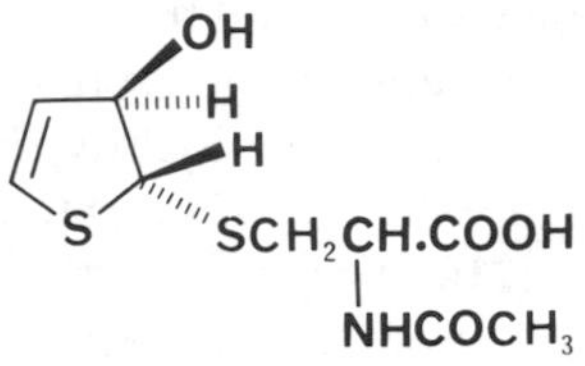   (26)

These metabolites presumably arise via conjugation of an intermediate thiophen 2,3-epoxide with glutathione, the premercapturic acid being analogous to metabolites formed by bromobenzene (Gillham and Young, 1968) and naphthalene (Boyland and Sims, 1958). Unchanged thiophen was virtually undetectable in rat urine and faeces although one-third of the dose was recovered unchanged in the exhaled air (Bray *et al.*, 1971). Overall, nearly 30% of the dose in the rat remains unaccounted for. Since the products of sulphoxidation are chemically unstable, studies with radiolabelled thiophen are required to establish the complete metabolic fate of the compound.

The proposed glutathione conjugation reaction described above for the parent compound has been reported recently for a thiophen derivative. Thus, Lynch *et al.* (1987) have described the *in vivo* and *in vitro* conversion of the anthelmintic compound morantel (27a), the metabolism of which is described in more detail below, to a mercaptan or mercapturic acid product (27b); *in vitro*, the metabolite was produced by the addition of glutathione and cytosol (a source of glutathione transferase) to the products formed after incubating [3H]morantel in rat and bovine liver microsomes. Its formation is therefore consistent with the pathway proposed for thiophen involving glutathione conjugation of an intermediate metabolite, in this case probably the 4,5-epoxide.

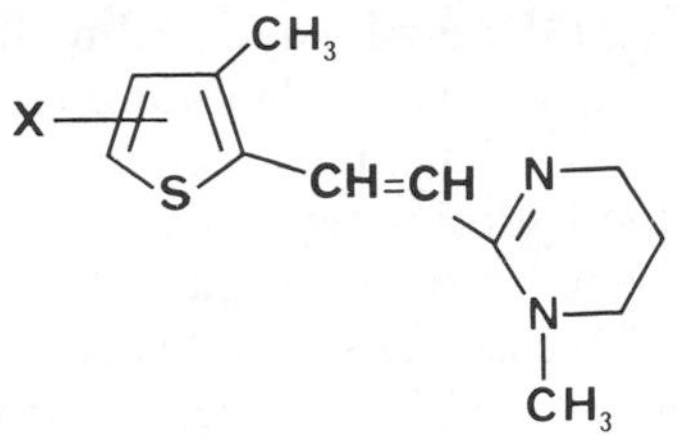

(27) **a**; X=H
     b; X=SH or SCH$_2$CH.COOH
                      |
                  NHCOCH$_3$

A metabolic reaction of greater importance for thiophen derivatives is ring hydroxylation. For 2-substituted derivatives, hydroxylation generally occurs at the C$_5$ position. 5-Hydroxylation is a major pathway of metabolism for the anti-inflammatory drug suprofen (**28**) in man, monkey, rat and mouse, although the route is unimportant in the guinea pig and apparently absent in the dog (Mori *et al.*, 1985). Similarly, the natriuretic tienilic acid (**29a**) is excreted in human or rat urine principally (up to 50% dose) as the 5-hydroxy derivative (**29b**) while dog is again anomalous in not producing this metabolite (Mansuy *et al.*, 1984); structural assignment was by comparison with a synthetic standard in contrast to that of other workers (Vinay *et al.*, 1980) who, probably erroneously, assigned the major urinary metabolite in man as the dihydroxylated compound (**30**).

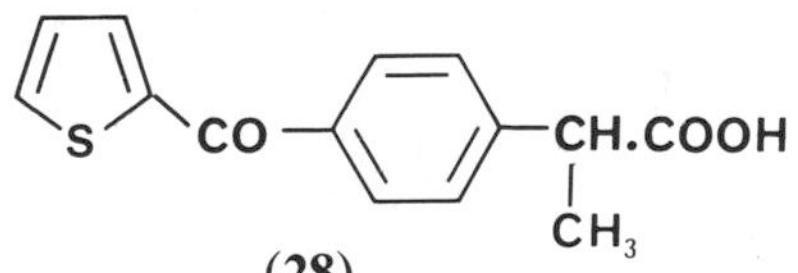

(**28**)

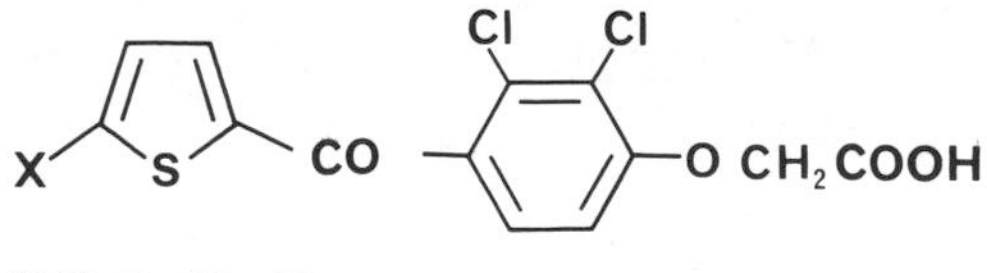

(**29**) **a**; X= H
        b; X = OH

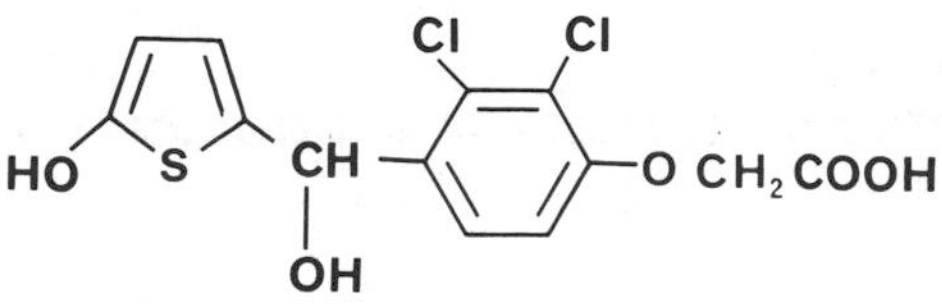

(**30**)

The anthelmintics pyrantel (**31**); R=H) and its 3-methyl analogue morantel also undergo 5-hydroxylation in farm animal and laboratory species, including the dog. In the case of pyrantel, this pathway (Scheme 3) is inferred (Faulkner *et al.*, 1972) from the presence in alkaline urine hydrolysates of keto acids (**34**)–(**36**) derived from a γ-thione acid (**33**), a ring-opened product of the 5-hydroxy metabolite (**32**). More recently, Lynch *et al.* (1987) have isolated the 5-hydroxy derivative of morantel and demonstrated that it is a major metabolite present in urine, plasma, liver and milk. The microbial conversion of thiophen-2-carboxylic acid (**37**) to 2-oxoglutarate (**38**) and inorganic sulphate follows a similar pathway (Scheme 4), reaction being initiated by 5-hydroxylation of the CoA ester (Cripps. 1973).

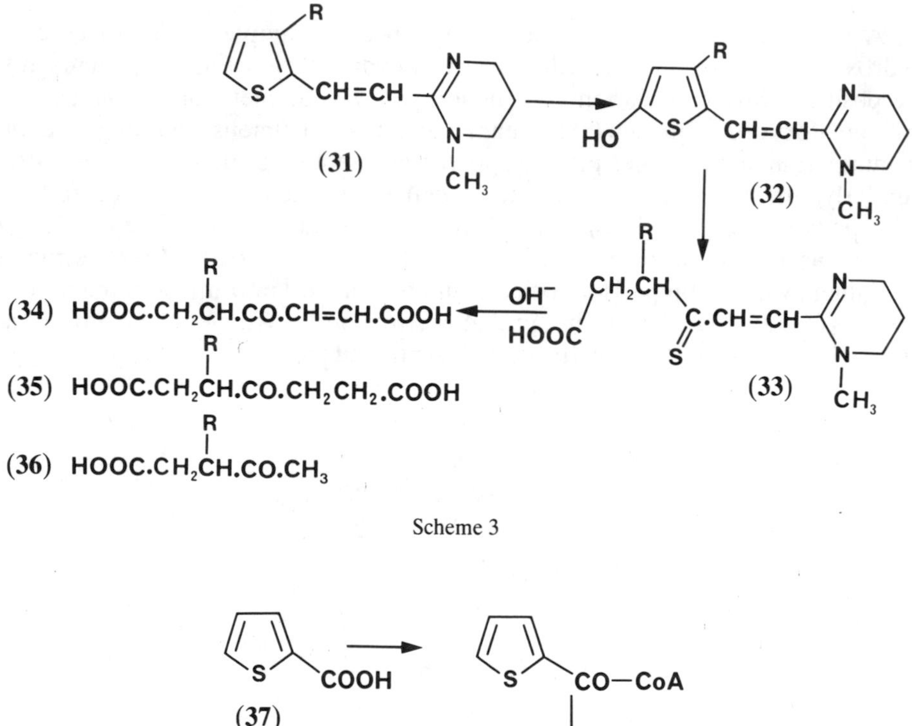

Scheme 3

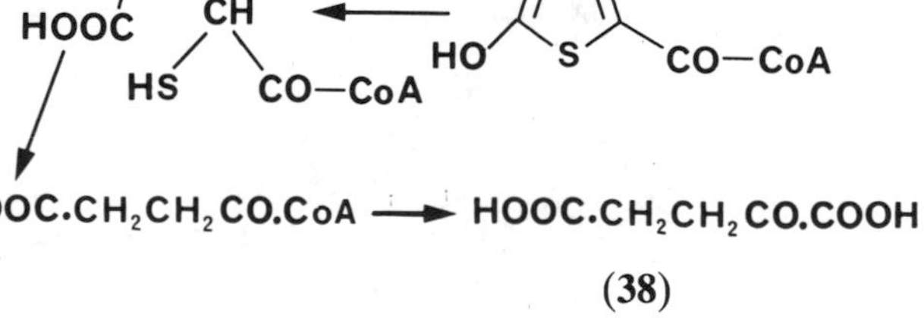

Scheme 4

An interesting example of ring hydroxylation occurs with the anticholinergic agent tiquizium bromide (39a) which contains two thiophen rings attached to the same sp^2 carbon atom of an olefinic bond. In dogs and rats, two 5-hydroxylated metabolites (39b,c) are produced (in comparable amounts), these being geometrical isomers about the double bond (Nishikawa *et al.*, 1985). Both metabolites are excreted in urine as glucuronide conjugates suggesting that they have hydroxy rather than ketone character; this behaviour is in contrast to the corresponding metabolites of suprofen and tienilic acid which are excreted in the free form.

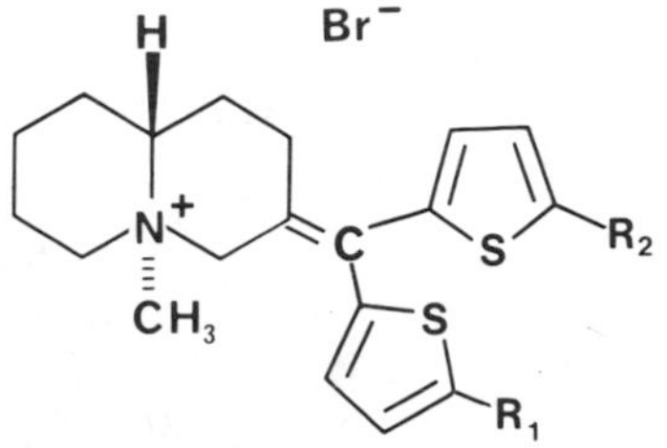

(39) a; $R_1 = R_2 = H$
b; $R_1 = OH$, $R_2 = H$
c; $R_1 = H$, $R_2 = OH$

Hydroxylation at the C_5 position of 2-substituted thiophens is consistent with oxidation by the cytochrome P-450 monooxygenases, this ring carbon being most activated towards electrophilic attack. Mansuy (1987) has recently confirmed the involvement of cytochrome P-450 in the *in vitro* hydroxylation of tienilic acid by human and rat liver microsomes and demonstrated that the reaction is mediated by a particular isozyme (cytochrome P -450–8) which is responsible for other aromatic hydroxylation reactions (e.g. *p*-hydroxylation of mephenytoin). This suggests that hydroxylation of the thiophen ring occurs via a 4,5-epoxide intermediate analogous to the arene oxide moiety implicated in aromatic carbocyclic hydroxylation. Metabolic hydroxylation of substituted thiophen ring systems has not been described at carbons other than C_5. Thus, Mansuy *et al.* (1984) showed that neither the 3-hydroxy nor the 4-hydroxy derivatives were present in the urine of human volunteers dosed with tienilic acid. Similarly, tiaprofenic acid (40), a 2,5-disubstituted thiophen derivative, has no reported metabolites in which hydroxylation of the heterocyclic ring has occurred (Pottier *et al.*, 1977).

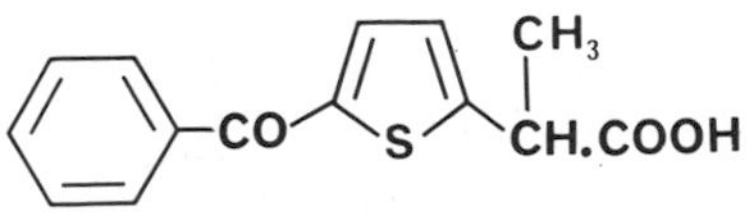

(40)

Detailed consideration of the metabolism of thiophen substituents is beyond the scope of this review. However, it is noteworthy that the aromatic nature of the thiophen ring means that an α-carbon is benzylic in character and hence easily oxidized; examples are the *in vivo* conversion of the diuretic azosemide (41) to the N-dealkylated product (42) (Asano *et al.*, 1984) and the *in vitro* formation of 2-hydroxymethylthiophen (44) and thiophen-2-carboxylic acid (37) from methapyrilene (43) (Kammerer and Schmitz, 1986).

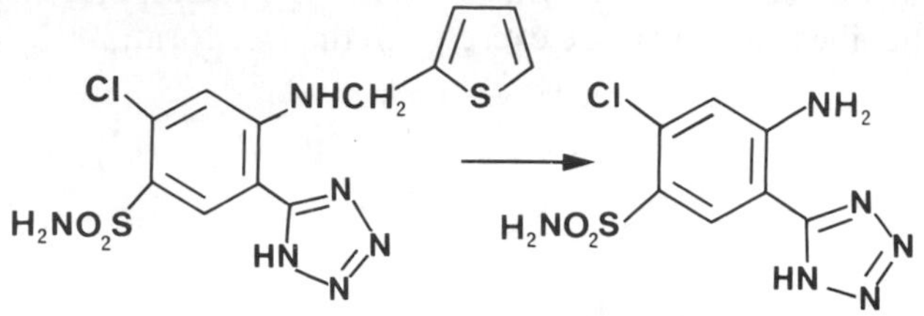

(41)   (42)

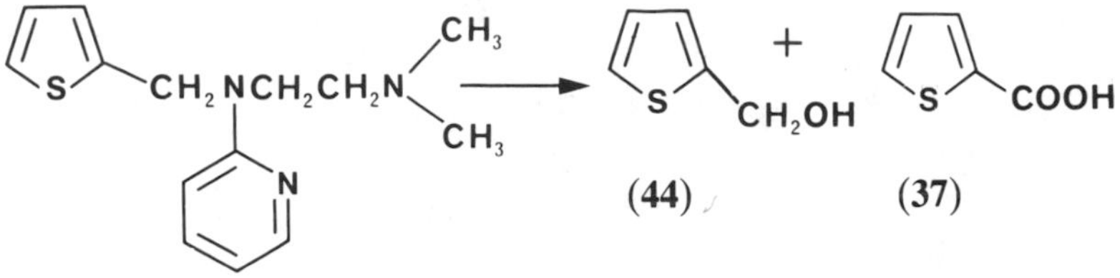

(43)   (44)   (37)

There is evidence that several thiophen-containing drugs — cephaloridine ((45); S. D. Nelson *et al.*, 1977a), suprofen and tienilic acid (Vaughan and Tucker, 1987) — produce renal toxicity; tienilic acid causes both kidney and liver damage on prolonged use. Studies by Fehring *et al.* (1984) and Mansuy (1987) suggest that this tissue necrosis is caused by covalent binding to tissue macromolecules by a reactive metabolite. In the case of tienilic acid, the hepatotoxicity, which occurs in only a small proportion of patients, appears to be of immunological origin and to be associated with production of circulatory antibodies to the specific cytochrome P-450 isozyme responsible for 5-hydroxylation (see above). From a mechanistic standpoint, the reactive metabolite would appear to be the 4,5-epoxide intermediate rather than a ring-opened reactive thiol suggested recently by Vaughan and Tucker (1987) to explain the toxicity of suprofen.

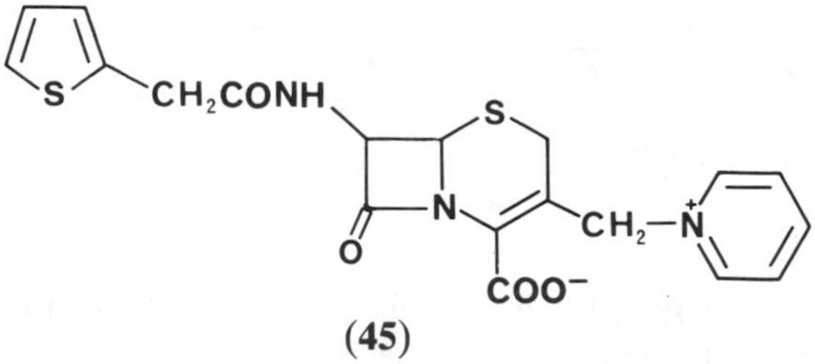

(45)

9.3.3 Metabolism of compounds containing a fused thiophen ring

Bohm (1941) characterized a urinary metabolite of benzothiophen in the rabbit as a glucuronide of the 2-hydroxy derivative (46); in the same species, Bray and Carpanini (1968) accounted for 80% of the benzothiophen dose in the urine as 'mercapturic acid-like compounds' (cf. (25) and (26)), one of which was tentatively identified as the 3-mercapturic acid (47). Despite the greater chemical stability (see above) of the S-oxidized derivatives compared with those of thiophen, no sulphoxidized metabolites of benzothiophen have been characterized. Clearly, as with thiophen, definitive studies with radiolabelled benzothiophen are required. Interestingly, the carbamate insecticide mobam (48), a derivative of benzothiophen, is excreted as a stable sulphoxide (49) in the urine of goats and cows (Robbins et al., 1970).

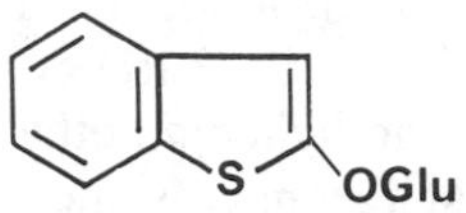

(46)

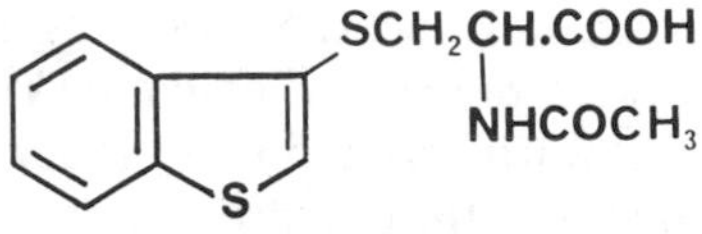

(47)

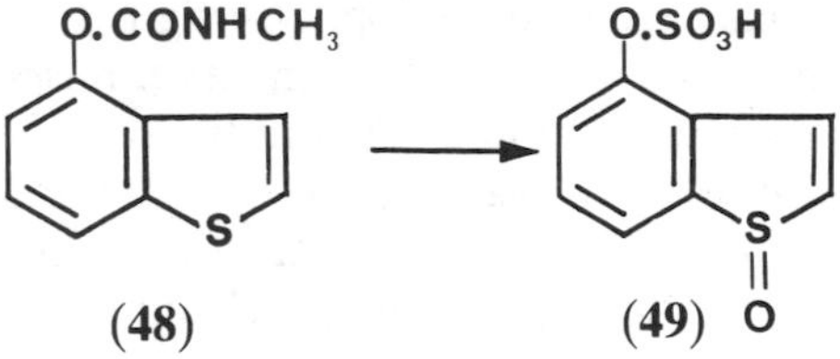

(48)

(49)

Ambaye et al. (1961) reported that dibenzothiophen forms the 1-hydroxy sulphone metabolite (50) in vivo in the rat and also a water-soluble sulphonic acid derivative. More recent in vivo and in vitro studies in the rat (Hoodi and Damani, 1984; Vignier et al., 1985) have demonstrated the conversion of dibenzothiophen to its sulphoxide (51) and sulphone (52) metabolites. Induction and inhibition studies indicated that both steps were mediated by cytochrome P-450, rather than the flavin-containing monooxygenase, and that the sulphoxidation reaction is sex dependent. The sulphoxidation of dibenzothiophen, which occurs as a marine pollutant from oil, has also been reported to occur in a marine model ecosystem (Lu et al., 1978).

The related dibenzothiophen derivative, benz[b]naphtho[2,3-d]thiophen (for structure see Table 4, chapter 5, this volume) and related carcinogenic thiaarenes have recently been reported to be oxidized by rat liver microsomes to the corresponding sulphones, apparently by a cytochrome P-450$_{mc}$ system (Jacob et al., 1986). It has been suggested that the sulphur in aromatic heterocyclic rings is oxidized exclusively by a cytochrome P-450 dependent system with negligible contribution from the flavin monooxygenase (Hoodi and Damani, 1984; Hoodi, 1986).

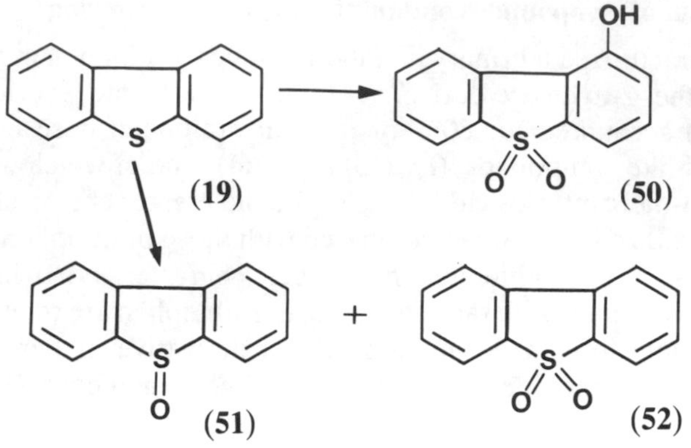

(19) → (50)

(51) + (52)

There are only two reports of ring hydroxylation of benzothiophen derivatives, in both of which the site of oxidation is C_6, the position in the carbocylic ring most activated towards electrophilic attack. Thus, 3-(2-dimethylamino ethyl)benzothiophen **(53a)** forms the 6-hydroxy metabolite **(53b)** in rat liver microsomes although the reaction appears to be of limited importance *in vivo* (Harrison *et al.*, 1974) However, the 6-hydroxy derivative **(54b)** of the antidiarrhoeal agent UK-59,354 **(54a)** was identified as a major metabolite in the mouse (Bowers *et al.*, 1987). In both cases, NMR was used to assign the position of hydroxylation in the molecule. In the antianxiety drug brotizolam **(55a)**, the thiophen ring is fused to a non-benzenoid system. Hydroxylation has been described in several species to occur in the benzodiazepine ring (giving **(55c)**) rather than in the thiophen ring (Bechtel *et al.*, 1986). In this case, the most activated sites in the thiophen ring are already substituted, unlike the situation in tenoxicam **(56a)**, an anti-inflammatory thienothiazine derivative, where a thiophen ring hydroxylated metabolite **(56b)** has been reported in the rat (Ichihara *et al.*, 1984).

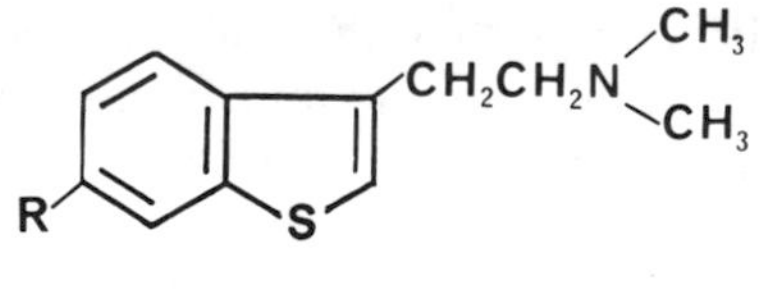

(53) a; R=H
b; R=OH

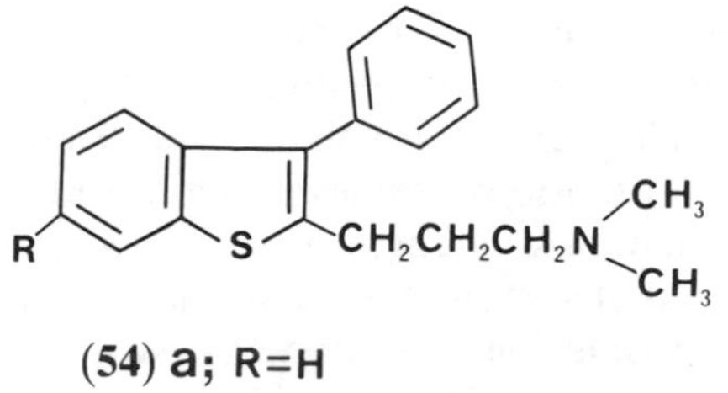

(54) a; R=H
b; R=OH

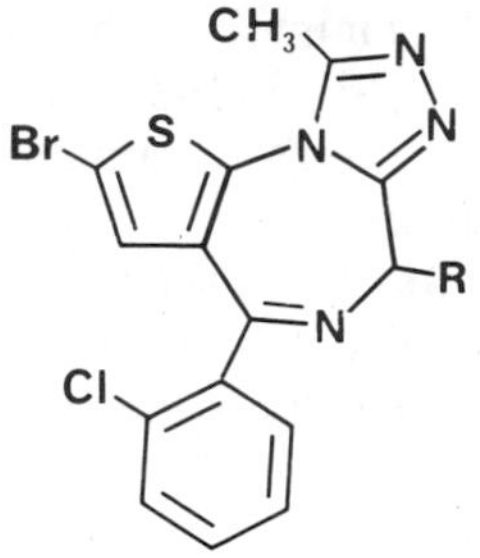

(55) **a**; R=H
   b; R=OH

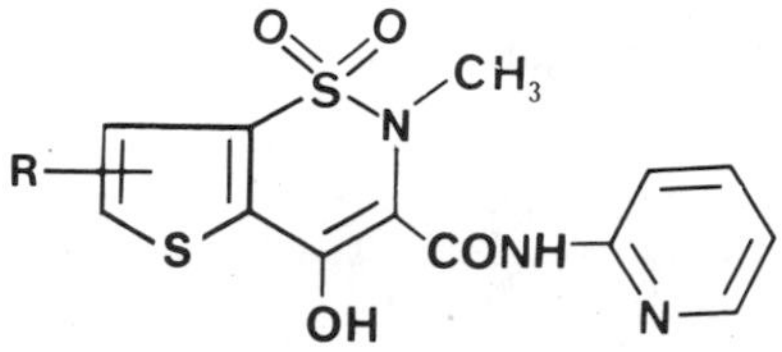

(56) **a**; R=H
   b; R=OH

For clotizaepam (57), a structural analogue of brotizolam, a major route of metabolism involves oxidation of the ethyl substituent to produce (58) (Arendt *et al.*, 1982); this is another example of the activation of α-carbon benzylic substituents on the thiophen ring.

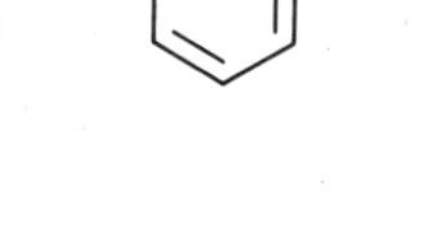

$R = CH_3CH_2-$; (57)

$R = CH_3CH$; (58)
      |
      OH

9.3.4 Metabolism of compounds containing a reduced thiophen ring

Tetrahydrothiophen (THTP; (20)) is non-aromatic, having properties typical of aliphatic thioethers, and as such forms a stable sulphoxide (59) and sulphone (60) (see Scheme 5); the latter is sometimes known as sulpholane.

Roberts and Warwick (1961) administered [^{35}S]THTP intraperitoneally to rats and identified 3-hydroxy-THTP sulphone (61) as a major urinary metabolite; the same metabolite was also obtained from animals dosed with THTP sulphone, S-β-L-alanyltetrahydrothiophenium ((62); as mesylate salt) and the antileukaemic drug busulfan ((63); myleran), suggesting common pathways of metabolism for these compounds. Recently, Hassan and Ehrsson (1987a) have perfused rat livers with [^{14}C]busulfan and identified γ-glutamyl-β-(S-tetrahydrothiophenium)alanylglycine ((64); the sulphonium ion of glutathione) as the major metabolite in the biliary perfusate, this sulphonium ion decomposing in the bile to produce THTP. The same authors (Hassan and Ehrsson, 1987b) also demonstrated that urinary metabolites of [^{14}C]busulfan in the rat included THTP sulphoxide and THTP sulphone, as well as the previously reported 3-hydroxy-THTP sulphone. They suggest that *in vivo* the sulphonium ion is degraded in the gut to THTP which is then reabsorbed and further oxidized. Hence, the overall metabolism of busulfan would appear to be as in Scheme 5; in this unusual scheme, it should be noted that the sulphur atom in the THTP metabolites emanates from glutathione and not from the mesylate functions of the original busulfan.

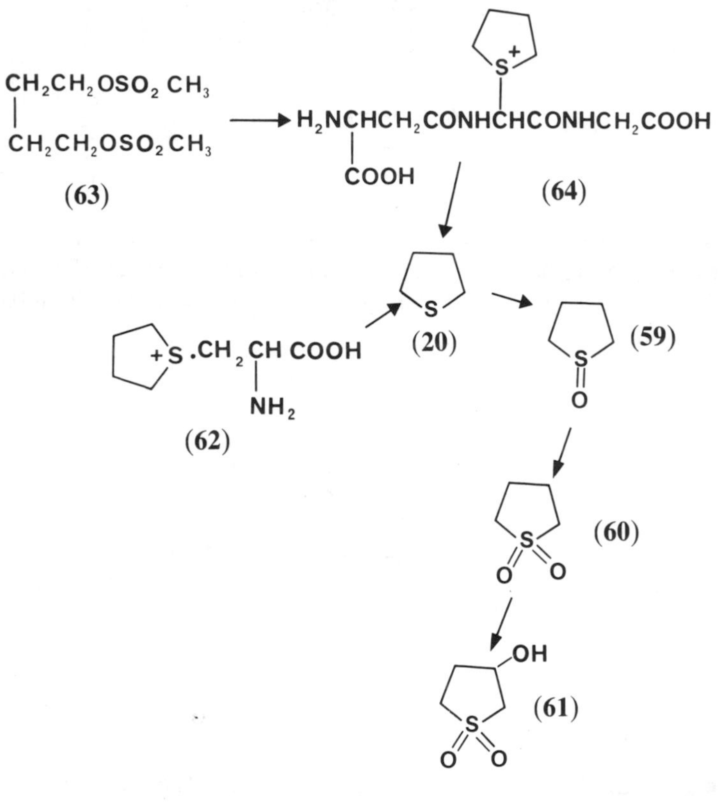

Scheme 5

Other studies (Hoodi and Damani, 1984; Hoodi, 1986) on the metabolic fate of THTP in the rat describe the excretion of the sulphoxide and sulphone metabolites in urine in amounts corresponding to 15–25% and 5% of the dose respectively, but these authors did not look for the 3-hydroxysulphone. Roberts and Warwick (1961) reported that there was no unchanged THTP in urine or exhaled air. It is therefore possible that the 3-hydroxy sulphone may constitute a large part of the dose unaccounted for by Damani and coworkers, but this remains to be verified.

Hoodi and Damani (1984) also examined the *in vitro* sulphoxidation of THTP in rat liver microsomes which results predominantly in the sulphoxide with only trace amounts of the sulphone. Experiments with purified enzymes and specific inhibitors and inducers suggested that the flavin-containing monooxygenase rather than cytochrome P-450 was responsible for sulphoxide formation, in contrast to the enzymology for the *S*-oxidation of aromatic sulphur heterocycles (see section 9.3.3 above).

Takata *et al.* (1980) have shown that cytochrome P-450, present either in rabbit liver microsomes or in a purified reconstituted form, is responsible for the sulphoxidation of 2-methyl-2,3-dihydrobenzothiophen (65) to give (66). This substrate is analogous to an acyclic alkylaryl sulphide and, since Waxman *et al.* (1982) have shown that such compounds can be metabolized by both cytochrome P-450 and the flavin monooxygenase, the oxidation of (65) and analogues by the latter enzyme would appear worthy of study. Of further interest is the fact that in sulphoxides such as (66) the sulphur atom is bonded to three different groups. When the remaining lone pair of electrons on the sulphur atom is taken into account, the molecule can be seen to be chiral and hence capable of existing as enantiomers. In the studies of Takata *et al.* (1980) the *trans* sulphoxide was the major product, suggesting stereospecific attack by the enzyme.

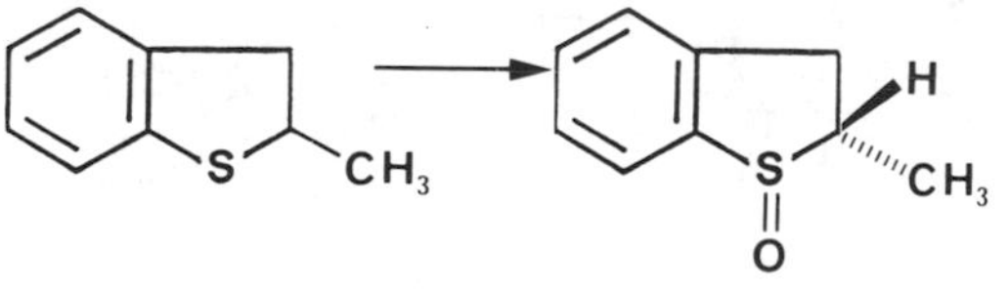

(65)   (66)

The isobenzothiophen derivative Lu-5-003 (67) affords sulphoxides (68a–c) as major metabolites in rat, dog and man (Overo *et al.*, 1970); there is some evidence in the rat for sulphone formation. In the case of (68a) and (68b), one isomer appears to be preferentially produced suggesting stereospecific binding of substrate to enzyme as discussed above.

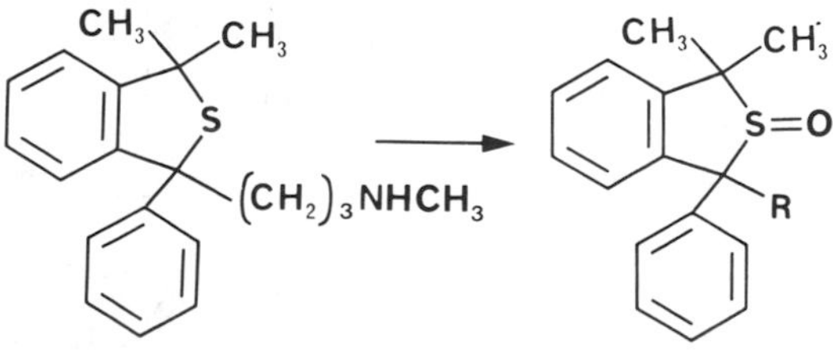

(67)   (68) **a**; R= $-(CH_2)_3NHCH_3$
                             b; R= $-(CH_2)_3NH_2$
                             c; R= $-(CH_2)_2COOH$

Sulphoxidation of reduced thiophens has also been described for non-mammalian organisms; biotin has been found to exist in bacteria as biotin sulphoxide (69); Maw, 1972).

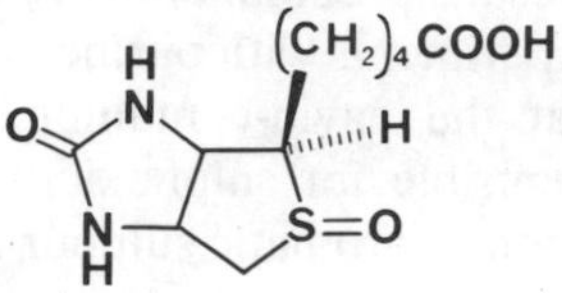

(69)

9.3.5 Metabolism of compounds containing two sulphur atoms

The only well-studied compound in this class is the antischistosomal agent oltipraz (70), a 1,2-dithiole-3-thione derivative, which has an interesting if somewhat complex metabolic scheme (Jolles, 1984; Bieder *et al.*, 1983). An abbreviated version relevant to the present discussion is given in Scheme 6. The main metabolic pathway in mouse, rat, monkey and man involves a complete rearrangement of the molecule which appears to be initiated by attack at C_5 by a sulphur nucleophile, possibly a

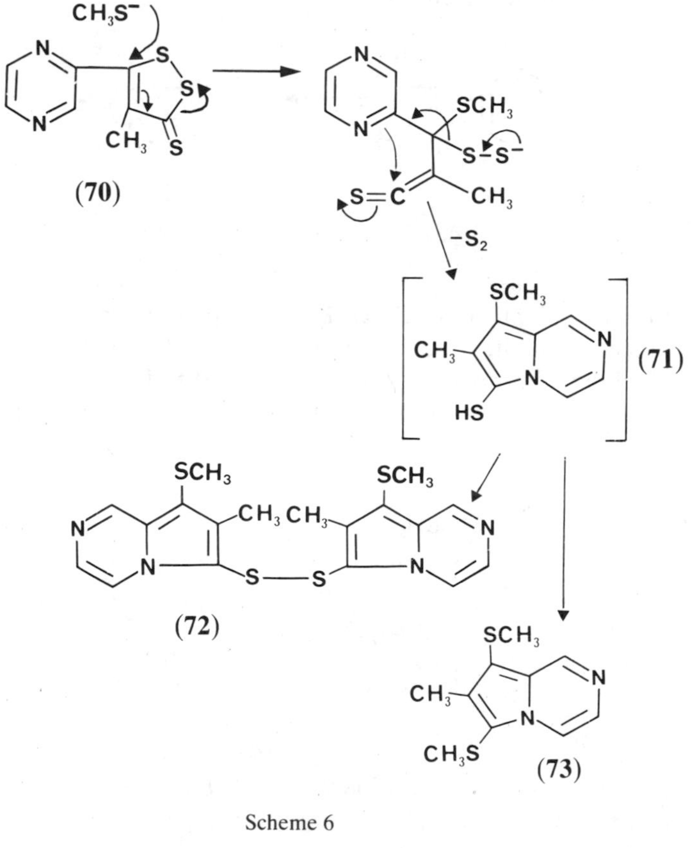

Scheme 6

methionine residue. This results in rupture of the 1,2-dithiole ring and a rearrangement reaction involving recyclization into an intermediate pyrrolo[1,2-a]pyrazine thiol (71) and elimination of two sulphur atoms. Subsequent disulphide bridge formation and the action of an *S*-methyltransferase enzyme results in the formation of the isolable metabolites (72) and (73). Metabolites from the rearrangement reaction are also present in the worms to be found in schistosome-infected mice, but it is not clear whether they are formed in the worm or taken up from the host (Heusse *et al.*, 1985).

The anticirrhotic drug malotilate (74), di-isopropyl-1,3-dithiol-2-ylidenemalonate, forms a 4,5-dihydroxy metabolite (75) which has been isolated from rat liver; this metabolite is thought to represent an intermediate stage in the loss of the dithiole ring leading to the major metabolites di-isopropylmalonate (76) and isopropyl hydrogen malonate ((77); Funayama *et al.*, 1983).

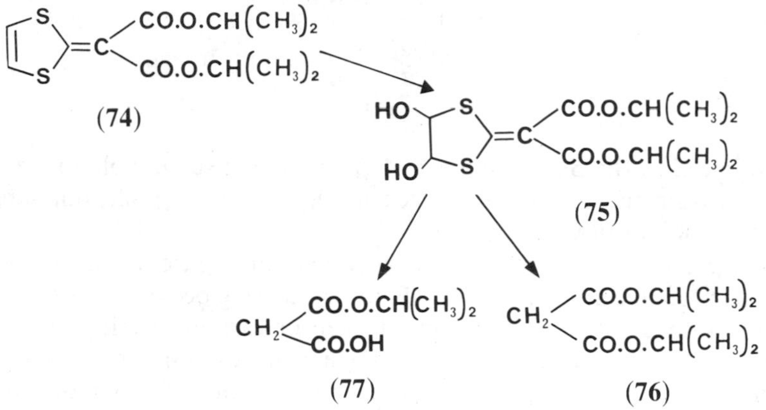

9.4 FIVE-MEMBERED SULPHUR HETEROCYCLES CONTAINING SULPHUR AND OTHER HETEROATOMS

9.4.1 Chemical considerations

Although five-membered heterocyclic rings containing sulphur and oxygen are known, the most studied being 1,3-oxathiole (78), the metabolic fate of these systems has not been described and consequently they will not be further considered. However, rings containing sulphur and nitrogen appear in many drugs and foreign compounds. The parent heterocycles thiazole (79), isothiazole (80) and the isomeric thiadiazoles (81)–(84) possess varying degrees of aromatic character and some useful reviews on their chemistry have been published (Campbell, 1979; Metzger, 1984).

In thiazoles, electrophilic reagents preferentially attack the lone electron pair on nitrogen rather than sulphur to form a quaternary thiazolium ion. The C_2 carbon of thiazoles is electron deficient and consequently substituents at this position are quite reactive and subject to nucleophilic displacement, particularly if the nitrogen atom is quaternized. Thiazoles do not readily undergo electrophilic substitution reactions but C_5 is the more favoured of the two sites. Benzothiazoles (85) preferentially undergo electrophilic substitution in the carbocyclic ring, principally at C_4 and C_6.

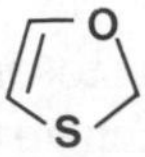

(78) 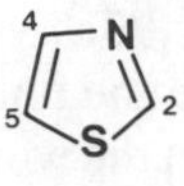 (79) 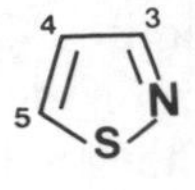 (80)

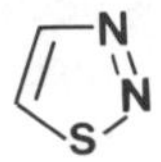

(81) 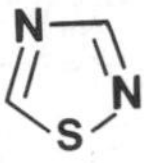 (82) 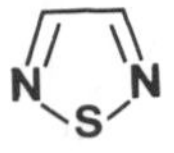 (83) 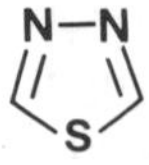 (84)

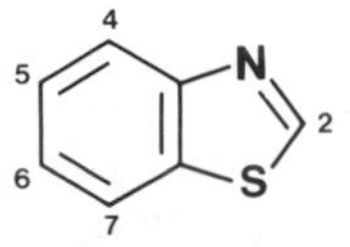

(85)

In isothiazoles, the S—N bond is ambiphilic, being susceptible to electrophilic attack at nitrogen and nucleophilic attack at sulphur. Electrophilic ring substitution occurs preferentially at C_4.

The 1,2,5-thiadiazole system (83) is very electron deficient and relatively inert towards electrophilic attack; when oxidation of the ring occurs, the site of attack is usually the sulphur atom rather than nitrogen or carbon, leading to sulphoxide, sulphone and ring-opened products. 1,3,4-Thiadiazoles (84) undergo a facile reaction with electrophiles (e.g. alkylating agents) on one or other of the annular nitrogen atoms; direct oxidation of the sulphur has not been reported and the lower electron density of the carbon atoms means that electrophilic substitution reactions do not normally occur. The chemistry of the 1,2,3- and the 1,2,4-thiadiazoles (81) and (82) will not be discussed here in view of the absence of data on the biotransformation of such compounds.

5-Hydroxythiazoles exist as tautomeric mixtures, e.g. ((86a) and (86b); cf. hydroxy-thiophens), predominantly in the keto form, although the position of equilibrium depends on the solvent and substitution patterns. Similarly, reduced tetrahydrothiazoles (thiazolidines) are normally in tautomeric equilibrium with an acyclic thiol (87a,b).

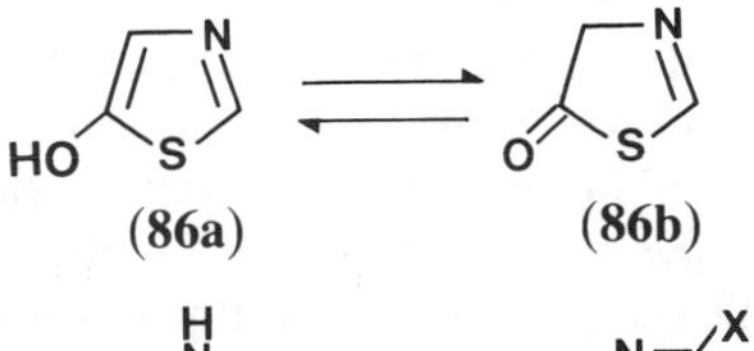

(86a) (86b)

(87a) (87b)

Dihydro- and tetrahydro-isothiazoles have been little studied; the 1,1-dioxide derivatives ((**88**); sultams) generally behave as cyclic sulphonamides with stable $S-C_5$ and $S-N$ bonds.

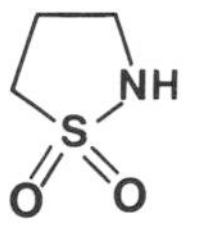

(**88**)

9.4.2   Metabolism of compounds containing an unfused ring

There are essentially two metabolic options available to these ring systems. These involve attack either at a heteroatom (sulphur or nitrogen) or at one of the ring carbon atoms. In the former case, the metabolites tend to be stable but in the latter a ring-opened product usually results.

Formation of *S*- and *N*-oxygenated metabolites is, in fact, rare probably because the lone pairs of electrons on the heteroatoms are, to some extent, involved in aromatic stabilization of the ring. Nevertheless, the beta-blocker timolol (**89**), a 1,2,5-thiadiazole derivative, forms a sulphoxide **90** as a minor metabolite in the rat but not mouse or man (Tocco *et al.*, 1980). Oxidation in this case occurs at sulphur rather than nitrogen which is consistent with the relative susceptibility of these heteroatoms towards electrophilic attack.

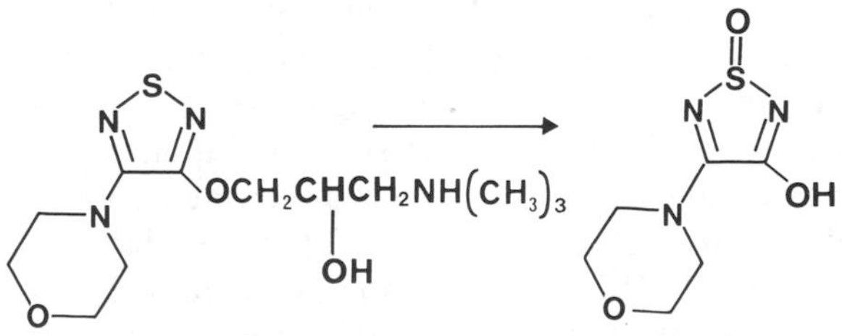

(**89**)   (**90**)

The sedative anticonvulsant chlormethiazole ((**91**), Scheme 7), a thiazole, gives rise to a stable 'di-oxide' metabolite (**93a**) in human urine, in which both the nitrogen and sulphur atoms have been oxidized (Offen *et al.*, 1985); the stability of this unusual metabolite may result from a resonance contribution from the canonical forms (**93b**) and (**93c**). The enzyme responsible for this sulphoxidation reaction has not been characterized but attack at sulphur by an electrophilic species such as cytochrome P-450 appears likely. The *N*-oxidation reaction may proceed via a quaternary oxygenated species (**92**). Such a species has been suggested as an intermediate in the production of other metabolites of this compound in man. Thus the metabolites 2-methylthiochlormethiazole (**94**) and 5-acetyl-4-methyl-2- mercaptothiazole (**95**) are thought to be formed by initial oxidative quaternization at

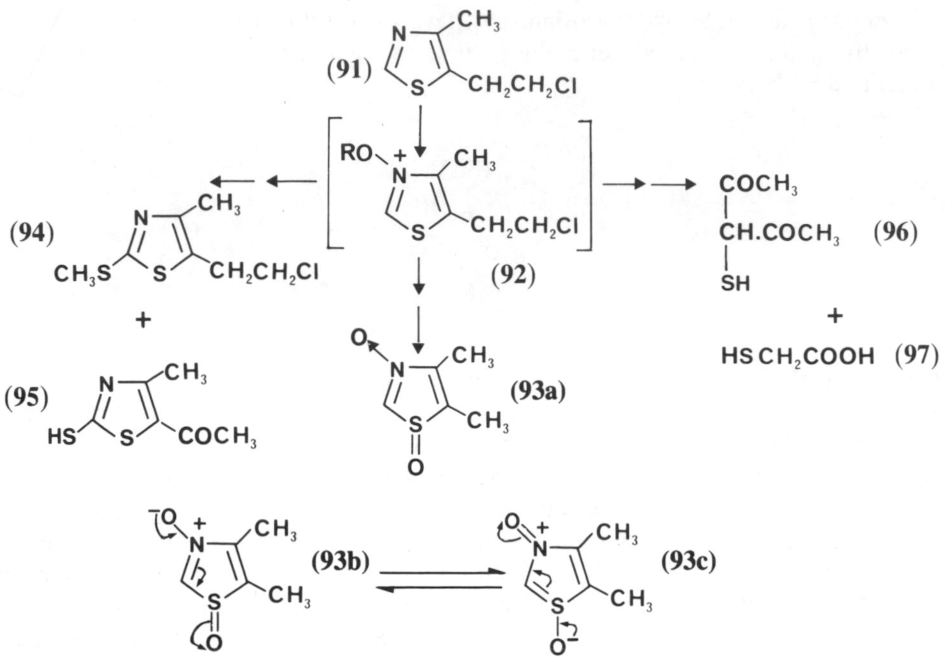

Scheme 7

nitrogen followed by glutathione attack with subsequent β-lyase cleavage and thiomethylation (Pal and Spiteller, 1982). Similarly, 2,4-pentadione-3-thiol (96) and mercaptoacetic acid (97) probably result from nucleophilic attack of hydroxide ion at C_2 of a quaternary intermediate, causing ring opening and loss of carbon dioxide (Grupe and Spiteller, 1982). Clearly, the suggested quaternary intermediate may play a pivotal role in regulating some of the complex metabolism of chlormethiazole and further study in this area is warranted. The metabolism of chlormethiazole is discussed in more detail by Wilson (see Chapter 6, Volume 3, Part A of this series).

An interesting quaternization reaction has been reported for the anticancer agent 2-amino-1,3,4-thiadiazole (98); ATDA; NSC 4728). Following administration of [5-^{14}C]ATDA to mice bearing L1210 cells, most of the dose is excreted as unchanged drug (El Dareer et al., 1978) but formation of an ATDA mononucleotide (99) was demonstrated in cells harvested from the peritoneal cavity (J. A. Nelson et al., 1977). The structure of the metabolite is analogous to that of nicotinamide mononucleotide but it is not reported which of the two nitrogen atoms (N_3 or N_4) is quaternized. Both ATDA mononucleotide and the ATDA adenine dinucleotide (which is not found in vivo) have been synthesized and are inhibitors of inosine monophosphate dehydrogenase, and the mononucleotide is thereby considered responsible for the cytotoxicity of ATDA (O'Dwyer et al., 1986). The antidiabetic glybuzole (100), also a 1,3,4-thiadiazole derivative, is claimed to be excreted in rat bile as an N-glucuronide. It is not reported whether conjugation occurs at an annular nitrogen giving a quaternary derivative (cf. ATDA) or at the sulphonamide moiety (Ishii et al., 1978).

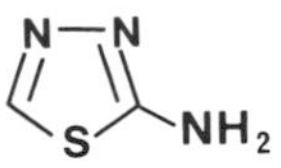

(98) 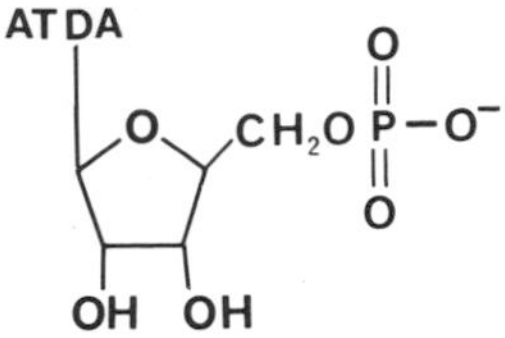   (99)

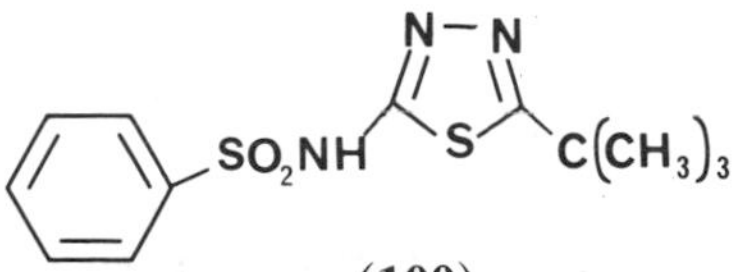

(100)

In contrast to the C_2−S ring cleavage of chlormethiazole (where the C_2 position is vacant) C_5−S scission appears to be a common biotransformation pathway for 2-substituted thiazoles. It is thus a major route of metabolism *in vivo* in the rat for some anti-inflammatory 4-substituted-2-acetamido thiazoles ((**101**); R=H, CH$_3$, C$_6$H$_5$). Products of ring opening are acetylthiohydantoic acids (**102**) and acetyl thiourea (**103**); reaction also occurs *in vitro* with the post-mitochondrial supernatant fraction of rat liver homogenates (Chatfield and Hunter, 1973a,b).

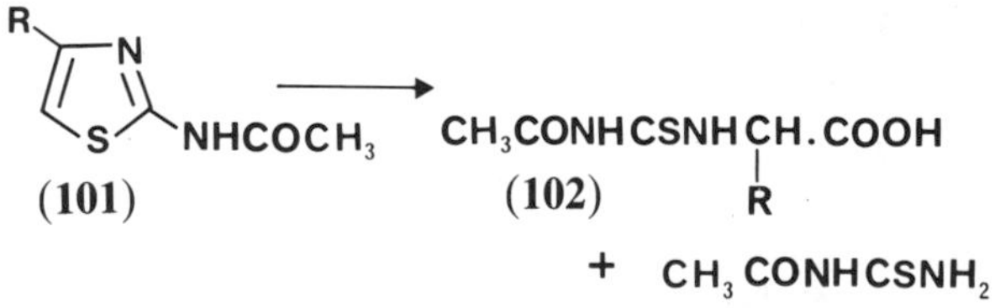

(101)   CH$_3$CONHCSNHCH.COOH
                        (102)   R

                        +   CH$_3$CONHCSNH$_2$

                        (103)

A similar ring opening occurs for the anti-inflammatory sudoxicam (**104**) in rat, dog and monkey to produce analogous products, (**106**) and (**107**), and there is some evidence for this pathway in man (Hobbs and Twomey, 1977); the authors suggest a mechanism (Scheme 8) in which the thiolactone tautomer (**105b**) of 5-hydroxy sudoxicam (**105a**) undergoes hydrolytic ring scission, a scheme which could also apply to (**101**). Another example is the immunoregulator fanetizole (**108**) which forms the α-carboxy thiourea derivative (**109**) in all species studied (Frame, 1987). 2-Aminothiazole (**110**) and 2-acetamido-5-nitrothiazole (**111**) are also reported to be extensively metabolized (Perrault and Bouet, 1946; Bousquet *et al.*, 1965) but the products have not been characterized.

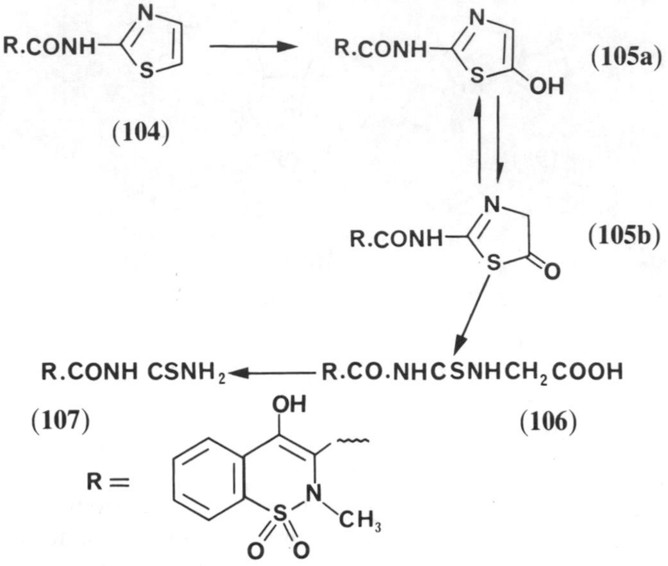

Scheme 8

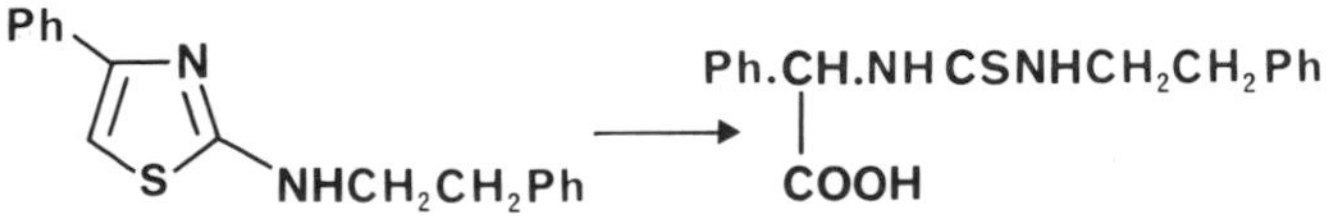

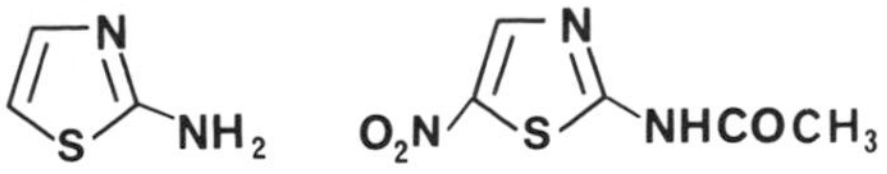

The antischistosomal drug niridazole (112), a 2,5-disubstituted thiazole, is metabolized by the gut flora of mammals to 1-thiocarbamoyl-2-imidazolidinone (TCI; (113)) which possesses immunosuppressive properties (Tracy et al., 1980; Tracy and Webster, 1981); TCI has also been identified as a niridazole metabolite in schistosomes (Catto et al., 1984). Although the product is analogous to the thiohydantoic acids described above, it is uncertain whether the mechanism is the same as that suggested for sudoxicam. There is some evidence (Tracy and Webster, 1981) that the bacterial production of TCI may be reductive, although an oxidative pathway involving a 4,5-epoxythiazole cannot be ruled out; the latter has been suggested to underlie the high dose carcinogenicity of niridazole (Blumer *et al.*, 1979).

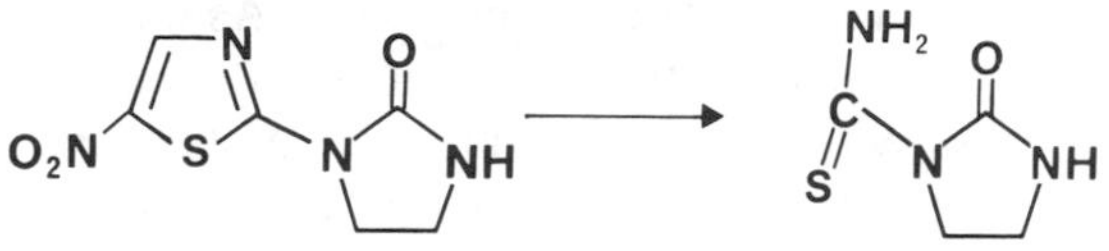

(112)   (113)

In addition to these examples, there are several drugs in which an S–N ring is apparently stable to metabolism by virtue of either the presence in the molecule of other preferentially metabolized functional groups or high polarity which promotes renal elimination. Thus, the H_2-antagonist nizatidine (114) is excreted predominantly unchanged in all species, although some sulphoxidation of the side chain may occur (Knadler *et al.*, 1986). Similarly, for fentiazac (115) the major routes of metabolism are glucuronidation and hydroxylation of the phenyl ring (Fumero *et al.*, 1980; Franklin *et al.*, 1984).

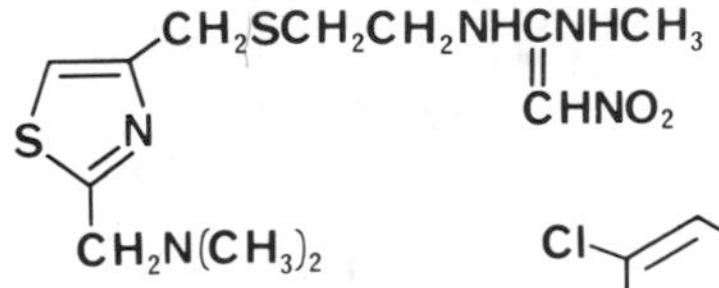

(114)

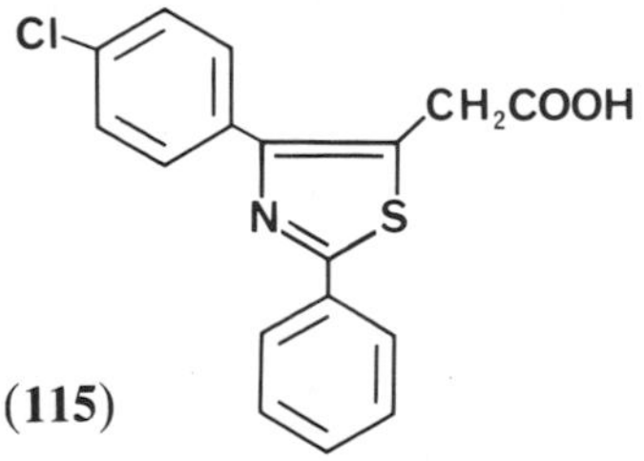

(115)

The isomeric phenyl thiazoles (116)–(118) are reported to be inhibitors of an *in vitro* epoxidase, the relative potency being 2-Ph ≫ 5-Ph > 4-Ph (Smith and Wilkinson, 1978); inhibition presumably takes place via coordination of the N_3 atom to haem iron, as has been reported for imidazoles and triazoles.

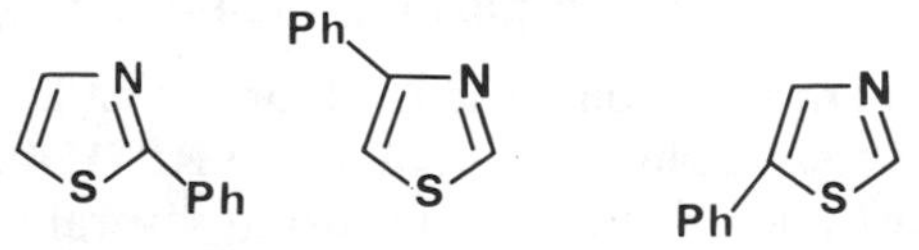

(116)   (117)   (118)

9.4.3   Metabolism of compounds containing a fused ring

Information on the metabolism of compounds in which an S–N heterocycle is fused to a benzene ring is very limited. The calcium channel blocker fostedil (119), a 2-substituted benzothiazole, undergoes ring hydroxylation in the dog, mainly at C_5, C_6 and C_7 (Thomas, 1984; Thomas and Bopp, 1984). As discussed above, the C_2 position of benzothiazoles is very electron deficient and therefore activated towards nucleophilic attack. This reactivity is presumably the basis of the unusual reaction (Scheme 9) shown by benzothiazole 2-sulphonamide (120) in which the sulphonamide group is displaced by glutathione with resulting formation of a mercapturic acid (121), thiol (122) and S-glucuronide (123) (see Crooks, chapter 7 of this volume and references cited therein).

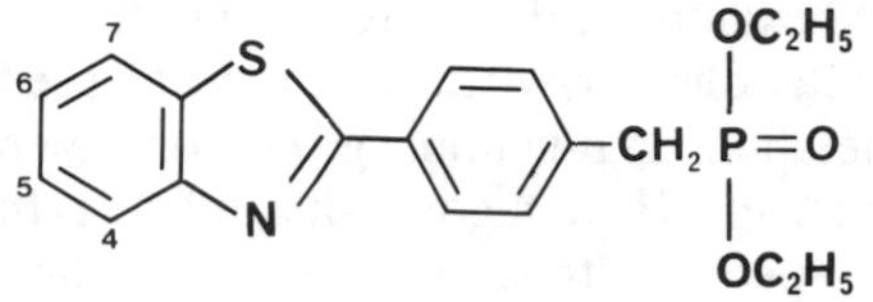

(119)

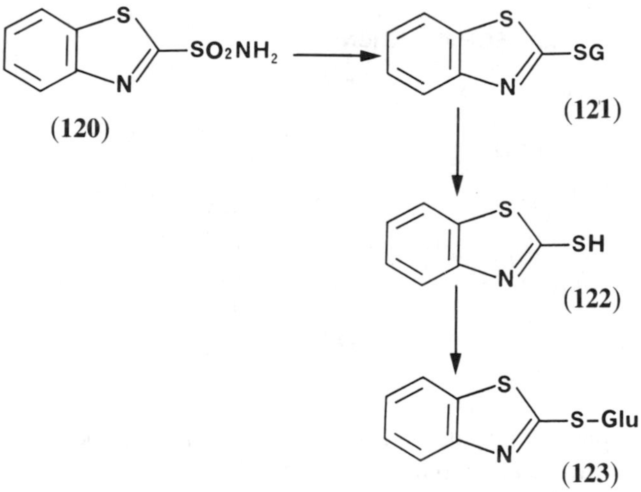

Scheme 9

9.4.4   Metabolism of compounds containing a reduced ring

The two biotransformation options described for aromatic S–N rings can also occur for their reduced forms. Sulphoxidation of 48740 RP (124), a reduced pyrrolothiazole, has recently been described (Scheme 10) in several species including man by Decouvelaere *et al.* (1986). This compound, which has antiplatelet aggregatory factor (anti-PAF) properties, forms two diastereomeric sulphoxides (125) and a sulphone **(126)**. In addition, 48740RP undergoes hydroxylation at C_1 to produce an alcohol which is then conjugated with glucuronic acid **(127)**. *N*-oxidation of reduced S–N rings has not been reported.

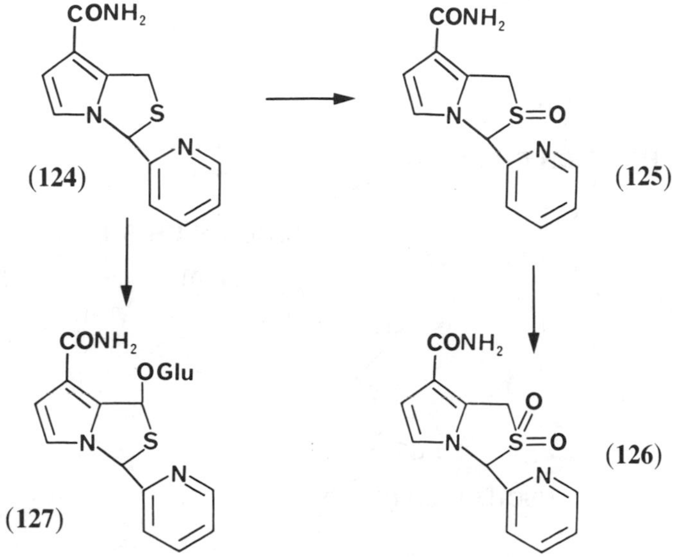

Scheme 10

Thiazolidine 4-carboxylic acid (thioproline; **(128)**) is metabolized *in vitro* by liver mitochondria to produce the ring-opened *N*-formylcysteine **(129)**; reaction is reported to be catalysed by a specific dehydrogenase (MacKenzie and Harris, 1957; Cavallini *et al.*, 1956). Oeriu *et al.* (1970) described the *in vivo* conversion of thioproline to cysteine **(130)**, the reaction presumably occurring via *N*-formylcys-teine. Similarly, MK-771 **(131)**, a tripeptide analogue of thyrotropin releasing hormone, is metabolized in the rat to thioproline (Scheme 11) and there is some evidence of further degradation to cysteine (Vickers *et al.*, 1983). After oral dosing of [^{35}S]thioproline to rats, radioactivity accumulates in liver, pancreas and the endocrine glands but the identity of these drug-derived components has not been established (Harmand and Blanquet, 1982).

Interestingly, the thiazolidine ring of the angiotensin converting enzyme (ACE) inhibitor fentiapril **(132)** appears stable to metabolism in the dog, renal excretion of parent compound being the major elimination process with some *S*-methylation of the side chain thiol and formation of a bis-disulphide (Abe *et al.*, 1984).

Similarly, the thiazolidine ring of pencillin antibiotics does not undergo *S*- or *N*-oxidation, and these compounds are normally eliminated either unchanged or as metabolites (e.g. penicilloic acids; Cole *et al.*, 1973) with an intact thiazolidine ring.

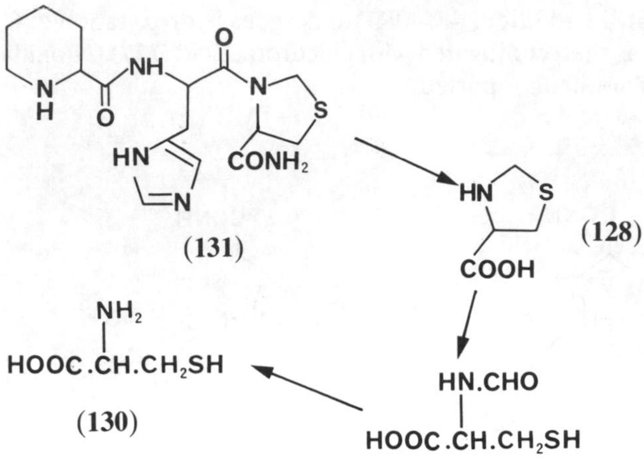

(131)

(130)

HOOC.CH.CH₂SH
|
NH₂

(128)

(129)

Scheme 11

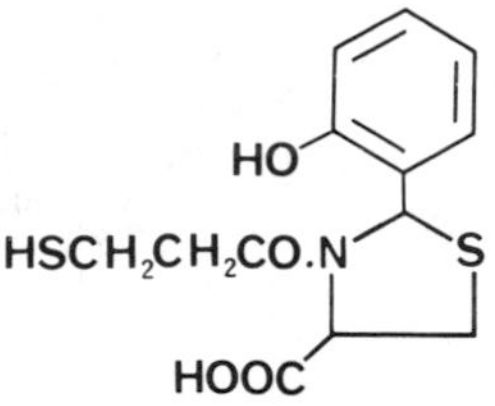

(132)

There have been some claims that the thiazolidine ring in the 5R,6R penicilloic acid (134) produced from ampicillin (133) undergoes metabolic C₂–S cleavage to form a penemaldic acid derivative ((136); Masada *et al.*, 1980; Uno *et al.*, 1981). However, on the basis of its spectroscopic properties, Bird *et al.* (1983) reassigned the ampicillin metabolite as the 5S,6R epimer (135) of the penicilloic acid, and the

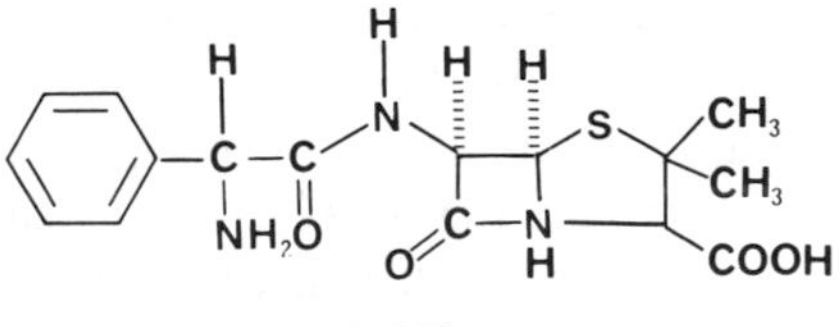

(133)

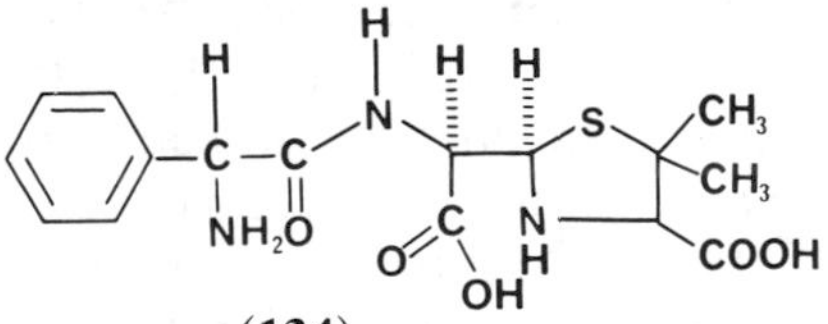

(134)

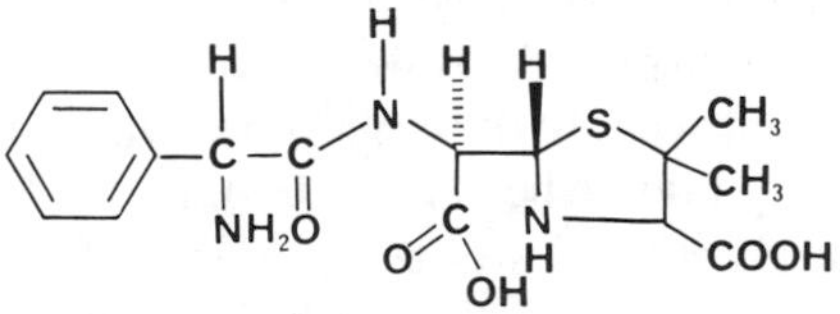

(135)

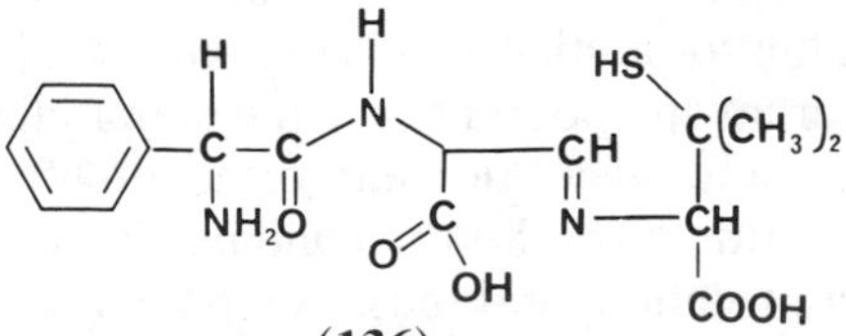

(136)

urinary excretion of this metabolite has recently been quantitated (Haginaka and Wakai, 1987). Although the epimerization reaction requires scission and subsequent reclosure of the thiazolidine ring, the process may be chemically rather than enzymatically mediated.

The antimetastatic agent Wy-13,876 (**137a**), a dihydrothiazolobenzimidazole, is converted in the rat to the thyrotoxic metabolite benzimidazole-2-thiol (**138**); Janssen *et al.*, 1979, 1981). The reaction is thought to occur via the ring-opened form of Wy-13,876 (**137b**), which predominates in solution at physiological pH, rather than by direct cleavage of the thiazolidine ring. Support for this view comes from the observation that the dehydrated analogue Wy-18,251 (**139**), which cannot easily ring open, does not produce the thiol and consequently is not thyrotoxic (Janssen *et al.*, 1982).

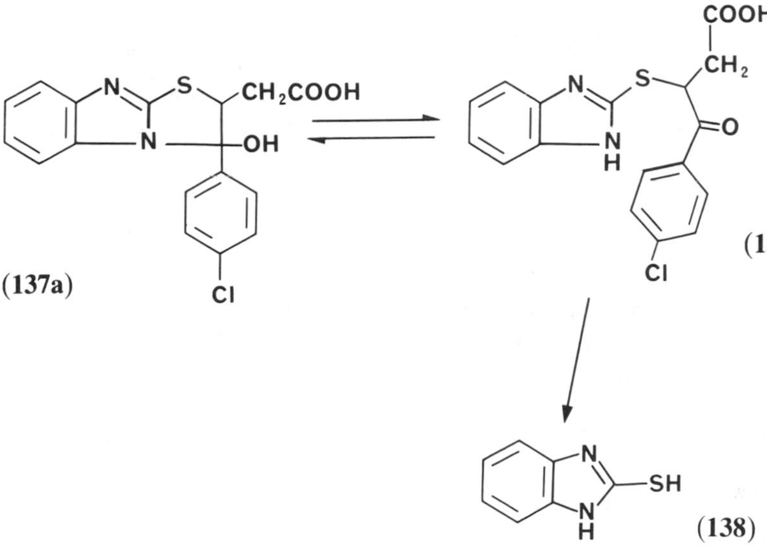

(137a)

(137b)

(138)

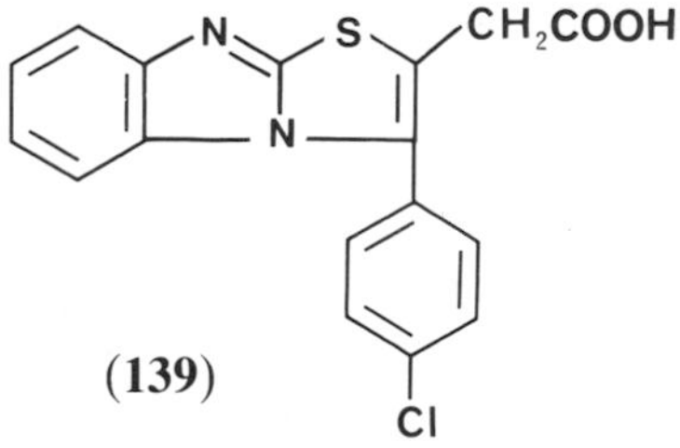

(139)

The antiviral nucleoside Acluracil (**140**) provides another example (*cf.* busulfan) where the metabolic involvement of glutathione leads to the generation of a metabolite containing heterocyclic sulphur from a compound initially without such a ring. Thus (Scheme 12), the reduced thiazolopyrimidine derivative (**143**) is formed as a minor urinary metabolite in the rabbit via a free thiol intermediate (**141**) which, after epoxidation of the allylic double bond to give (**142**), undergoes ring closure (Kaul *et al.*, 1982).

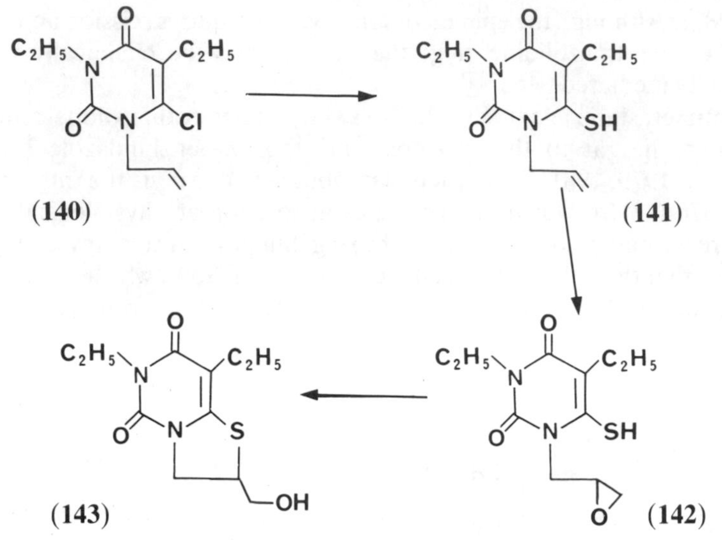

Scheme 12

Available information on reduced isothiazoles is limited to the 1,1-dioxide derivatives, which behave as cyclic sulphonamides. Thus, the major route of metabolism in rat and dog for the cardiac stimulant UK-35,493 (**144**) is N–C_3 cleavage to produce a β-carboxy sulphonamide (**145**) (Atkins and Rance, 1987). Renwick and Williams (1978) have shown that the fused analogue benzisothiazoline-1,1-dioxide (BIT; (**147**)), an impurity in saccharin (**151**), is similarly metabolized in the rat to 2-sulphamoylbenzoic acid (**150**) and 2-sulphamoylbenzyl alcohol (**146**) as well as saccharin itself and an unidentified labile conjugate. The authors suggest (Scheme 13) that an intermediate 3-hydroxy metabolite (**149**) is in equilibrium with ring-opened 2-sulphamoylbenzaldehyde (**148**) from which (**146**) and (**150**) arise.

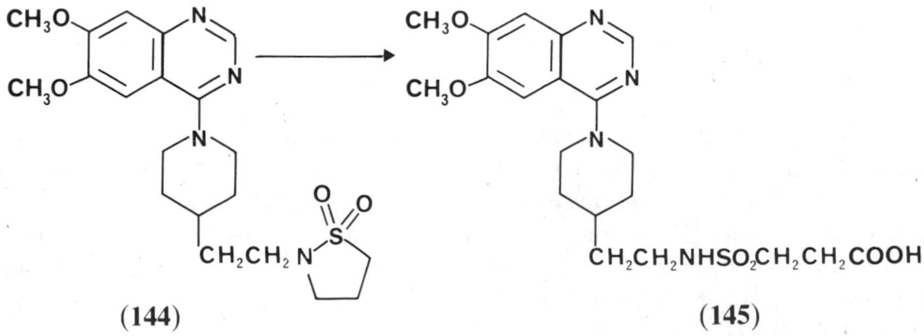

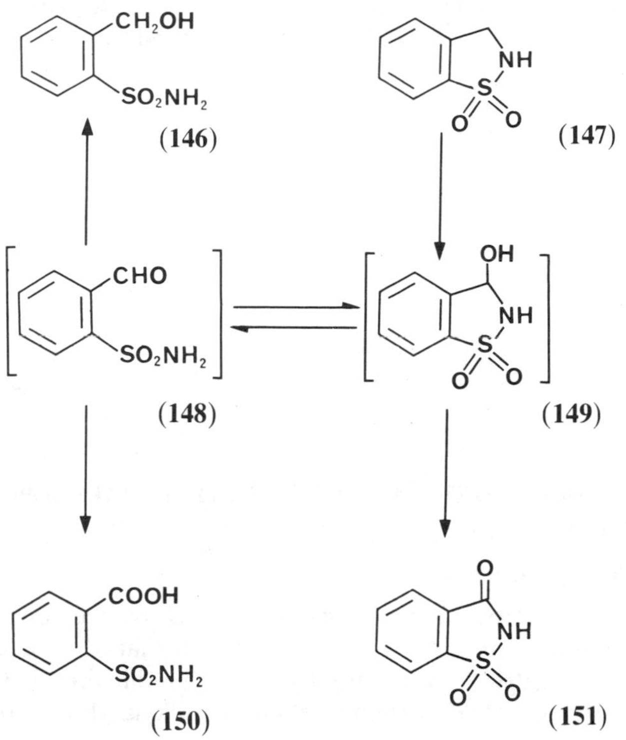

Scheme 13

In view of its wide usage as a sweetener and its purported tumorigenicity, the *in vivo* fate of saccharin has been a subject of much debate. The most definitive studies (Ball *et al.*, 1977) concluded that saccharin is not metabolized *in vivo* or *in vitro* in any species, possibly because of its polarity, and that earlier claims for the existence of metabolites (Pitkin *et al.*, 1971, Kennedy *et al.*, 1972; Lethco and Wallace, 1975) were based on the use of impure saccharin or inappropriate analytical procedures. In addition, Renwick (1978) has shown that the 3-amino and 5-chloro analogues ((152) and (153) respectively), which are also saccharin impurities, are essentially stable to metabolism in the rat.

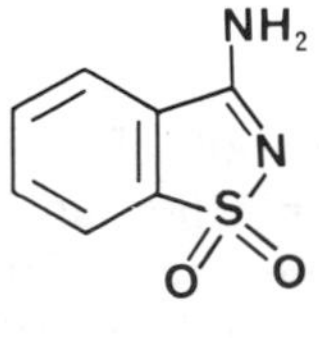

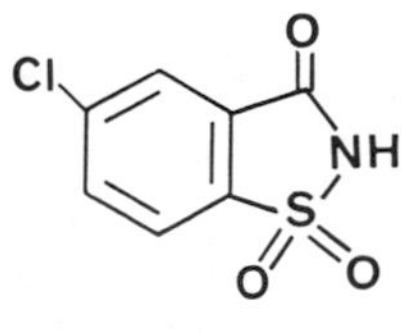

The tranquilliser supidimide (154), structurally related to saccharin, is reported to undergo non-enzymic hydrolytic ring opening in man to give (155), a reaction which does not occur in rat where carbon oxidation of the piperidone ring is the predominant process (Becker *et al.*, 1982).

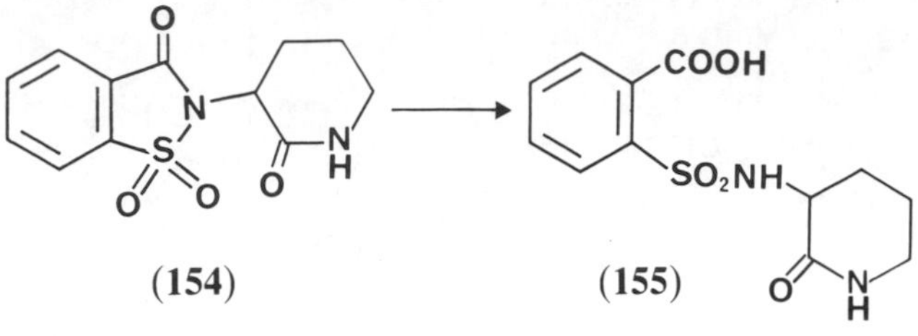

(154)   (155)

9.5   SIX-MEMBERED HETEROCYCLES CONTAINING SULPHUR AS THE ONLY HETEROATOM

9.5.1   Chemical considerations

Tetrahydro-2H-thiin (156) and its simple substituted derivatives are to be regarded as cyclic sulphides and their chemistry is analogous to that of their acyclic aliphatic counterparts (Ingall, 1984). By the same token, the 3,4-dihydro-2H-benzothiin (157) and 9H-thioxanthene (158) systems behave as alkylaryl and diaryl sulphides respectively.

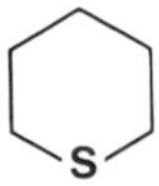

(156) 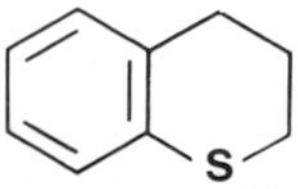   (157)

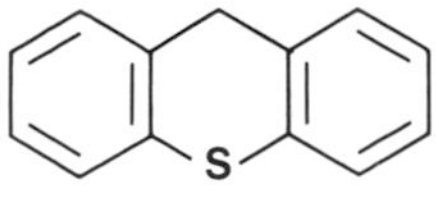

(158)

9.5.2   Metabolism of fused and unfused systems

Takata *et al.* (1980) have studied the sulphoxidation of 4-(*p*-chlorophenyl)-tetra-hydro-2H-thiin (159) and some alicyclic and benzothiin analogues by rabbit liver microsomes and by purified reconstituted cytochrome P-450 from the same preparations. In each case, the *trans* sulphoxide was the major metabolic product (*cf.* the dihydrobenzothiophen (65)) in contrast to chemical oxidations which yielded comparable amounts of the *cis* and *trans* isomers. An additional steric effect observed for the 2-substituted benzothiins (160) was that product yield decreased as the size of the R substituent increased. It is interesting that although (159) and (160), like THTP, are analogues of aliphatic sulphides (e.g. diethyl sulphides (DES)), they were substrates for cytochrome P-450, in contrast to THTP and DES which are exclusively metabolized by the flavin-containing monooxygenase (Hoodi and Damani, 1984). The possibility that this series of compounds may be substrates for both of these oxidative enzymes has already been discussed.

(159) 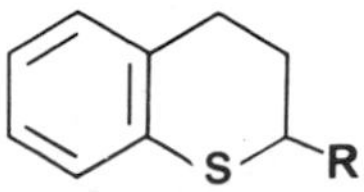   (160)

The metabolism of the dibenzo-fused thiins, the 9H-thioxanthenes, has been reasonably well studied in view of the neuroleptic and antidepressant properties of several derivatives. As with the structurally similar phenothiazines, sulphoxide formation is common in all species, often in conjunction with other metabolic changes in the molecule; unlike the phenothiazines, the sulphone metabolites have not been reported but their formation should not be ruled out. Early literature for thiothixene ((162); Hobbs, 1968), flupenthixol ((161); Jorgensen et al., 1969) and clopenthixol ((163); Khan, 1969) suggested that thioxanthenes differed from phenothiazines in not undergoing ring hydroxylation. However, more recently Breyer-Pfaff et al. (1985) have identified phenolic metabolites of chlorprothixene (164) in urine of man and dog, largely in the conjugated form. In both species, 6- and 7-hydroxylation was observed but dog differed from man in the preferential formation of 5-hydroxy compounds which were absent from human urine.

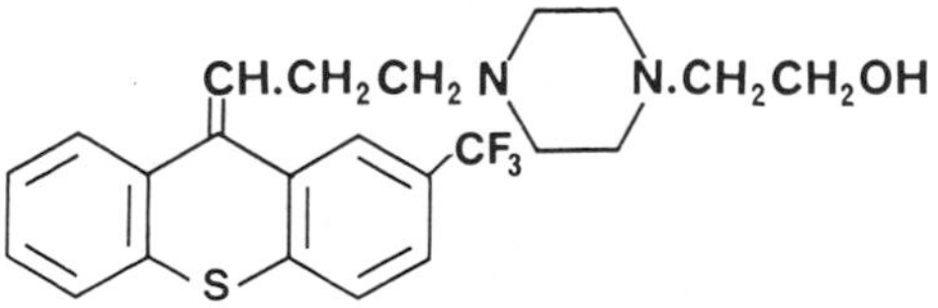

(161) 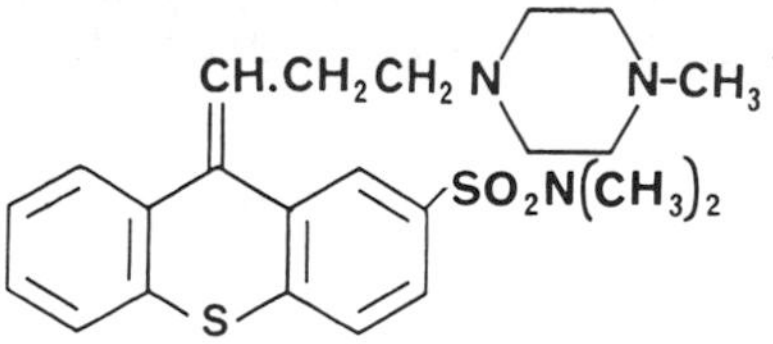   (162)

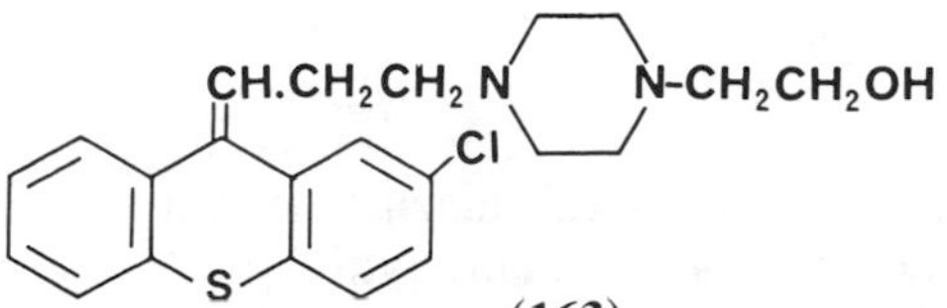

(163) 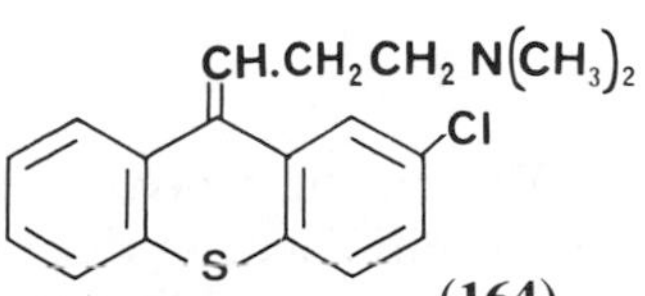   (164)

The metabolism of compounds containing two sulphur atoms in a six-membered ring has received little study; the 1,3-dithiane tetraoxide ring of the calcium antagonist tiapamil (165) is apparently resistant to metabolism, possibly by virtue of the sulphone groups that reduce the susceptibility of the ring carbons, particularly C_2, towards electrophilic attack (Wendt, 1982).

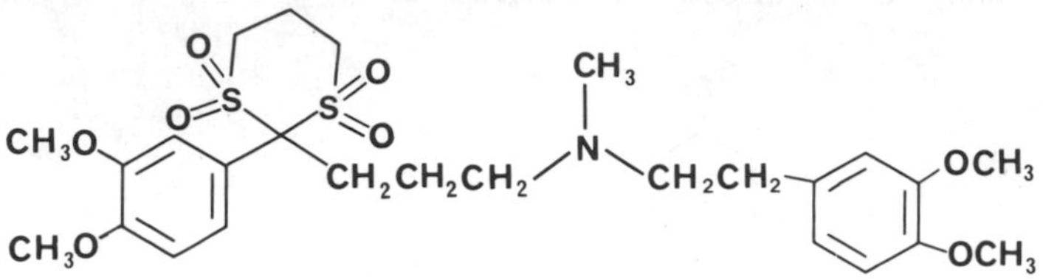

(165)

9.6 SIX-MEMBERED HETEROCYCLES CONTAINING SULPHUR AND OTHER HETEROATOMS

9.6.1 Chemical considerations

The ring systems to be covered in the section are fairly diverse and therefore relevant aspects of their chemistry are included in the context of their metabolism.

9.6.2 Metabolism of compounds containing an unfused ring

Two sulphoxide metabolites of the fungicide carboxin (vitavax) (166), a 1,4-oxathiin, were present in urine after oral dosing to rats and rabbits (Waring, 1973). In both metabolites, hydroxylation of the phenyl ring had also occurred, the sulphoxide of the parent compound being absent. This is in apparent contrast to the situation in dog where vitavax sulphoxide was described as the major metabolite by Chin et al. (1969), although Waring has pointed out that these workers may have used inappropriate analytical procedures; for this reason, the claims by Chin et al. (1970) that sulphoxide and sulphone metabolites of vitavax are present in barley and wheat plants should be treated with some caution.

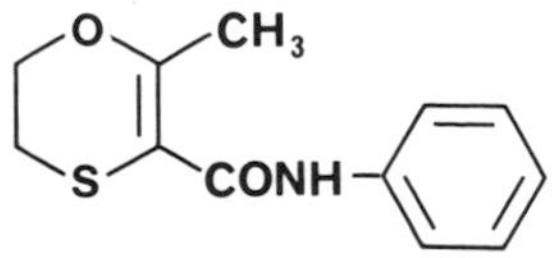

(166)

The muscle relaxant chlormezanone (167), a 1,3-thiazine derivative, shows a species difference in ring opening (Scheme 14). In man, hydrolytic N–C_4 scission predominates, yielding the carboxylic acid (169) which is the major component in plasma (Koppel *et al.*, 1986); the compound may in fact be formed non-enzymatically in the stomach. In mice and rats oxidative S–C_2 cleavage occurs, producing the sulphonic acid (168) which undergoes further metabolism (Hakusui *et al.*, 1978); since the secondary metabolites (170) and (171) may be produced from either (168) or (169), the relative importance of the two ring cleavage processes in rodents is unclear.

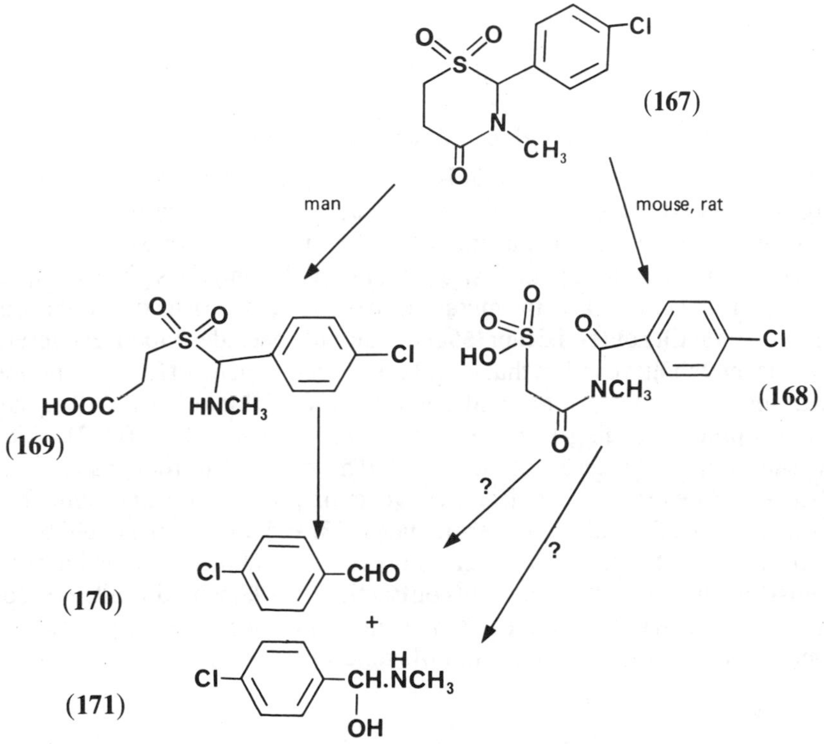

Scheme 14

The tetrahydro-1,4-thiazine (thiomorpholine) ring of the antitrypanosomal compound (172) is metabolized to the sulphoxide and sulphone in the mouse (Chatfield, 1976); there was no evidence of oxidative scission of the thiazole ring.

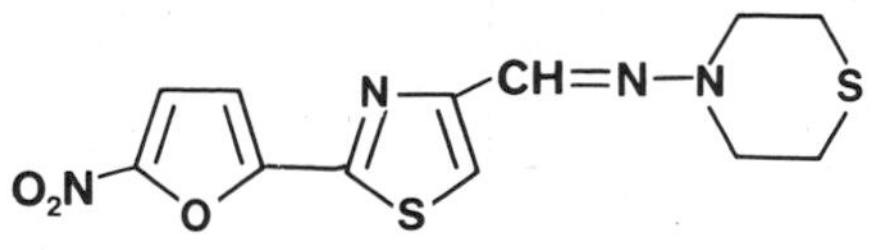

(172)

Limited studies on the metabolism of the chemotherapeutic agent taurolin (173) have suggested that the thiadiazine ring is cleaved at C_3; thus, when drug radiolabelled with ^{14}C at each of the three methylene bridges between pairs of nitrogen atoms was administered to rats, most of the radioactivity was excreted as exhaled carbon dioxide (Ganz *et al.*, 1980; Steinbach-Lebbin *et al.*, 1982).

(173); *indicates position of* ^{14}C

9.6.3 Metabolism of compounds containing fused rings

Two important groups of compounds to be considered in this section are the 4-hydroxy-2-methyl-2H-1,2-benzothiazine-3-carboxamide-1,1-dioxide family (the oxicams) and the dibenzo-1,4-thiazine family (the phenothiazines).

In drug metabolism terms, the oxicam family **(174)** comprises piroxicam, sudoxicam, isoxicam and tenoxicam; the metabolism of the thiazole ring of sudoxicam **(104)** and the thiophen ring of tenoxicam **(56a)** has already been described. An interesting and unusual metabolic reaction that is apparently common to all four compounds but with varying importance (Hobbs and Twomey, 1981; Ichihara *et al.*, 1984) is a ring contraction process leading to formation of *N*-methyl saccharin ((**177**); thieno-*N*-methyl saccharin in the case of tenoxicam). The mechanism remains to be clearly established. However, for tenoxicam, the reaction probably occurs (Ichihara *et al.*, 1984) directly via a C_3 oxidation intermediate **(175)** and this scheme could be general for all compounds (Scheme 15). The authors cite several pieces of evidence for this mechanism including the fact that ring contraction of **(175)** would yield, as secondary products, oxamic acid derivatives **(176)** which have been isolated in the case of tenoxicam and isoxicam (Borondy and Michiliewicz, 1984).

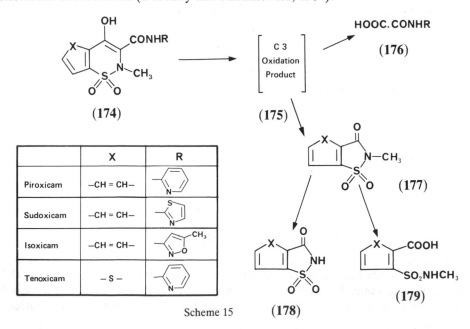

	X	R
Piroxicam	—CH = CH—	
Sudoxicam	—CH = CH—	
Isoxicam	—CH = CH—	CH₃
Tenoxicam	— S —	

Scheme 15   **(178)**

Both of the saccharin derivatives (177) and (178) were reported as metabolites of tenoxicam and piroxicam (Hobbs and Twomey, 1981); these authors also described the presence of (178) as a minor metabolite of isoxicam and sudoxicam. Borondy and Michiliewicz (1984) do not mention formation of (177) or (178) from isoxicam, and nor do Hobbs and Twomey (1981) in the original work on sudoxicam, but in both of these studies the saccharin metabolites would not have been radiolabelled and hence would only have been detected by specific assay. The o-(methylsulphamoyl) aromatic carboxylic acid (179) was reported for tenoxicam and isoxicam. In the case of piroxicam two additional metabolites (180) and (181) were present in the excreta of rhesus monkey, resulting from amide hydrolysis and decarboxylation of the parent drug; oxidation of (181) would provide an alternative route to the N-methyl saccharin.

In addition to the ring contraction reaction, rat (but not dog or rhesus monkey) formed two metabolites of piroxicam in which hydroxylation of the benzene ring had occurred (Hobbs and Twomey, 1981); deuterium labelling experiments suggested that these arose from hydroxylation at two of the three positions C_5, C_6 and C_7 but not at C_8. In all species, but particularly in dog, a cyclodehydrated metabolite was also formed and assigned the tetracyclic structure (182).

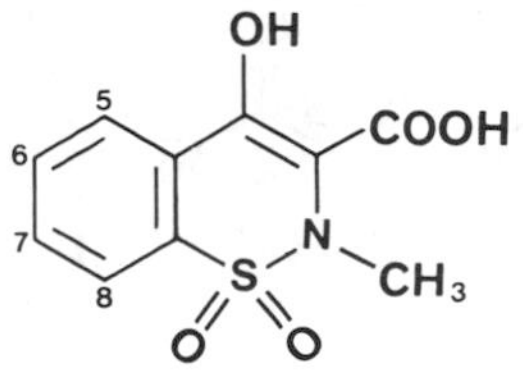

(180) 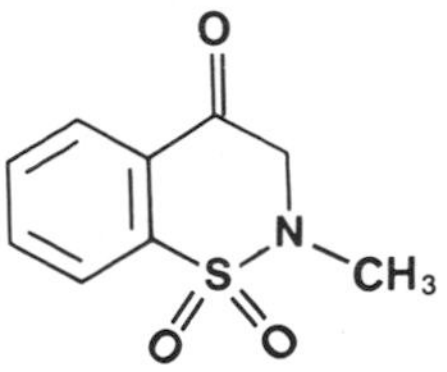   (181)

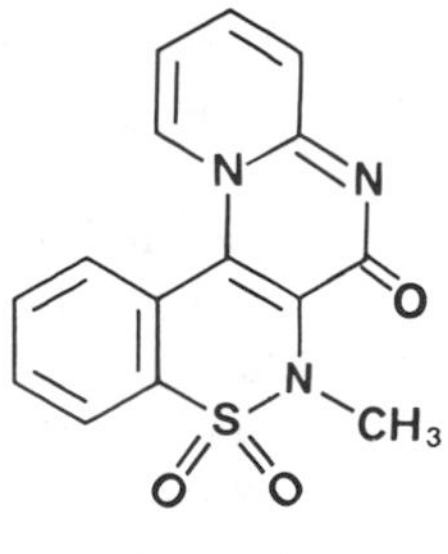

(182)

Despite the qualitative similarities in metabolic pathways for the oxicams, it is important to appreciate that quantitative differences exist and that each compound shows species variations. Moreover, in terms of the human metabolism of piroxicam, sudoxicam and isoxicam, biotransformation processes in which the oxicam ring is not involved play an important part, these being hydroxylation of the pyridine ring, opening of the thiazole ring (see above), and hydroxylation of the methyl group on the isoxazole ring respectively.

The metabolism of phenothiazines (e.g. chlorpromazine; (184)) has been widely studied in view of the fact that many members of the series are potent tranquillizers. Chemically they may be regarded as diaryl sulphides. Phenothiazine itself (183), which has anthelmintic properties, is metabolized to the sulphoxide in farm animals and man (Clare, 1947; Mitchell, 1982) and this biotransformation, together with the subsequent conversion to the sulphone, is almost universal for compounds with the phenothiazine ring system; the parent compound also undergoes N-glucuronidation (Mitchell and Waring, 1979) and N-oxidation *in vitro* (Beckett *et al.*, 1975) but the *in vivo* occurrence of the latter reaction is uncertain.

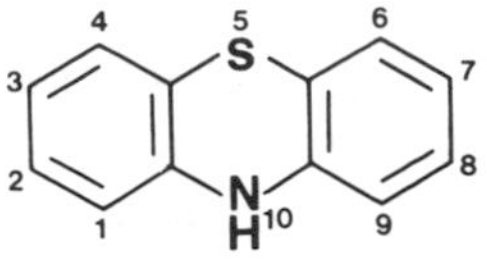

(183)

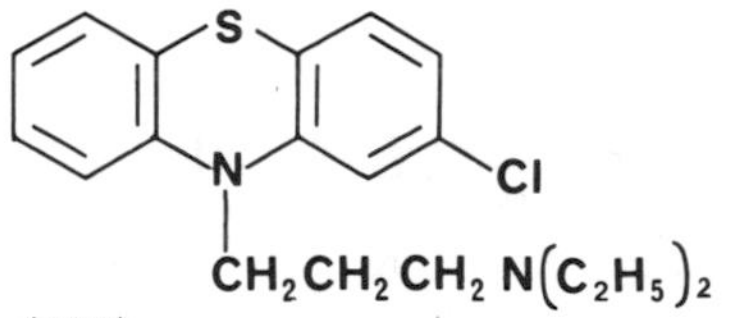

(184)

The detailed metabolism of individual phenothiazine derivatives will not be discussed here and the reader is referred to the chapter by Mitchell in Volume 3 of this series. It is sufficient to say that the sulphoxide rather than the sulphone metabolites generally predominate, often in conjunction with ring hydroxylation and oxidation of the N_{10} side chain.

Other than for phenothiazine itself, the sulphoxides are chiral and the presence of an asymmetric carbon in the molecule means that diastereomeric metabolites may be produced. This is apparent in the metabolism of thioridazine (Scheme 16) which forms diastereomeric pairs for the 5-sulphoxide derivatives of parent compound (186), the N-demethylated metabolite (188) and the 2-sulphone (187); the presence of (187) demonstrates the ability of thioridazine to undergo oxidation at both sulphur functions (Papadopoulos and Crammer, 1986).

Studies in rat and man (Watkins *et al.*, 1986) suggest that, for (186) and (188), the sulphoxide diastereomers are produced in approximately equal amounts. Moreover, separate administration of the diastereomers of (186) showed they had similar patterns of disposition and elimination (Hale *et al.*, 1985). These results suggest that, unlike the 2-substituted benzothiin and dihydrobenzothiophen analogues discussed above, the sulphoxidation of thioridazine is not subject to marked steric control, possibly because the enzyme can approach equally well from either face of the molecule.

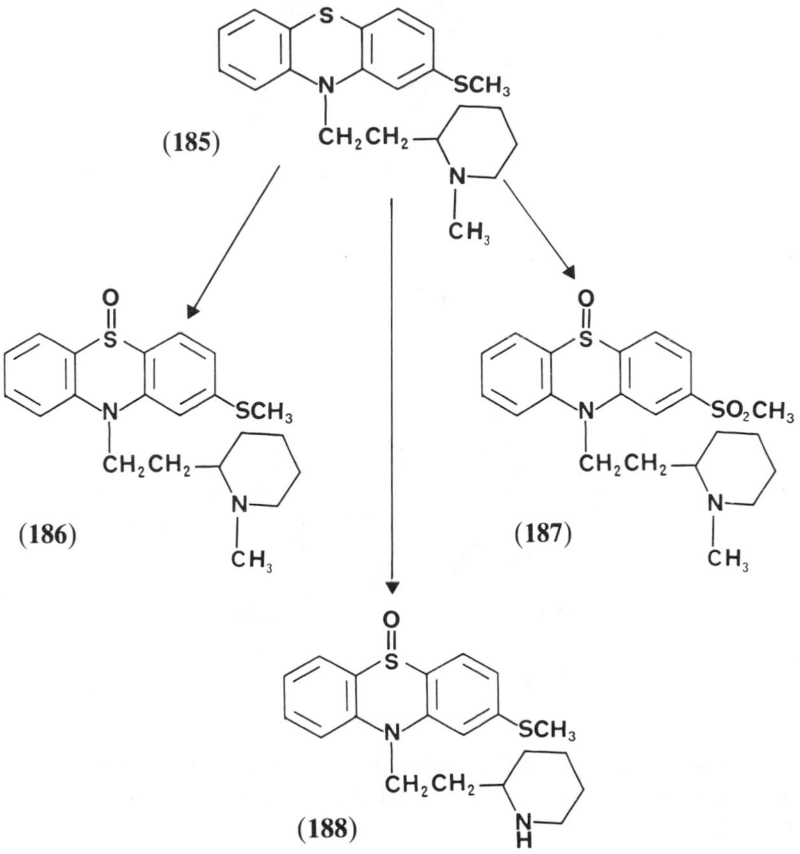

Scheme 16

In terms of enzymology, both cytochrome P-450 (Breyer, 1971) and the flavin-containing monooxygenase (Prema and Gopinathan, 1976) have been reported to catalyse sulphoxidation of chlorpromazine (**184**). Liver and extrahepatic tissues (e.g. intestine; Knoll *et al.*, 1977) have the ability to produce chlorpromazine sulphoxide; lung homogenates from rats dosed with thioridazine contain large amounts of the sulphoxide metabolites (Prema and Gopinathan, 1976). Enzyme activity is present in non-mammalian species such as the cestode *M. expansa* and nematode *A. Suum,* and the helminths also possess a sulphoxide reductase enzyme which produces chlorpromazine from its sulphoxide metabolite (Douch and Buchanan, 1979). The same reduction reaction can be effected *in vitro* by the enzyme aldehyde oxidase present in mammalian liver cytosol (Tatsumi *et al.*, 1984) (also, see chapter 5, this volume).

The anti-inflammatory drug SQ 11,579 (**189**), a mono-fused 1,4-benzothiazine, is converted to sulphoxide metabolites in rat, dog and monkey; *in vitro*, the reaction is inhibited by SKF 525-A, a known inhibitor of cytochrome P-450 (Lan *et al.*, 1973, 1976). Similarly tomizine (**190**), in which the 1,4-thiazine ring is fused to a pyrimidine system, is reported to form metabolites (**191**) and (**192**) where either the sulphur or the nitrogen of the thiazine ring has undergone oxidation (Linberg *et al.*, 1979).

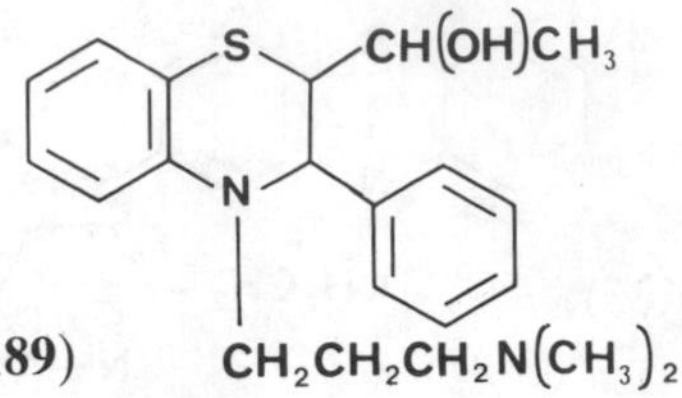

(189) $CH_2CH_2CH_2N(CH_3)_2$

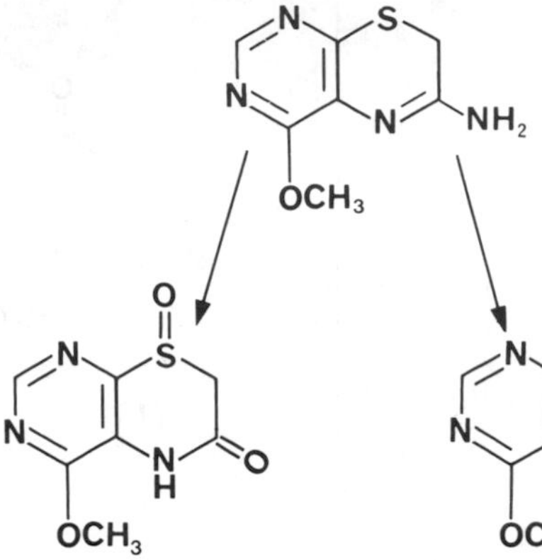

(190)

(191)   (192)

Finally, Chasseaud *et al.* (1972) suggest that a minor metabolite in the rat of the herbicide bentazon (193) is the N_1-glucuronide, the N–H moiety of the 2,1,3-benzothiadiazine ring system being slightly acidic.

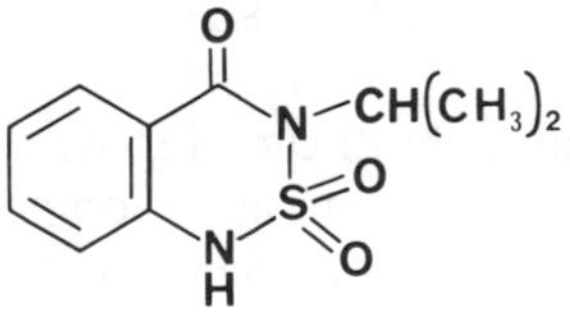

(193)

9.7   SEVEN-MEMBERED HETEROCYCLES CONTAINING SULPHUR

9.7.1   Chemical considerations

In terms of relevance to drug metabolism, the two principal ring systems where sulphur is present in a seven-membered ring are the isomeric [b,f] and [b,e] dihydro dibenzothiepins (194) and (195) respectively. The numbering system is such that the sulphur atom is at the 5 position in both cases. (194) should be regarded as a diaryl sulphide and hence may be expected to show properties similar to the thioxanthenes and phenothiazines. (195) is analogous to an aryl benzyl sulphide and so will have similarities to both aromatic and alicyclic thioethers.

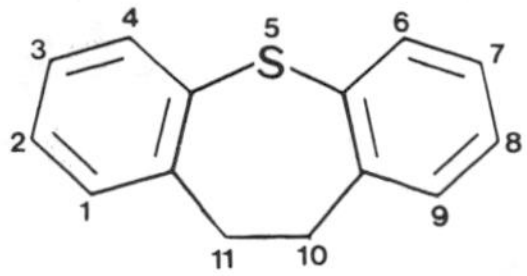

(194) 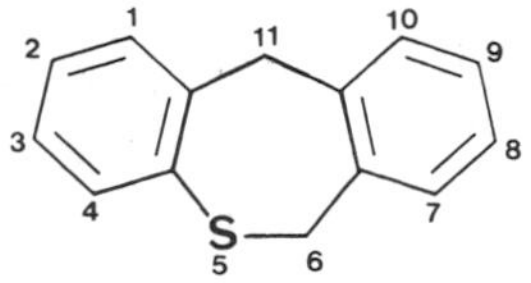   (195)

9.7.2   Metabolism of compounds containing a seven-membered sulphur ring

There are several centrally acting dibenzothiepin derivatives of both the [b,f] and the [b,e] type, and their metabolism has been studied in reasonable detail. An example of the [b,f] type is isofloxythepine (196) which forms five sulphoxide metabolites in rat, in some of which ring hydroxylation also occurred (Polakova *et al.*, 1986); (196) also formed a 10,11-dehydrogenated metabolite which was excreted in faeces. The structural analogues methiothepin (197) and oxyprotrepine (198), which like thioridazine possess two sulphur functions, are metabolized to the 5,8-disulphoxide and other 8-sulphoxidized products (Eschenhof *et al.*, 1976; Helia and Paulikova, 1986); for oxyprotepine, reaction occurred in mouse, rat or rabbit liver microsomes prepared from animals pretreated with phenobarbital. Sulphoxidation was also a reported route of metabolism for the neuroleptics zotepine ((199); Noda *et al.*,1979) and clorotepine ((200); octoclothepin; Jindra and Jindrova, 1979); the latter also undergoes hydroxylation at C_3 in hamster liver microsomes.

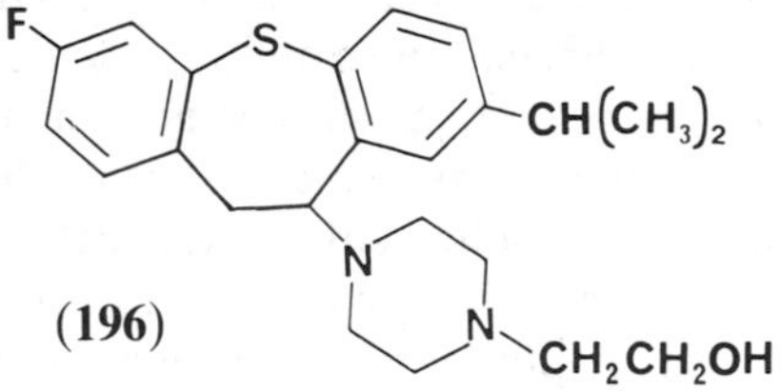

(196)

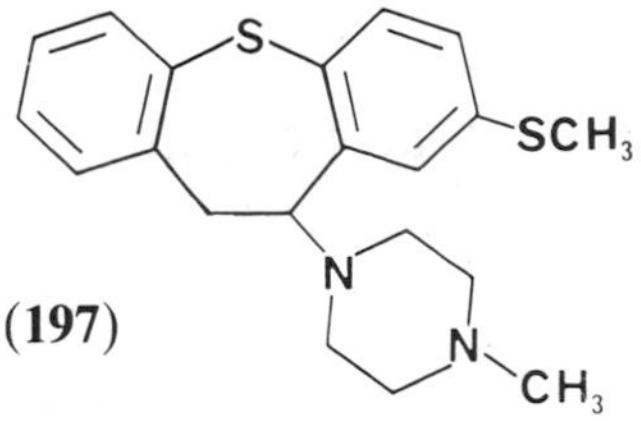

(197)

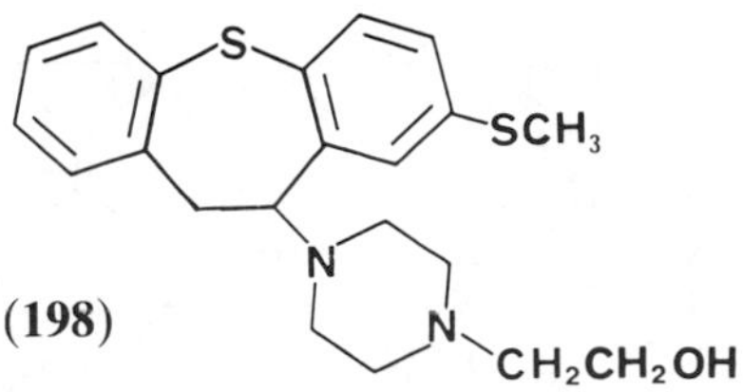

(198)

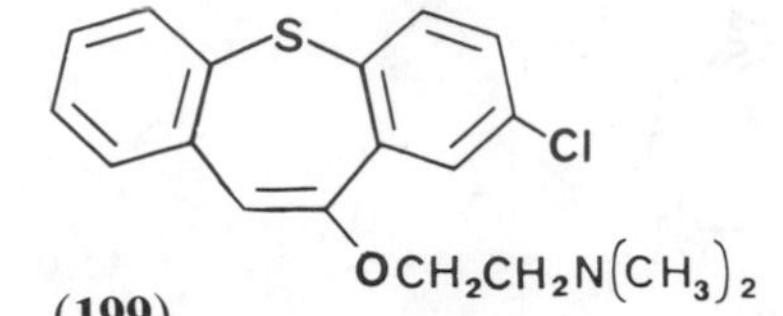

(199)

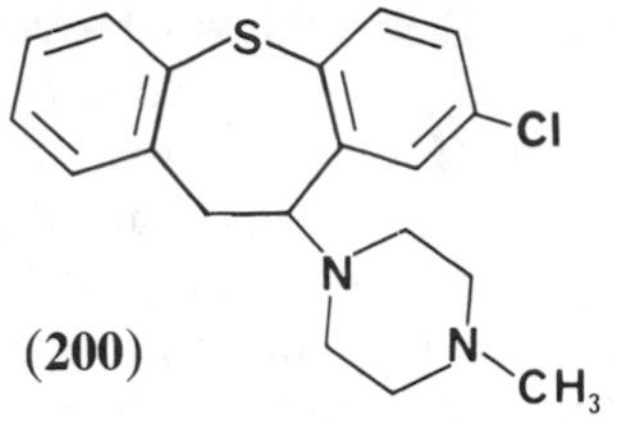

(200)

A well-studied dibenzo[b,e]thiepin derivative is the antidepressive dothiepin (201). Sulphoxidation occurs in all species studied including man and also takes place *in vitro* in rat liver 10 000 g supernatant fraction (Crampton *et al.*, 1978); the sulphoxides of parent compound and the mono-*N*-demethylated derivatives were significant metabolites in human urine (Rees, 1981). The structural analogue tropatepine (202) also forms sulphoxide metabolites in rat and man (Arnoux *et al.*, 1986). Similar pathways of metabolism occur in the rat for dithiadene (203) in which a thiophen replaces one of the benzene rings of dothiepin. Sulphoxidation apparently takes place only on the S_{10} sulphur atom and not on the thiophen sulphur (S_1) (Queisnerova *et al.*, 1971); the analogous thienobenzothiepin, pipethiaden (204), also forms a thiepin- but not a thiophen-sulphoxide metabolite (Lapka *et al.*, 1985). Dithiadene undergoes further metabolism to the sulphone and there are also reports of the formation of inorganic sulphate in urine after administration of [10-^{35}S]dithiadene to rats, which would suggest metabolic cleavage of the thiepin ring (Franc *et al.*, 1973).

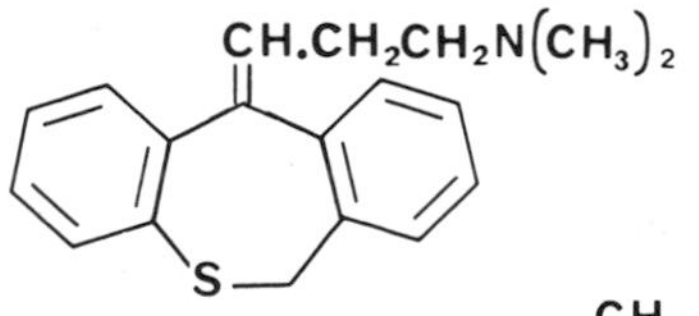

(201)

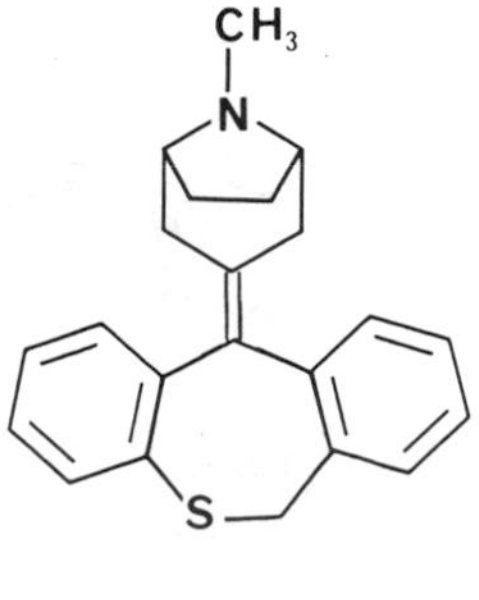

(202)

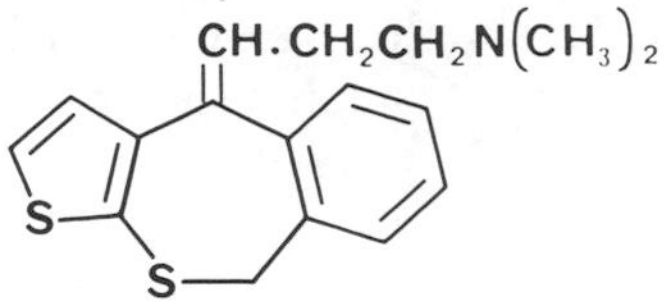

$CH.CH_2CH_2N(CH_3)_2$

(203)

CH_3

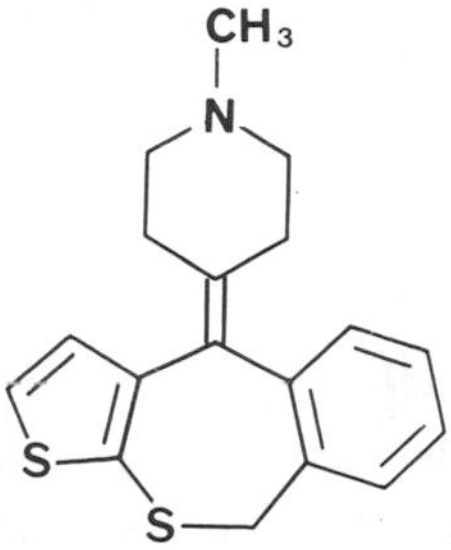

(204)

Delocalization of sulphur electrons might be expected to be greater for the [b,f] compared with the [b,e] isomer of dibenzothiepin. Hence, in view of the differing roles played by cytochrome P-450 and the flavin-containing monooxygenase in the oxidation of sulphur heterocycles, the comparative enzymology of sulphoxidation for the isomeric thiepins seems an area worthy of further study.

The antipsychotic clothiapine (205), a dibenzo[b,f]thiazepine which features a seven-membered ring containing sulphur and nitrogen, forms sulphoxide metabolites in dog and man, and a trace of clothiapine sulphone in man (Gauch and Lehner, 1969; Gauch et al., 1969).

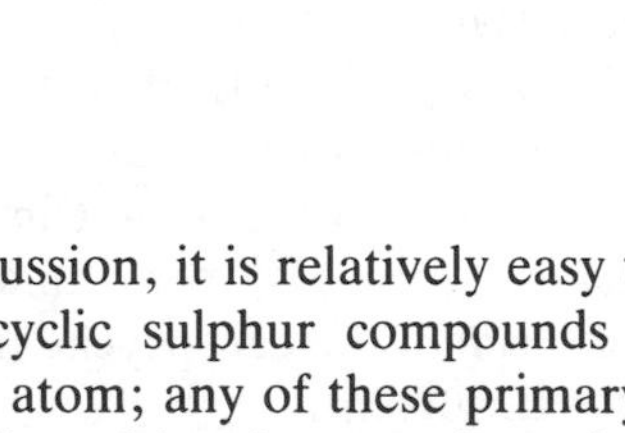

(205)

9.8 CONCLUSIONS

In the light of the above discussion, it is relatively easy to classify the metabolism options available to heterocyclic sulphur compounds as attack at sulphur, at nitrogen, or at a ring carbon atom; any of these primary reactions, but especially attack at carbon, may be followed by ring opening, yielding acyclic products. In chemical terms, the initial attack is normally viewed as being electrophilic but there are examples (e.g. oltipraz (70)) of nucleophilic attack.

However, it is a much more difficult task to formulate general rules which can be used to predict the metabolic fate of a given compound. Despite the numerous examples described, it should be recognized that there are still relatively few within each type, and consequently broad generalizations are unwise. Nevertheless, some general conclusions are appropriate and these are conveniently discussed in terms of the four aspects of the chemistry of sulphur heterocycles referred to in the Introduction.

9.8.1 Ring size

Although a convenient basis for classification, ring size is not a major determinant of metabolism for sulphur heterocycles except for three- and four-membered rings for which ring strain is the basis of their reactivity. For example, the phenothiazines (183) and the dibenzo[b,f]thiepins (194), which may be regarded as six- and seven-membered ring homologues, show generally similar routes of metabolism.

9.8.2 Degree of saturation

The degree of ring saturation is an important factor determining routes of metabolism for sulphur heterocycles, particularly for the S-oxygenation reaction. Thus, formation of sulphoxides assumes a greater importance for saturated systems than for their aromatic counterparts, this being apparent for both thiophens and thiazoles; by the same token, pipethiaden (204) undergoes S-oxygenation in the thiepin but not the thiophen ring. Nevertheless, S-oxygenation of simple aromatic heterocycles should not be ruled out since a significant proportion of the dose remained unaccounted for in early metabolism studies on thiophen derivatives, and therefore chemically unstable sulphoxide metabolites may have escaped detection. Differences in the enzymology of S-oxygenation for the aromatic dibenzothiophen (19) and the saturated tetrahydrothiophen (20) have been attributed (Damani, 1987; Hoodi and Damani, 1984) to substrate nucleophilicity as reflected in the ability of the sulphur atom to donate electrons to the oxidizing enzyme. More generally, it is possible that the extent to which sulphur electrons are involved in aromatic stabilization will determine both the ease and the enzymology of S-oxygenation.

9.8.3 Ring fusion

Fusion of an S-heterocycle to another ring system usually increases the number of metabolism options open to the molecule. Limited data for hydroxylation of fused thiophens are generally consistent with the view that enzymic attack preferentially occurs in the heterocyclic ring (e.g. benzothiophen (18)), as expected on electronic grounds, unless this ring is already substituted in which case reaction occurs in the other ring (e.g. UK-59,354 (54a) and brotizolam (55a)); in the case of tenoxicam (56a) both the thiophen and the thiazine rings are metabolized.

Available data do not permit corresponding comparisons for thiazole systems. However, purely on chemical grounds, benzothiazoles would be expected to undergo preferential metabolism in the carbocyclic ring, as occurs in fostedil (119), even when the C_2 position is unsubstituted.

9.8.4 Heteroatom replacement

In some circumstances, replacement of a ring carbon by a nitrogen atom has a marked effect on the metabolism of sulphur heterocycles. Thus, the incidence of ring-opened products from thiazoles appears to be more common than from thiophen derivatives. 5-Hydroxy metabolites of both ring systems are in tautomeric equilibrium with lactone forms, but the position of the equilibrium and the hydrolytic stability of the lactone tautomers are presumably such that ring opening is more favoured for thiazoles than for thiophens.

In contrast, thioxanthenes and phenothiazines, two series of compounds related by substituion of a nitrogen for a carbon atom γ to the sulphur generally undergo similar metabolism reactions at sulphur and carbon. In this case, however, the heteroatom replacement takes place at a site more distant from the sulphur, the replacement does not perturb greatly the electron density in the carbocyclic rings, and the overall symmetry of the ring system is retained.

9.8.5 Sulphoxides vs. sulphones

Some mention should be made as to the relative occurrence of sulphoxide and sulphone metabolites of sulphur heterocycles. Although end-product analysis can sometimes be misleading, there is a body of evidence to suggest that sulphone metabolites assume less quantitative importance *in vivo* than the corresponding sulphoxides; compounds for which this is the case include THTP (**20**), Lu-5-003 (**67**), the thioxanthenes and the phenothiazines. One possible reason for this trend is the high polarity of sulphoxide metabolites which would be expected to promote their renal elimination and to limit access to the membrane-bound enzymes which mediate sulphone formation. In addition, steric factors presumably determine, at least in part, the ease with which a sulphoxide can fit into the active site of a metabolizing enzyme. It is interesting in this respect that sulphone formation at the 2-thiomethyl substituent of thioridazine (**185**) appears to occur more readily than at the S_5 atom of the phenothiazine ring (see Scheme 16). It may be that the steric constraints associated with an acyclic sulphoxide are less than for a cyclic analogue, possibly, in the case of thioridazine, owing to the close proximity of the carbocyclic rings to the sulphur atom.

9.8.6 Areas for further study

Two other aspects of the metabolism of sulphur heterocycles are particularly worthy of future study. Thus, the enzymology of S-oxygenation reactions is currently unclear, particularly the relative importance of pathways mediated by cytochrome P-450 and the flavin-containing monooxygenase and the mechanistic studies of Hoodi and Damani (1984) should be extended to a wider range of sulphur heterocycles. Finally, in the light of the chirality of many sulphoxides, more attention should be given to stereochemical aspects of the formation and disposition of these metabolites; the increasing availability of purified isozymes and the development of analytical methodology for the separation of enantiomers should facilitate progress in this field in the next few years.

ACKNOWLEDGEMENTS
The assistance and advice of many Pfizer colleagues is gratefully acknowledged; particular thanks are due to Dr M. J. Randall, Mrs V. Addis and Mrs C. Deeley.

REFERENCES
Abe, Y., Okahara, T., Miura, K., Yukimara, T., Takada, T., Iwatani, T., Iso, T., and Yamamoto, K. (1984). Renal excretion of an ACE inhibitor (SA-446) in dogs. *Naunyn-Schmiedeberg's Arch. Pharmacol.*, **325**, 356–359.

Ambaye, R. Y., Panse, T. B., and Tilak, B. D. (1961). Metabolism of thioesters of carcinogenic hydrocarbons; metabolism of dibenzothiophene. *Proc. Indian Acad. Sci.*, **53**, 149–156.

Arendt, R., Ochs, H. R., and Greenblatt, D. J. (1982). Electron capture GLC analysis of the thienodiazepine Clotiazepam. *Arzneim. Forsch.*, **32**, 453–455.

Arnoux, P., Placidi, M., Aubert, C., and Cano, J. P. (1986). Simultaneous determination of tropatepine and its major metabolites by hplc–ms identification: application to metabolic and kinetic studies. *J. Chromatogr.*, **381**, 75–82.

Asano, T., Inoue, T., and Kurono, M. (1984). Disposition of azosemide. *Yakugaku Zasshi*, **104**, 1181–1190.

Atkins, L. C. and Rance, D. J. (1987). Unpublished results.

Ball, L. M., Renwick, A. G., and Williams, R. T. (1977). The fate of [^{14}C] saccharin in man, rat and rabbit and of 2-sulphamoyl-[^{14}C] benzoic acid in the rat. *Xenobiotica*, **7**, 189–203.

Bechtel, W. D., Mierau, J., Brandt, K., Förster, H. J., and Pook, K. H. (1986). Metabolic fate of [^{14}C]-brotizolam in the rat, dog, monkey and man. *Arzneim. Forsch.*, **36**, 578–586.

Becker, R., Frankus, E., Graudums, I., Günzler, W. A., Helm, F. C., and Flohe, L. (1982). The metabolic fate of supidimide in the rat. *Arzneim. Forsch.*, **32**, 1101–1111.

Beckett, A. H., Al-Sarraj, S., and Essien, E. E. (1975). The metabolism of chlorpromazine and promethazine to give new 'pink spots'. *Xenobiotica*, **5**, 325–355.

Bieder, A., Decouvelaere, B., Gaillard, C., Depaire, H., Heusse, D., Ledoux, C., Lemar, M., Le Roy, J. P., Raynaud, L., Snozzi, C., and Gregoire, J. (1983). Comparison of the metabolism of oltipraz in the mouse, rat and monkey and in man. *Arzneim. Forsch.*, **33**, 1289–1297.

Bird, A. E., Cutmore, E. A., Jennings, K. R., and Marshall, A. C. (1983). Structure reassignment of a metabolite of ampicillin and amoxycillin and epimerisation of their penicilloic acids. *J. Pharm. Pharmacol.*, **35**, 138–143.

Block, E. (1984). Thietanes, thietes and fused-ring derivatives. In W. Lwowski (ed.), *Comprehensive Heterocyclic Chemistry*, Vol. 7, Pergamon, Oxford, pp. 403–447.

Blumer, J. L., Novak, R. F., Lucas, S. V., Simpson, J. M., and Webster, L. T. (1979). Aerobic metabolism of niridazole by rat liver microsomes. *Mol. Pharmacol.*, **16**, 1019–1030.

Bohm, F. (1941). In welcher form scheiden kaninchen substanzen aus die einen mit einen benzolring gekuppelten funfring enthalten und dem indoltypus nahe stehen? *Hoppe-Seyler's Zeitschrift Physiol. Chemie*, **269**, 17–28.

Borondy, P. E. and Michiliewicz, B. M. (1984). Metabolic disposition of isoxicam in man, monkey, dog and rat. *Drug Metab. Disp.*, **12**, 444–451.

Bousquet, W. F., Rogler, J. C., Knevel, A. M., Spahr, J. L., and Christian, J. E. (1965). Studies on the tissue clearance and metabolism of 2-acetamido-5-nitro-thiazole-C^{14} in turkeys. *J. Agr. Food Chem.*, **13**, 571–573.

Bowers, G. D., Kaye, B., Rance, D. J., and Waring, L. (1987). Unpublished results.

Boyland, E. and Sims, P. (1958). An acid-labile precursor of 1-naphthylmercapturic acid and naphthol: an N-acetyl-S-(1:2 dihydrohydroxynaphthyl)-L-cysteine. *Biochem. J.*, **68**, 440–447.

Bray, H. G. and Carpanini, F. M. B. (1968). The metabolism of thiophen and benzo[b]thiophen. *Biochem. J.*, **109**, 11P.

Bray, H. G., Carpanini, F. M. B., and Waters, B. D. (1971). The metabolism of thiophen in the rabbit and rat. *Xenobiotica*, **1**, 157–168.

Breyer, U. (1971). Metabolism of the phenothiazine drug perazine by liver and lung microsomes from various species. *Biochem. Pharmacol.*, **20**, 3341–3351.

Breyer-Pfaff, U., Wiest, E., Prox, A., Wachsmuth, H., Protiva, M., Sindelar, K., Friebolin, H., Krauss, D., and Kunzelman, P. (1985). Phenolic metabolites of chlorprothixene in man and dog. *Drug Metab. Disp.*, **13**, 479–489.

Campbell, M. M. (1979). Five-membered ring systems. In P. G. Sammes (ed.), *Comprehensive Organic Chemistry*, Vol. 4, Pergamon, Oxford, pp. 961–1050.

Catto, B. A., Tracy, J. W., and Webster, L. T. (1984). 1-thiocarbamoyl-2-imidazolidinone: a metabolite of niridazole in *S. mansoni. Mol. Biochem. Parasitol.*, **10**, 111–120.

Cavallini, D., De Marco, C., Mordovi, B., and Trasarti, F. (1956). Studies of the metabolism of thiazolidine carboxylic acid by rat liver homogenate. *Biochim. Biophys. Acta*, **22**, 558–564.

Chasseaud, L. F., Hawkins, D. R., Cameron, B. D., Fry, B. J., and Saggers, V. H. (1972). The metabolic fate of bentazon in the rat. *Xenobiotica*, **2**, 269–276.

Chatfield, D. H. (1976). Disposition and metabolism of some nitrofurylthiazoles possessing antiparasitic activity. *Xenobiotica*, **6**, 509–520.

Chatfield, D. H. and Hunter, W. H. (1973a). The metabolism of acetamido thiazoles in the rat: 2-acetamido-, 2-acetamido-4-methyl- and 2-acetamido-4-phenyl-thiazole. *Biochem. J.*, **134**, 869–878.

Chatfield, D. H. and Hunter, W. H. (1973b). The metabolism of acetamido thiazoles in the rat: 2-acetamido-4-chloromethyl thiazole. *Biochem. J.*, **134**, 879–884.

Chin, W. T., Stone, G. M., Smith, A. E., and Von Schmeling, B. (1969). Fate of carboxin in soil, plants and animals. *Proc. Br. Insect. Fungic. Conf.*, **2**, 322–327.

Chin, W. T., Stone, G. M., and Smith, A. E. (1970). Metabolism of carboxin (vitavax) by barley and wheat plants. *J. Agric. Food Chem.*, **18**, 709–712.

Clare, N. T. (1947). The metabolism of phenothiazine in ruminants. *Aust. Vet. J.*, **23**, 340–344.

Cole, M., Kenig, M. D., and Hewitt, V. A. (1973). Metabolism of penicillins to penicilloic acids and 6-amino-penicillanic acid and its significance in assessing penicillin absorption. *Antimicrob. Agents Chemother.*, **3**, 463–468.

Crampton, E. L., Dickinson, W., Haran, G., Marchant, B., and Risdall, P. C. (1978). The metabolism of dothiepin hydrochloride *in vivo* and *in vitro*. *Brit. J. Pharmacol.*, **64**, 405P.

Cripps, R. E. (1973). The microbial metabolism of thiophen-2-carboxylate. *Biochem. J.*, **134**, 353–366.

Damani, L. A. (1987). Metabolism of sulphur-containing drugs. In *Drug Metabolism — From Molecules to Man*, D. J. Benford, J. W. Bridges, and G. G. Gibson, (eds.), Taylor and Francis, London, pp. 581–603.

Damani, L. A. and Case, D. E. (1984). Metabolism of heterocycles. In O. Meth-Cohn (ed.), *Comprehensive Heterocyclic Chemistry*, Vol 1, Pergamon, Oxford, pp. 223–246.

Decouvelaere, B., Gaillard, C., Kettler, J . P., Martin, S., Gires, P., and Gaillot, J.
    (1986). Biotransformation of 48740 RP. Poster. *10th European Drug Metabo-
    lism Workshop, Guildford.*
Dittmer, D. C. (1984). Thiiranes and thiirenes. In W. Lwowski (ed.), *Comprehen-
    sive Heterocyclic Chemistry,* Vol. 7, Pergamon, Oxford, pp. 131–184.
Douch, P. G. C. and Buchanan, L. L. (1979). Some properties of the sulphoxidases
    and sulphoxide reductases of the cestode *M. expansa,* the nematode *A. suum*
    and mouse liver. *Xenobiotica,* **9**, 675–679.
El Dareer, S. M., Tillery, K. F., and Hill, D. L. (1978). Distribution and metabolism
    of 2-amino-1,3,4-thiadiazole in mice, dogs and monkeys. *Cancer Treat. Rep.,*
    62, 75–83.
Eschenhof, E., Meister, W., Oesterhelt, G., and Vetter, W. (1976). On metabolism
    of methiothepin in rat, dog and man. *Arzneim. Forsch.,* **26**, 262–271.
Faulkner, J. K., Figdor, S. K., Monro, A. M., Schach von Wittenau, M., Stopher,
    D. A. and Wood, B. A. (1972). The comparative metabolism of pyrantel in five
    species. *J. Sci. Fd. Agric.,* **23**, 79–91.
Fehring, S. I., Henderson, T., Ahokas, J. T., Ham, K., Ravenscroft, P. J., and
    Emmerson, B. T. (1984). Is tienilic acid toxicity mediated by metabolic activa-
    tion? *Clin. Exp. Pharmacol. Physiol., Suppl. 8,* 75–76.
Figdor, S. K. and Caldwell, J. (1987). Unpublished results.
Frame, G. M. (1987). Personal communication.
Franc, Z., Smolik, S., Raz, K., Lipovska, M., Horesovsky, O., and Francova, V.
    (1973). Pharmacokinetics of dithiadene (10-^{35}S). *Ceskoslov. Farm.,* **22**,
    306–309.
Franklin, R. A., Norris, R., Shepherd, N. W., and Rhenius, S. T. (1984). Prelimi-
    nary studies on the fate of ^{14}C-fentiazac in man. *Xenobiotica,* **14**, 955–960.
Fumero, S., Mondino, A., Silvestri, S., Zanolo, G., De Marchi, G., and Pedrazzini,
    S. (1980). Metabolism of fentiazac. *Arzneim. Forsch.,* **30**, 1253–1256.
Funayama, S., Murakami, N., Izawa, Y., Suzuki, T., Uchida, M., Tsuchiya, K.,
    Sugimoto, T., Nambara, Y., Suzuki, M., and Kitagawa, H. (1983). Studies on
    the fate of malotilate (NKK-105), a new drug for chronic hepatitis and liver
    cirrhosis in rats. *J. Pharm. Dyn.,* **6**, S-85.
Ganz, A. J., Steinbach-Lebbin, C., and Waser, P. G. (1980). Pharmacokinetics of
    ^{14}C-tauroline, a novel chemotherapuetic, in rats. *Experientia,* **36**, 707.
Gauch, R. and Lehner, H. (1969). The metabolism of clotiapine: metabolism and
    excretion in the human. *Farmaco. Ed. Prat.,* **24**, 100–109.
Gauch, R., Hunziker, F., Lehner, H., Michaelis, W., and Schindler, O. (1969). The
    metabolism of clotiapine; metabolism and excretion in the dog. *Farmaco. Ed.
    Prat.,* **24**, 92–99.
Gillham, B. and Young, L. (1968). The isolation of premercapturic acids from the
    urine of animals dosed with chlorobenzene and bromobenzene. *Biochem. J.,*
    109, 143–147.
Gronowitz, S. and Hörnfeldt, A-B. (1986). Synthesis, physical properties and
    reactions of compounds containing thiophene–oxygen bonds. In S. Gronowitz
    (ed.), *Thiophene and its Derivatives,* Part 3, Interscience, New York, pp. 1–133.
Grupe, A. and Spiteller, G. (1982). Unexpected metabolites produced from clo-
    methiazole. *J. Chromatogr.,* **230**, 335–344.

Haginaka, J. and Wakai, J. (1987). LC determination of ampicillin and its metabolites in human urine by post-column alkaline degradation. *J. Pharm. Pharmacol.*, **39**, 5–8.

Hakusui, H., Tachizawa, H., and Sano, M. (1978). Biotransformation of chlormezanone, a muscle-relaxing and tranquilising agent: the effect of combination with aspirin on its metabolic fate in rats and mice. *Xenobiotica*, **8**, 229–238.

Hale, P. W., Melethil, S., and Poklis, A. (1985). The disposition and elimination of stereoisomeric pairs of thioridazine 5-sulphoxide in the rat. *Eur. J. Drug Met. Pharmacokinet.*, **10**, 333–341.

Harmand, M. F. and Blanquet, P. (1982). Pharmacokinetics and metabolism of [^{35}S]-thiazolidine carboxylic acid in the rat. *Eur. J. Drug Metab. Pharmacokinet.*, **7**, 323–327.

Harrison, S. D., Bosin, T. R., and Maickel, R. P. (1974). Metabolism of 3-(2-dimethylaminoethyl)benzo[b]thiophene *in vitro* and *in vivo* in the rat. *Drug Metab. Disp.*, **2**, 228–236.

Hassan, M. and Ehrsson, H. (1987a). Metabolism of ^{14}C-busulfan in isolated perfused rat liver. *Eur. J. Drug Metab. Pharmacokinet.*, **12**, 71–76.

Hassan, M., and Ehrsson, H. (1987b). Urinary metabolites of busulfan in the rat. *Drug Metab. Disp.*, **15**, 399–402.

Heffter, A. (1886) Über das Verhalten des Thiophens im Thierkörper. *Pflugers Arch. Ges. Physiol.*, **39**, 420–425.

Helia, O. and Paulikova, I. (1986). Species differences in oxyprotepine metabolism *in vitro*. *Ceskoslov. Farm.*, **35**, 274–275.

Heusse, D., Marlard, M., Bredenbac, J., Decouvelaere, B., Leroy, J. P., Bieder, A., and Jumeau, H. (1985). Disposition of ^{14}C-oltipraz in animals. *Arzneim. Forsch.*, **35**, 1431–1436.

Hobbs, D. C. (1968). Metabolism of thiothixene. *J. Pharm. Sci.*, **57**, 105–111.

Hobbs, D. C. and Twomey, T. M. (1977). Metabolism of sudoxicam by the rat, dog and monkey. *Drug Metab. Disp.*, **5**, 75–81.

Hobbs, D. C. and Twomey, T. M. (1981). Metabolism of piroxicam by laboratory animals. *Drug Metab. Disp.*, **9**, 114–118.

Hoodi, A. A. (1986). *Ph.D. Thesis,* University of Manchester.

Hoodi, A. A. and Damani, L. A. (1984). Cytochrome P-450 and non-P-450 sulphoxidations. *J. Pharm. Pharmacol.*, **36**, 62P.

Ichihara, S., Tsuyuki, Y., Tomisawa, H., Fukazawa, H., Nakayama, N., Tateishi, M., and Joly, R. (1984). Metabolism of tenoxicam in rats. *Xenobiotica*, **14**, 727–739.

Ingall, A. H. (1984). Thiopyrans and fused thiopyrans. In A. J. Boulton and A. McKillop (eds.) *Comprehensive Heterocyclic Chemistry*, Vol. 3, Pergamon, Oxford, pp. 885–942.

Ishii, A., Oishi, T., Endo, M., and Mineura, K. (1978). Metabolism of glybuzole in rats after pretreatment of phenobarbital, SKF 525A and carbon tetrachloride. *Yakugaku Zasshi*, **98**, 1208–1214.

Jacob, J ., Schmoldt, A., and Grimmer, G. (1986). The predominant role of S-oxidation in rat liver metabolism of thiaarenes. *Cancer Lett.*, **32**, 107–116.

Janssen, F. W., Sharma, R. N., Young, E. M., Kirkman, S. K., and Ruelius, H. W.

(1979). Metabolism and thyroid toxicity of an immunomodulator, Wy-13,876. *Pharmacologist,* **21**, 193.

Janssen, F. W., Young, E. M., Kirkman, S. K., Sharma, R. N., and Ruelius, H. W. (1981). Biotransformation of the immunomodulator, 3-(p-chlorophenyl)-2,3-dihydro-3-hydroxythiazolo[3,2a]benzimidazole-2-acetic acid, and its relationship to thyroid toxicity. *Toxicol. Appl. Pharmacol.,* **59**, 355–363.

Janssen, F. W., Kirkman, S. K., Fenselau, C., Stogniew, M., Hofman, B. R., Young, E. M., and Ruelius, H. W. (1982). Metabolic formation of N- and O-glucuronides of 3-(p-chlorophenyl)thiazolo[3,2-a]benzimidazole-2-acetic acid. *Drug Metab. Disp.,* **10**, 599–604.

Jindra, A. and Jindrova, N. (1979). Oxidative biotransformation of octoclothepin in liver microsomes. *Pharmazie,* **34**, 12.

Jolles, G. (1984). Pharmacokinetics and mechanism of action of oltipraz in animals and man. *Report of W.H.O. Scientific Working Group on Biochemistry and Chemotherapy of Schistosomiasis,* Geneva.

Jorgensen, A., Hansen, V., Larsen, U. D., and Khan, A. R. (1969). Metabolism, distribution and excretion of flupenthixol. *Acta Pharmacol. Toxicol.,* **27**, 301–313.

Kamataki, T., Lee Lin, M. C. M., Belcher, D. H., and Neal, R. A. (1976). Studies of the metabolism of parathion with an apparently homogeneous preparation of rabbit liver cytochrome P-450. *Drug Metab. Disp.,* **4**, 180–189.

Kammerer, R. C. and Schmitz, D. A. (1986). Metabolism of methapyriline by rat-liver homogenate. *Xenobiotica,* **16**, 671–680.

Kaul, R., Hempel, B., and Kiefer, G. (1982). Structure of a novel sulphur-containing metabolite of Acluracil. *Xenobiotica,* **12**, 495–498.

Kennedy, G., Fancher, O. E., Calandra, J. C., and Keller, R. E. (1972). Metabolic fate of saccharin in the albino rat, *Food Cosmet. Toxicol.,* **10**, 143–149.

Khan, A. R. (1969). Some aspects of clopenthixol metabolism in rats and humans. *Acta Pharmacol. Toxicol.,* **27**, 202–212.

Knadler, M. P., Bergstrom, R. F., Callaghan, J. T., and Rubin, A. (1986). Nizatidine, an H_2-blocker: its metabolism and disposition in man. *Drug Metab. Disp.,* **14**, 175–182.

Knoll, R., Christ, W., Mueller Oerlinghausen, B., and Coper, H. (1977). Formation of chlorpromazine sulphoxide and monodesmethyl chlorpromazine by microsomes of small intestine. *Arch. Pharmacol.,* **297**, 195–200.

Koga, N., Inskeep, P. B., Harris, T. M., and Guengerich, F. P. (1986). S-[2-(N^7-guanyl)ethyl]glutathione, the major DNA adduct formed from 1,2-dibromoethane. *Biochemistry,* **25**, 2192–2198.

Koppel, C., Tenczer, J., and Wagemann, A. (1986). Metabolism of chlormezanone in man. *Arzneim. Forsch.,* **36**, 1116–1118.

Lan, S. J., Chando, T. J., Cohen, A. I., Weliky, I., and Schreiber, E. C. (1973). Metabolism of SQ 11,579 under *in vitro* and *in vivo* conditions by rats, dogs and monkeys. *Drug Metab. Disp.,* **1**, 619–627.

Lan, S. J., Dean, A. V., Walker, B. D., and Schreiber, E. C. (1976). Metabolism of SQ 11,579 by the intact rat, isolated perfused rat liver and rat-liver microsomes. *Xenobiotica,* **6**, 171–183.

Lapka, R., Franc, Z., and Smolik, S. (1985). Pharmacokinetics of ^{3}H-pipethiaden after single oral and intravenous administration in rats. *Arzneim. Forsch.*, **35**, 486–488.

Lethco, E. J. and Wallace, W. C. (1975). Metabolism of saccharin in animals. *Toxicology*, **3**, 287–300.

Linberg, L. F., Yadrovskaya, V. A., Safonova, T. S., and Sheinker, Y. N. (1979). Study of the metabolism of tomizine labelled with ^{14}C and deuterium. *Khim. Farmatsevt. Zh.*, **13**, 24–31.

Lu, P. Y., Metcalf, R. L., and Carlson, E. M. (1978). Environmental fate of 5 radiolabelled coal conversion by-products evaluated in a laboratory model ecosystem. *Envir. Hlth. Perspect.*, **24**, 201–208.

Lynch, M. J., Mosher, F. R., Levesque, W. R., and Newby, T. J. (1987). The *in vitro* and *in vivo* metabolism of morantel in cattle and toxicology species. *Drug Metab. Rev.*, **18**, 253–288.

MacKenzie, C. G. and Harris, J. (1957). N-formyl cysteine synthesis in mitochondria from formaldehyde and L-cysteine via thiazolidine-carboxylic acid. *J. Biol. Chem.*, **227**, 393–406.

Mansuy, D. (1987). A chemical approach to reactive metabolites. In D. J. Benford, J. W. Bridges, and G. G. Gibson (eds.), *Drug Metabolism — from Molecules to Man*, Taylor and Francis, London, pp. 669–678.

Mansuy, D., Dansette, P. M., Foures, C., and Jaouen, M. (1984). Metabolic hydroxylation of the thiophene ring. *Biochem. Pharmacol.*, **33**, 1429–1435.

Masada, M., Kuroda, Y., Nakagawa, T., and Uno, T. (1980). Structural investigation of new metabolites of aminopenicillins excreted in human urine. *Chem. Pharm. Bull.*, **28**, 3527–3536.

Maw, G. A. (1972). In A. Senning (ed.), *Sulphur in Organic and Inorganic Chemistry*, Vol. 2, Dekker, New York, p. 113.

Meth-Cohn, O. (1979). Thiophens. In P. G. Sammes (ed.), *Comprehensive Organic Chemistry*, Vol. 4, Pergamon, Oxford, pp. 789–838.

Metzger, J. (1984). Thiazoles and their benzo derivatives. In K. T. Potts (ed.), *Comprehensive Heterocyclic Chemistry*, Vol. 6, Pergamon, Oxford, pp. 235–332.

Mitchell, S. C. (1982). Mammalian metabolism of orally administered phenothiazine. *Drug Met. Rev.*, **13**, 319–343.

Mitchell, S. C. and Waring, R. H. (1979). Metabolism of phenothiazine in the guinea-pig. *Drug Metab. Disp.*, **7**, 399–403.

Mori, Y., Kuroda, N., Sakai, Y., Yokoya, F., Toyoshi, K., and Baba, S. (1985). Species differences in the metabolism of suprofen in laboratory animals and man. *Drug Metab. Disp.*, **13**, 239–245.

Nelson, J. A., Rose, L. M., and Bennett, L. L. (1977). Mechanism of action of NSC 4728. *Cancer Res.*, **37**, 182–187.

Nelson, S. D., Boyd, M. R., and Mitchell, J. R. (1977). Role of metabolic activation in chemical-induced tissue injury. In D. M. Jerina and R. F. Gould (eds.), *Drug Metabolism Concepts*, American Chemical Society, Washington, p. 155.

Nishikawa, T., Nagata, O., Tanbo, K., Yamada, T., Takahara, Y., and Kato, H. (1985). Absorption, excretion and metabolism of tiquizium bromide in dogs. *Xenobiotica*, **15**, 1053–1060.

Noda, K., Suzuki, A., Okui, M., Noguchi, H., Nishiura, M., and Nishiura, N. (1979). Pharmacokinetics and metabolism of zotepine in rat, mouse, dog and man. *Arzneim. Forsch.*, **29**, 1595–1600.

O'Dwyer, P. J., Wagner, B. H., Stewart, J. A., and Leyland-Jones, B. (1986). Aminothiadiazole: an antineoplastic thiadiazole. *Cancer Treat. Rep.*, **70**, 885–889.

Oeriu, S., Winter, D., Dobre, V., and Vilanescu, T. (1970). Pharmacokinetic data on ^{35}S labelled thiazolidine-carboxylic acid in the rat. *Stud. Cercet. Fiziol.*, **15**, 107–118.

Offen, C. P., Frearson, M. J., Wilson, K., and Burnett, D. (1985). 4,5-dimethylthia-zole-N-oxide-S-oxide: a metabolite of chlormethiazole in man. *Xenobiotica*, **15**, 503–511.

Overo, K. F., Jorgensen, A., and Hansen, V. (1970). Metabolism, distribution and excretion of the thiophthalane Lu-5-003, a bicyclic thymoleptic. *Acta Pharmacol. Toxicol.*, **28**, 81–96.

Pal, R. and Spiteller, G. (1982). Thiomethylation and thiohydroxylation — a new pathway of metabolism of heterocyclic compounds. *Xenobiotica*, **12**, 813–820.

Papadopoulos, A. S. and Crammer, J. L. (1986). Sulphoxide metabolites of thioridazine in man. *Xenobiotica*, **16**, 1097–1107.

Perrault, M. and Bovet, D. (1946). Aminothiazole in the treatment of thyrotoxicosis. *Lancet,* May 18, 731–734.

Pitkin, R. M., Anderson, D. W., Reynolds, W. A., and Filer, L. J. (1971). Saccharin metabolism in *Macaca mulatta*. *Proc. Soc. Exp. Biol. Med.*, **137**, 803–806.

Polakova, L., Roubal, Z., Metysova, J., Kleinerova, E., Svatek, E., Koruna, I., and Ryska, M. (1986). A metabolic study of isofloxythepine in rats. *Ceskoslov. Farm.*, **35**, 255–260.

Pottier, J., Berlin, D., and Raynaud, J. P. (1977). Pharmacokinetics of the anti-inflammatory tiaprofenic acid in humans, mice, rats, rabbits and dogs. *J. Pharm. Sci.*, **66**, 1030–1036.

Prema, K. and Gopinathan, K. P. (1976). Distribution, induction and purification of a monooxygenase catalysing sulphoxidation of drugs. *Biochem. Pharmacol.*, **25**, 1299–1303.

Quiesnerova, M., Metys, J., and Svatek, E. (1971). Biotransformation of dithiadene in rats. *Ceskoslov. Farm.*, **20**, 379–382.

Rajappa, S. (1984). Thiophenes and their benzo derivatives: (ii) reactivity. In C. W. Bird and G. W. H. Cheeseman (eds.), *Comprehensive Heterocyclic Chemistry*, Vol. 4, Pergamon, Oxford, pp. 741–861.

Rees, J. A. (1981). Clinical interpretation of pharmacokinetic data on dothiepin hydrochloride. *J. Int. Med. Res.*, **9**, 98–102.

Renwick, A. G. (1978). The fate of saccharin impurities: the metabolism and excretion of 3-amino[3-^{14}C]benz[d]isothiazole-1,1-dioxide and 5-chloro-saccharin in the rat. *Xenobiotica*, **8**, 487–495.

Renwick, A. G. and Williams, R. T. (1978). The fate of saccharin impurities: the excretion and metabolism of [3-^{14}C]-BIT in man and rat. *Xenobiotica*, **8**, 475–486.

Robbins, J. D., Bakke, J. E. and Feil, V. J. (1970). Metabolism of mobam in dairy goats and a lactating cow. *J. Agric. Food Chem.*, **18**, 130–134.

Roberts, J. J. and Warwick, G. P. (1961). The formation of 3-hydroxytetrahydroth-iophene-1:1-dioxide from myleran, S-β-L-alanyltetrahydothiophenium mesy-late, tetrahydrothiophene, and tetrahydrothiophene-1:1-dioxide in the rat, rabbit and mouse. *Biochem. Pharmacol.*, **6**, 217–227.

Smith, L. R. and Wilkinson, C. F. (1978). Influence of steric factors on the interaction of isomeric phenyloxazoles and phenylthiazoles with microsomal oxidation. *Biochem. Pharmacol.*, **27**, 2466–2467.

Steinbach-Lebbin, C., Ganz, A. J., Chang, A., and Waser, P. G. (1982). Uber die pharmacokinetik von taurolin. *Arzneim. Forsch.*, **32**, 1542–1546.

Takata, T., Yamazaki, M., Fujimori, K., Kim, Y. H., Oae, S., and Iyanagi, T. (1980). Stereochemistry of sulphoxides by enzymatic oxygenation of sulphides with rabbit liver microsomal cytochrome P-450. *Chem. Lett.*, 1441–1444.

Tatsumi, K., Kitamura, S., and Yamada, H. (1984). Sulphoxide reductase activity of liver aldehyde oxidase and a new electron transfer system containing the enzyme as a component. *J. Pharm. Dyn.*, **7**, S-86.

Thomas, E. W. (1984). Hplc determination of a new calcium antagonist, fostedil, in plasma and urine using fluorescence detection. *J. Chromatogr.*, **305**, 233–238.

Thomas, E. W. and Bopp, B. A. (1984). Pharmacokinetics of fostedil in beagle dogs following oral and intravenous administration. *J. Pharm. Sci.*, **73**, 1400–1403.

Tocco, D. J., Duncan, A. E. W., Deluna, F. A., Smith, J. L., Walker, R. W., and Vandenheuvel, W. J. A. (1980). Timolol metabolism in man and laboratory animals. *Drug Metab. Disp.*, **8**, 236–240.

Tracy, J. W. and Webster, L. T. (1981). The formation of 1-thiocarbamoyl-2-imidazolidinone from niridazole in mouse intestine. *J. Pharmacol. Exp. Ther.*, **217**, 363–368.

Tracy, J. W., Fairchild, E. H., Lucas, S. V., and Webster, L. T. (1980). Isolation, characterisation and synthesis of an immunoregulatory metabolite of niridazole. *Mol. Pharmacol.*, **18**, 313–319.

Uno, T., Masada, M., Yamaoka, K., and Nakagawa, T. (1981). Hplc determination and pharmacokinetic investigation of aminopenicillins and their metabolites in man. *Chem. Pharm. Bull.*, **29**, 1957–1968.

Vaughan, D. P. and Tucker, G. F. (1987). Suprofen toxicity. *Pharmaceutical J.*, 9 May, 577–578.

Vickers, S., Duncan, C. A. H., Arison, B. H., Ramjit, H. G., Rosegay, A., Nutt, R. F. and Veber, D. F. (1983). Metabolism of MK-771 in gut and brain tissue of rats: the implications for its bioavailability. *Drug Metab. Disp.*, **11**, 147–151.

Vignier, V., Berthou, F., Dreano, Y., and Floch, H. H. (1985). Dibenzothiophene sulphoxidation: a new and fast hplc assay of mixed-function oxidation. *Xenobiotica*, **15**, 991–999.

Vinay, P., Paquin, J., Lemieux, G., Gougoux, A., and Bertrand, M. (1980). Metabolism of tienilic acid in man. *Eur. J. Drug Metab. Pharmocokinet.*, **5**, 113–125.

Walshe, N. D. A. (1979). Other sulphur systems. In P. G. Sammes (ed.), *Comprehensive Organic Chemistry*, Vol. 4, Pergamon, Oxford, pp. 845–852.

Waring, R. H. (1973). The metabolism of vitavax by rats and rabbits. *Xenobiotica*, **3**, 65–71.

Watkins, G. M., Whelpton, R., Buckley, D. G., and Curry, S. H. (1986). Chromato-graphic separation of thioridazine sulphoxide and N-oxide diastereoisomers: identification as metabolites in the rat. *J. Pharm. Pharmacol.*, **38**, 506–509.

Waxman, D. J., Light, D. R., and Walsh, C. (1982). Chiral sulphoxidations catalysed by rat liver cytochromes P-450. *Biochemistry*, **21**, 2499–2507.

Wendt, G. (1982). Pharmacokinetics and metabolism of tiapamil. *Cardiology*, **69** (Suppl. 1), 68–78.

10

Glucosinolates, alliins and cyclic disulphides: sulphur-containing secondary metabolites

G. R. Fenwick and A. B. Hanley
AFRC Institute of Food Research, Norwich Laboratory, Colney Lane, Norwich
NR4 7UA, UK

SUMMARY

1. This review describes the distribution and nature of sulphur-containing secondary metabolites in plants.
2. Emphasis is mainly on three examples of such compounds in foods and feeding stuffs; glucosinolates in brassicas, the amino acid derivatives of Allium species and the cyclic disulphides in asparagus.
3. The chemistry and biochemistry of such compounds, their physiological properties and those of their chemical or metabolic breakdown products is described.

10.1 INTRODUCTION

While not as ubiquitous as their oxygen- and nitrogen-containing counterparts, secondary metabolites containing sulphur embrace a considerable range of structural types and exhibit a broad spectrum of biological activity. In certain cases, most notably the alliums (onion, garlic) and crucifers (mustard, cabbage), these compounds are prized by consumers for their contribution to flavour and pungency and, therefore, are also of concern to food processors and technologists.

In comparison with nitrogen-containing metabolites, the extent of their study has been, until quite recently, rather limited. While significant advances in the identification, and synthesis, of many alkaloids occurred through the 'golden age' of natural product chemistry, 1870–1940, it is only really since that period that naturally occurring sulphur compounds have been actively studied. Thus, while at first it might seem that parallels do exist between the isolation, structural elucidation and synthesis of, say, strychnine (1818, 1946 and 1954, respectively) and of the glucosinolates of the crucifers (first isolated in crystalline form in 1830, the structure elucidated and synthesized in 1956), this is really not so. While the former had been the basis of extensive study in various laboratories, especially those of Leuchs and Wieland since before the end of the last century, with the noticeable exception of the detailed work of Gadamer, the latter occasioned relatively little interest (Fenwick *et al.*, 1983).

It is tempting to suggest reasons for these differences. Whereas alkaloids and other natural products often possessed very obvious biological properties (frequently being associated with acute toxicity) which attracted the interests of research groups closely associated with pharmaceutical companies, this was not generally so for the sulphur compounds. Indeed, it was exactly the identification of antibiotic activity in garlic extracts which led Cavallito and coworkers and Seebeck and Stoll during the 1940s into research programmes which eventually established the structures, biological activities and relationships between alliin, allicin and volatile sulphides (see below). These compounds, like the glucosinolates, were also readily available from 'common' plants, even weeds, which did not perhaps have the glamour of some of the more exotic oriental or equatorial alkaloid-containing species.

There are chemical reasons too which might be cited — the availability of many alkaloids as crystalline, and therefore readily purified, salts or derivatives, the development during the second half of the last century of many colour reactions and degradative processes (including structure-specific rearrangements and eliminations) which were primarily applicable to studies on alkaloids. In contrast to the alkaloids, both glucosinolates and alliins are readily broken down enzymically on tissue disruption and this introduced additional problems in handling, separation and purification. However, in the end, it was perhaps the social problem — the lingering, unpleasant and intensely strong aromas of onion, garlic or cabbage — which discouraged these very early studies. In the absence of fume cupboards, extractors, pumps and leak-proof equipment, the steam distillations and other chemical procedures which were a necessary part of researches into sulphur-containing natural products were viewed with considerable disfavour by individual scientists, their colleagues and, very probably, their families. Once such researches began, however, progress was rapid — thus in the period 1950–1965 over 70 different glucosinolates were identified and their chemical, physiological and biochemical properties determined.

The purpose of the present chapter is to describe and discuss the distribution and nature of sulphur-containing secondary metabolites in plants, with emphasis on three examples occurring in foods and feedingstuffs, the glucosinolates in brassicas, the amino acid derivatives of allium species and the cyclic disulphides in asparagus. The chemistry and biochemistry of such compounds, their physiological properties and those of their metabolic and breakdown products will be discussed.

10.2 GLUCOSINOLATES

10.2.1 Occurrence

It is now commonly accepted that the distribution of glucosinolates lies within certain families of dicotyledous angiosperms, predominantly within the order Capparales *sensu* Cronquist or Taktajan, embracing the Capparaceae, Cruciferae, Moringaceae, Resedaceae and Tovariaceae (Kjaer, 1974). The usefulness of glucosinolate composition as a chemotaxanomic index of botanical origins and relationships has been elegantly elaborated by, for example, Rodman (1978, 1981) and Al-Shehbaz (1985). In the context of human foods and animal feedingstuffs it is primarily members of the Cruciferae which are of interest. No authenticated member of this family has been found to be devoid of glucosinolates or (when examined) the associated enzyme, myrosinase. Claims have been made for the presence of glucosinolates in onions and cocoa beans (see Heaney and Fenwick, 1987) but more recent work has not supported such occurrences. The presence of trace amounts of glucosinolates in mushrooms has also been claimed, but in view of the above the authors are of the opinion that further evidence is needed to substantiate this.

10.2.2 Structure

The common structural skeleton for glucosinolates, now accepted following the studies of Ettlinger and Lundeen (1956), crystallographic analysis and total synthesis, is shown in Fig. 1. The sugar moiety is, with a single exception, glucose but, as techniques for isolation and chemical characterization develop, it would not be at all surprising if acylated analogues were to be identified. One such compound, having a sinapoyl residue attached to glucose (Linscheid et al., 1980) has already been found in radish seed and there have been claims that other, 'bound' forms of glucosinolates occur — however, these have not been substantiated.

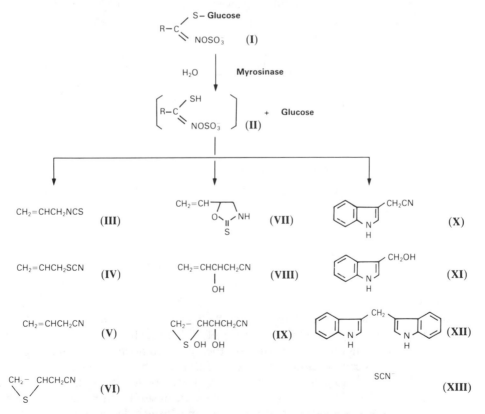

Fig. 1 — Glucosinolate structure and products of enzymic hydrolysis.

More than 100 different glucosinolates have currently been identified, possessing a range of different side chains (R, I, Fig. 1), comprising aliphatic, aromatic and indolic groups. The most common substituents are hydroxyl and terminal methythio (or oxidized analogues of) groups. The *anti* configuration between R and the sulphonated oxime moiety is of great importance in the enzymatic cleavage and rearrangement of glucosinolates, which is discussed below, where reference is made to many of the structures in Fig. 1.

10.2.3 Biosynthesis

Glucosinolates are derived from amino acids according to the pathway shown in Fig. 2. This involves N-hydroxylation, aldoxime formation by oxidative decarboxylation, oxidation to the nitro compound, insertion of sulphur, S-glucosylation and, finally, sulphonation. The evidence which has been assembled in support of such a pathway has been discussed in detail by Underhill and Wetter (1973) and Underhill (1980).

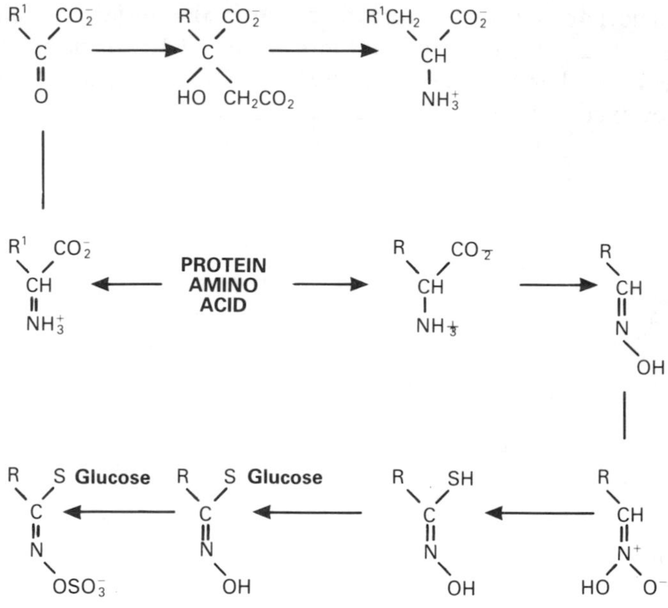

Fig. 2 — Biosynthesis of glucosinolates.

In some cases the reaction sequence proceeds directly and the amino acid is elaborated into the glucosinolate, so that the side chain is identical in both cases ($R=CH_3$ for alanine and glucocapparin; $R=p$-hydroxybenzyl for tyrosine and glucosinalbin), but in the majority of cases the initial amino acid side chain requires modification. This can occur at various points along the biosynthetic pathway; thus occasionally the amino acid itself may be modified so that both the non-protein amino acid and the corresponding glucosinolate are observed (for example, the co-occurrence of 2-amino-4-phenylbutyric acid and 2-phenylethyl glucosinolate in watercress). More generally, modification of the side chain occurs after the initial incorporation of the amino acid substrate. Glucosinolates containing terminal methylthio, methylsulphinyl or methylsulphonyl groups are derived from methionine, with subsequent chain elongation via the acetate pathway (and oxidation, if necessary). Elimination of CH_3SH from such compounds produces a family of terminally unsaturated glucosinolates, the lower members (prop-2-enyl, but-3-enyl, pent-4-enyl glucosinolates) of which are common amongst edible brassicas. The introduction of a β-hydroxyl group, forming glucosinolates which yield potent goitrogens on hydrolysis, is also important, although little is known about the biosynthetic nature of such insertions. Occasionally, intact glucosinolates may be further modified; thus labelling studies suggest that 3-indolylmethyl glucosinolate is converted to its 1-methoxy or 1-sulpho analogue in woad (*Isatis tinctoria*). A degree of lack of specificity in the enzymes involved in the biosynthesis of the indole glucosinolates is indicated in the observation that ring-halogenated tryptamines can be incorporated to form the analogous halogeno-indole glucosinolates (Goetz and Schraudolf, 1983).

As a result of considerable plant breeding activity, primarily centred on improving the rapeseed crop, there is some evidence that the biosynthesis of indole glucosinolates diverges from the more general pathway at a relatively early stage. Given the biological significance of these compounds and their ubiquity, it is surprising that this particular area has not been more fully explored.

10.2.4 Catabolism

Much of the significance of glucosinolates lies in the chemical properties and biological activities of their enzymically derived products. The enzyme responsible for catalysing the breakdown of glucosinolates, a thioglucoside glucohydrolase (EC 3.2.3.1), is commonly referred to as myrosinase. As far as is known all plants which contain glucosinolates also contain myrosinase, which may exist in a variety of isoenzymic forms. These differ in, for example, their molecular weights and in their requirement for ascorbate. Hence the myrosinase of *Wasabia japonica* (580 000 daltons) is much larger than that of mustard (120 000–150 000) or rapeseed (135 000). The enzymes appear to be glycoproteins, with the active site(s) containing an –SH grouping. Recently it has been shown that myrosinase also catalyses the breakdown of seleno-glucosinolates (Kjaer and Skrydstrup, 1987). In the intact tissue, enzyme and substrate are separated, although the exact nature of the compartmentalization is unclear. Disruption of cellular integrity is a prerequisite for enzymatic cleavage of glucosinolates, which therefore follows mechanical or other damage, fungal attack or food preparation. In passing, attention should be drawn to the fact that 'myrosinase activity' has also been found in bacteria, fungi and in the digestive tracts of animals and man (see Fenwick *et al.*, 1983). The latter may be of particular significance since it means that the glucosinolates in a variety of cooked or processed foods may be subject to breakdown *in vivo* even though the endogenous myrosinase has been inactivated during the earlier stages of food preparation.

The function of the enzyme is to cleave the C–S bond on the distal side of the side chain R, thus generating the aglycone (**II**, Fig. 1). This product is unstable and reacts further to produce a variety of secondary chemical compounds (Fig. 1), the exact nature and amounts of which depend on such factors as the structure of the side chain of the original glucosinolate, the conditions of the hydrolysis (e.g. temperature, pH) and the presence of cofactors. Amongst the latter the most important are ferrous ion and an inactive polypeptide, epithiospecifier protein (ESP).

The pungent flavours and physiological activities of the products of myrosinase activity have attracted study for many years (MacLeod, 1976) but, despite this, many questions about the product distribution remain unanswered. The aglycone, referred to above, is considered to expel a sulphate grouping associated with either rearrange- ments or desulphation process. Under normal conditions, pH 6–7, the former is greatly preferred in most cases. Indeed the rearrangement process (analogous to the Lossen rearrangement in organic chemistry) has particular structural requirements which are fulfilled in glucosinolates, i.e. the *anti* conformation of leaving group and migrating group (R, structure **I**). In support of this, desulphoglucosinolates (which may be readily produced by the action of sulphatase on the parent glucosinolate) are much more slowly broken down by myrosinase, if at all.

For simplicity, the authors have found it convenient to consider glucosinolates in three groups, according to the chemical structures of their main enzyme-induced breakdown products. The first group, which is also by far the largest, comprises those glucosinolates yielding isothiocyanates (**III**), commonly characterized by their pungency and antibiotic properties. While isothiocyanates are the major products at pH 5–7, below pH 3–4 it is the nitriles which become increasingly important (see below). This trend is also noted amongst members of the second group, which possess a β- (or less commonly γ-) hydroxyl group in the side chain. Such substituted isothiocyanates are unstable and cyclize, producing in the former case substituted oxazolidine-2-thiones (**VII**); at lower pH hydroxynitriles (**VIII**) are observed. A small number, currently six, of indole glucosinolates constitute the final group, which at pH 6–7 produced indole-3-carbinols (e.g. **XI**) and thiocyanate ion (**XIII**). The former may react further to yield diindolylmethane (**XII**) or ascorbigen (**XIV**). Indole-3-acetonitriles (e.g. **X**) are the main products obtained at pH 3–4.

(**XIV**)

It has, however, long been recognized that substantial amounts of nitriles are produced during the autolysis of cruciferous plant tissue, even under conditions when the pH would be expected to favour isothiocyanates (see Tookey *et al.*, 1980).

Recent studies have cast further light on this situation (Uda *et al.*, 1986a,b). Thus while ferrous ion significantly reduced myrosinase-induced isothiocyanate formation under acid conditions it had little effect at pH 7.5. This was taken as evidence that the ferrous ion was interacting with the aglycone, rather than with the parent glucosinolate, since the rate of release of glucose was found to be independent of ferrous ion concentration. Furthermore, other studies indicated that a combination of ferrous ion and thiols inhibited isothiocyanate formation of neutral pH. These findings led to the suggestion that, together, ferrous ion and thiols facilitate desulphation of the aglycone (to produce nitriles) at the expense of the isothiocyanate-yielding Lossen rearrangement. The involvement of thiols in this process may, at least partially, explain the reduction in nitrile production which is known to follow the heating of cruciferous tissue or extracts.

Ferrous ion is also a necessary cofactor, in combination with ESP, for the myrosinase-induced addition of sulphur across a terminal double bond to yield the episulphides (e.g. **VI**, **IX**, Fig. 1) (Tookey *et al.*, 1980). Not all cruciferous plants contain ESP and, in the absence of this, nitriles are produced as described above. It has been suggested that ESP acts as a non-competitive inhibitor of myrosinase (Petroski and Kwolek, 1985), while the sulphur is transferred across the terminal, unsaturated, site by a substantially intramolecular process (Brocker and Benn, 1983).

The formation of organic thiocyanates, rather than isothiocyanates, is known to occur in certain species. The factors affecting this product formation are little understood, although it appears to be competitive with the Lossen rearrangement; thus low temperatures and the absence of ascorbate (both of which slow down the enzymic hydrolysis) facilitate thiocyanate formation. It has been suggested that a Z→E isomerase may be operating and it is from the E form of the glucosinolate aglucose that thiocyanates are formed. The investigation of synthetic E glucosinolates might thus be expected to shed light on this proposal.

10.2.5 Levels and intakes

While there are many reports of the levels of glucosinolates in animal feedingstuffs such as rapeseed, their study in other plants, especially those utilized by animals and man, are much fewer. Before such levels, and the factors affecting their variation, can be considered, brief mention should be made of the way in which glucosinolates are measured.

Historically glucosinolates were first measured indirectly, i.e. as their products following myrosinase treatment. The main reason was the concentration in early work on mustard where such treatment released volatile isothiocyanates which could be determined by gravimetry, titration and much later by gas chromatography (GC). With the greatly increased interest in rapeseed, it was realized that other products, notably oxazolidine-2-thiones, needed to be measured if a reliable figure for total content was to be obtained; even then for a variety of reasons (discussed in detail by McGregor *et al.*, 1983) such methods underestimated the total figure. However, while the side-chain containing fragments differ greatly in their chemical composition, reactivity and chromatographic characteristics, the other products, i.e. glucose and sulphate, are stable and, more significantly, are considered to be produced stoichiometrically from the original glucosinolate molecule. While both have been used to determine total glucosinolate content, the availability of clinical glucose-detecting kits has resulted in enzymic glucose release being the most popular and widely used index of glucosinolate content.

Although valuable, such a method does not provide any information as to the nature of glucosinolates present, which may vary greatly depending on plant species, cultivar, part examined, maturity, etc. Consequently chromatographic methods using GC (both conventional and capillary) and HPLC have been developed in recent years and applied to the analysis of both oilseeds and other edible crucifers (Daun, 1986).

While the present authors, amongst others, have argued for systematic nomenclature to be used in glucosinolate research, many glucosinolates retain their original trivial names and are referred to as such in the literature. These early names were derived from the host botanical species, e.g. glucotropaeolin (benzyl glucosinolate) from *Tropaelous* spp. and sinigrin (prop-2-enyl glucosinolate) from *Sinapis nigra*. Indeed it was the latter compound, first isolated in 1830, that gave the family its name, glucosinolate. Table 1 thus contains both trivial and systematic nomenclature. Levels of total and major individual glucosinolates in a range of animal feedingstuffs and human foods shown in Tables 2 and 3.

Table 1 — Trivial nomenclature and structures of main glucosinolates occurring in edible plants

Prop-2-enyl	Sinigrin
But-3-enyl	Gluconapin
Pent-4-enyl	Glucobrassicanapin
2-hydroxybut-3-enyl	Progoitrin, *epi*-progoitrin
2-hydroxypent-4-enyl	Gluconapoleiferin
3-methylthiopropyl	Glucoiberverin
4-methylthiobutyl	Glucoerucin
5-methylthiopentyl	Glucoberteroin
3-methylsulphinylpropyl	Glucoiberin
4-methylsulphinylbutyl	Glucoraphanin
4-methylsulphinylbut-3-enyl	Glucoraphenin
3-methylsulphonylpropyl	Glucocheirolin
4-methylsulphonylbutyl	Glucoerysolin
Benzyl	Glucotropaeolin
2-phenylethyl	Gluconasturtiin
4-hydroxybenzyl	Glucosinalbin
3-indolylmethyl	Glucobrassicin
1-methoxy-3-indolylmethyl	Neoglucobrassicin
4-methoxy-3-indolylmethyl	—
1-hydroxy-3-indolylmethyl	—

Table 2 — Total glucosinolate content $(mg(100\ g)^{-1})$ of agriculturally important plants

Species	Glucosinolate content		Reference
	Range	Mean	
Cabbage			
White	26–106	53	VanEtten *et al.* (1980)
	42–156		Sones *et al.* (1984b)
Red	41–109	76	VanEtten *et al.* (1980)
Savoy	47–124	77	VanEtten *et al.* (1980)
	121–296		Sones *et al.* (1984b)
Chinese cabbage	17–1326	54	Daxenbichler *et al.* (1979)
Pe-tsai	10– 34	20	Lewis and Fenwick (1988)
Pak-choi	39– 70	51	Lewis and Fenwick (1988)
Brussels sprouts	60–390	200	Heaney and Fenwick (1980)
Cauliflower	61–114		Sones *et al.* (1984c)
Calabrese	42– 95	62	Lewis and Fenwick (1987)
Swede (peeled)	20–109	55	Mullin *et al.* (1980)
	113–231		Carlson *et al.* (1981)
Turnip (peeled)	21– 60	42	Mullin *et al.* (1980)
	97–227		Carlson *et al.* (1981)
Radish			
European	34– 57	45	Carlson *et al.* (1985)
European–American red	42–117	68	Carlson *et al.* (1985)
European–American white	57–119	77	Carlson *et al.* (1985)
Japanese daikon	66–253	139	Carlson *et al.* (1985)
Horseradish	332–354		Kojima *et al.* (1986)
Mustard			
White	220–520		Hälva *et al.* (1986)
Brown	<440		Hälva *et al.* (1986)
Black	180–450		Hälva *et al.* (1986)
Rapeseed			
High glucosinolate	>420		Röbbelen and Thies (1980)
Intermediate glucosino-late		250	
Low glucosinolate		130	

Table 3 — Content of major individual glucosinolates (mg (100 g)$^{-1}$) in agriculturally important crops

	A	B	C	D	E	F	G	H	I	J
White cabbage		4–59 7–41				5–27 2–89		0–14		
Red cabbage		1–10	2–5			2–14		15–39		
Savoy cabbage		0–16 13–65	2–6 3–50			7–42 33–129				
Brussels sprouts		1–156								
Cauliflower		6–63			0–30	1–42				
Calabrese					0–7		15–27		1–33	
Swede	0–24					4–38		1–14		2–22
Turnip	0–18		1–50	1–25						
Radish										
European									18–34	
European–American									29–107	
Japanese									50–223	
Chinese cabbage			0–25	1–28						
Horseradish		2700–2900	3300–6900							
Mustard										
White		32–800		0–32						
Brown		1800–4500								
Black										
Rapeseed										
High			880–960	220–270						
Low			50–70	20–30						

Table 3 – *Continued*

	K	L	M	N	O	P	Q	R	S
White cabbage						5–51[a]			
Red cabbage						1–8			
Savoy cabbage						16–33[a]			
Brussels sprouts						30–53[a]			
Cauliflower						2–13			
Calabrese		2–8				4.99			
Swede	0–11	22–67			1–47	0–10	27–49		
Turnip		0–45	0–19		3–55	19–36	7–57[a]		
Radish									
European							6–56[a]		
European–American									
Japanese	1–26	1–19							
Chinese cabbage					2–32			10–55[a]	
Horseradish					420–720				
Mustard									
White				2200–5200					
Brown									
Black									
Rapeseed									
High		2600–2800	130–160			<10			160–190
Low		120–170	<10			<20			120–1600

A, 1-methylpropyl glucosinolate; B, prop-2-enyl glucosinolate; C, but-3-enyl-glucosinolate; D, pent-4-enyl glucosinolate; E, 3-methylthiopropyl glucosinolate; F, 3-methylsulphinylpropyl glucosinolate; G, 4-methylthiobutyl glucosinolate; H, 4-methylsulphinylbutyl glucosinolate; I, 4-methylsulphinylbut-3-enyl glucosinolate; J, 5-methylthiopentyl glucosinolate; K, 5-methylsulphinylpentyl glucosinolate; L, 2-hydroxybut-3-enyl glucosinolate; M, 2-hydroxypent-4-enyl glucosinolate; N, 4-hydroxybenzyl glucosinolate; O, 2-phenylethyl glucosinolate; P, 3-indolylmethyl glucosinolate; Q, 1-methoxy-3-indolymethyl glucosinolate; S, 4-hydroxy-3-indolmethyl glucosinolate. References as for table 2.

[a] Indole glucosinolates (P–S) analysed as a group.

A variety of factors influence the glucosinolate content of a plant (Fenwick *et al.*, 1983), e.g. its genetic origin, ontogenic state and the part of the plant examined. Levels are also influenced by the environment, soil type, stress (e.g. lack of moisture, close plant spacing) and agronomic factors (fertiliser application).

After harvesting, a number of additional factors affect the glucosinolate content of the processed feedingstuff or food, notably the degree of tissue disruption, the extent of any heating processing (especially when combined with water) and chemical or physical treatment. Cooking, whether domestic or industrial, leaches out glucosinolates and initially causes cells to break down and glucosinolates to come into contact with myrosinase but thereafter inactivates this enzyme.

In recent years there have been a number of studies calculating the human intake of glucosinolates. All are subject to certain criticisms but the important fact is the nature of the figure obtained, rather than its exact value. Fenwick *et al.* (1987) have recently examined the available date and found likely mean intakes in North America and Europe to be approximately 15 mg person^{-1} day^{-1}. In contrast the UK figure is 46 mg person^{-1} day^{-1} (Sones *et al.*, 1984a), as a result of both the high intakes of cruciferous vegetables generally and that of glucosinolate-rich Brussels sprouts in particular (Table 4).

Table 4 — Mean daily intakes of intact glucosinolates (mg person^{-1} day^{-1})

	UC		Canada	USA[a]	The Netherlands
Cabbage	19.4	(14.0)[b]	3.4	8.5[c]	nd
Brussels sprouts	17.2	(9.4)	1.5	0.6[d]	(6.5)
Cauliflower	6.4	(4.4)	1.2	2.7[e]	(1.2)
Broccoli			1.4	3.3[f]	nd
Swede/turnip	3.1	(1.6)	5.8		nd
Total	46.1	(29.4)	13.5	15.1	(7.7)

UK data from Sones *et al.* (1984a; Canadian data adapted from Mullin and Sahasrabudhe (1978); USA data calculated from relevant food tables; Dutch data from Gramberg *et al.* (1986).
[a]Calculated.
[b]Values in parentheses refer to calculations based on cooked vegetable data.
[c]Including 1.0 mg from canned sauerkraut.
[d]Including 0.5 mg from frozen Brussels sprouts.
[e]Including 0.6 mg from frozen cauliflower.
[f]Including 1.4 mg from frozen broccoli.

When dietary, agronomic and cultural variations are taken into account, it has been estimated that 5% of the UK population (95% percentile) may regularly consume in excess of 300 mg person^{-1} day^{-1}. This, it should be reminded, relates to chemicals (albeit natural) whose long-term physiological and pharmacological effects in man have yet to be acertained.

10.2.6 Metabolism

Relatively little is known about the metabolism of glucosinolates in either animals or man. In early studies, it was reported that 2-hydroxybut-3-enyl glucosinolate was converted, in part, to 5-vinyloxazolidine-2-thione in both the rat and man, and bacteria were isolated from the human digestive tract which possessed the ability to bring about this change (Langer and Greer, 1977). Other studies, referred to below, have been interpreted as indicating that the breakdown with glucosinolates *in vivo*, at least partially, parallels that obtained with plant myrosinase.

Recently both Macholz *et al.* (1986) and Nugen-Boudon and Szylit (1987) have observed reduced breakdown of intact glucosinolates in germ-free, as compared with normal, rats and chickens. While some 5-vinyloxazolidine-2-thione was formed from 2-hydroxybut-3-enyl glucosinolate, the detailed investigation of product nature and distribution remains to be undertaken.

Incubation of the same glucosinolate with rumen liquor produced a series of products, including the above oxazolidine-2-thione, 5-vinylthiazolidine-2-one and polymeric material; these products were also found when the 5-oxazolidine-2-thione itself was incubated in the same manner (Lanzani *et al.*, 1974).

5-vinyloxazolidine-2-thione has been shown to yield the 4-hydroxy metabolite in rats (Langer and Michajlovskij, 1969). In common with other thioamides, this compound retained the antithyroid activity of the parent compound. As a result of detailed studies, Görler and coworkers (Brüsewitz *et al.*, 1977; Görler *et al.*, 1982a,b; Mennicke *et al.*, 1983) have shown species-dependent differences for the breakdown of prop-2-enyl and benzyl isothiocyanates; the main metabolic products of the latter were the hippuric acid (in the dog), 4-hydroxy-4-carboxy-3-benzylthiazolidine-2-thione (guinea pig, rabbit) and N-acetyl-S-(N-benzylthiocarbamoyl)-L-cysteine (rat, man). Given the large glucosinolate intakes in certain human populations, a more detailed understanding of the metabolism of glucosinolates and their products is urgently required.

10.2.7 Biological effects

To facilitate understanding, the biological effects of glucosinolates and their breakdown products will be considered separately, with further distinctions being made in the latter between isothiocyanates, oxazolidine-2-thiones and thiocyanate ion. In addition, nitriles and indoles will also be discussed since, although not sulphur containing, they are derived from glucosinolates and possess biological properties which are currently attracting considerable attention.

10.2.7.1 Intact glucosinolates

VanEtten *et al.* (1969) have examined the effect of 2-hydroxybut-3-enyl glucosinolate in rats. When levels corresponding to 5, 8.5, 15 and 26 g kg^{-1} diet were fed for 90 days, decreased growth occurred, the animals receiving the highest levels not surviving the treatment. Kidneys, liver and thyroid were all enlarged and showed pathological lesions. The feeding of the same glucosinolate to laying hens caused inhibition of hepatic trimethylamine oxidase (and associated tainted egg production), while this effect was absent in *in vivo* studies using liver homogenates (Butler and Fenwick, 1984).

Both the short- and longer-term effects of feeding a variety of intact glucosinolates to rats have been examined. Bille *et al.* (1983) fed four glucosinolates, at three levels (0.2, 1 and 5 g kg^{-1}), for 5 days. The diets were compared with and without the addition of myrosinase. At the highest level the glucosinolates caused palatability problems, reduced protein utilization and affected the size of internal organs, irrespective of whether the enzyme was present. Prop-2-enyl glucosinolate increased the weights of kidneys and, especially in the presence of myrosinase, decreased the size of the suprarenal glands. 2-Hydroxybut-3-enyl and 2-hydroxy-2-phenylethyl glucosinolates increased the size of both liver and kidneys while, the latter compound, in the presence of myrosinase, also decreased the weight of testicles and suprarenal glands. 4-Hydroxybenzyl glucosinolate increased kidney weight and, together with myrosinase, decreased the size of the suprarenal glands. When fed at the highest levels, all four glucosinolates caused doubling or trebling of the size of the thyroid gland.

In a recent study Vermorel *et al.* (1986) examined the effects of feeding six glucosinolates to rats for 29 days. Inclusion rates ranged from 0.5 to 3 g kg^{-1} diet and no adverse effects on feed intake of growth rate were observed. The feeding of 2-hydroxybut-3-enyl glucosinolate (3 g kg^{-1} diet) caused increases in the weights of liver and thyroid of 15% and 28% respectively, but the latter was not reflected in a reduction in thyroid hormone levels.

When the effects of 2-hydroxybut-3-enyl glucosinolate, as an isolated product or present in rapeseed meal, were compared with and without added myrosinase, it was apparent that thyroid metabolism was adversely affected (Vermorel *et al.*, 1988). However, the effect of the isolated glucosinolate (increasing thyroid size by 38% in the presence of myrosinase) was much less than that of the rapeseed (415% increase, with myrosinase). These findings, together with the observed effects on feed intake, weight gain and kidney and liver weights suggested that the extent and nature of glucosinolate breakdown was markedly affected by other dietary components.

10.2.7.2   *Individual hydrolysis products*

Isothiocyanates

The biological activities of isothiocyanates have been known for many centuries. Indeed, the earliest reasons for cultivating many brassicas were medicinal rather than culinary (Fenwick *et al.*, 1983). Extracts, decoctions and juices — all rich in isothiocyanates and, hence, pungent — were readily applied both internally and externally, the mustard poultice being a well-known example of the latter.

Volatile isothiocyanates exhibit a range of antibacterial, antifungal and antimicrobial activities. Most commonly the effect of the readily available prop-2-enyl isothiocyanate has been examined although, partly as a result of its volatility, it is not necessarily the most active. In most cases, the activity of the isothiocyanates is ascribed to their ready chemical binding to proteins, enzymes and biologically active –SH groups. The volatile oils of a number of crucifers have natural insecticidal activity and since Lichtenstein *et al.* (1964) have identified 2-phenylethyl isothiocyanate as a potent insecticide it is likely that other isothiocyanates behave similarly. Isothiocyanates are generally considered to be responsible for the high incidence of irritant and allergenic contact dermatitis experienced amongst vegetable handlers (Mitchell, 1974; Mitchell and Jordan, 1974).

Reports about the mutagenicity of prop-2-enyl isothiocyanate are contradictory. Thus Yamaguchi (1980) found this to be the most active of five isothiocyanates tested against *S. typhimurium* TAS100, but it possessed only weak activity against TA98 and was without effect in TA1535, 1636, 1537 and 1538 systems. In contrast, Eder *et al.* (1980) found the compound to be inactive when tested against *E. coli* WP2, *B. subtilis* H17 and *S. typhimurium* TA100, with and without metabolic activation. However, according to Neudecker and Henschler (1985), prop-2-enyl isothiocyanate was mutagenic in the TA100 system provided that sufficiently long pre-incubation periods were employed. It was suggested that this was necessary for the isothiocyanate to be metabolized to an active mutagen. A number of isothiocyanates are cytotoxic and while there is evidence for isothiocyanates' reducing foetal weights and causing embryonal death, there is no evidence that they are teratogenic. The only significant finding following a 2 year study of prop-2-enyl isothiocyanate in mice and rats was an increase in transition cell papillomas in the urinary tracts of male rats which were later related to a metabolite identified as N-acetyl-S-(N-(prop-2-enyl)-thiocarbamoyl)-L-cysteine (section 10.2.6 above). The LD_{50} figures for isothiocyanates have been summarized by Fenwick *et al.* (1983).

Isothiocyanates, possibly in combination with intact glucosinolates, have been identified as feeding and ovipositing stimuli in a number of insect species and as feeding deterrents in others. The relationship between insect and host plant has been discussed elsewhere (see Fenwick *et al.*, 1983).

Oxazolidine-2-thione

The short-term feeding of this compound to rats led to liver enlargement, associated with an increased number of hepatocytes. When fed to rats for 90 days at 2 g kg^{-1} diet, mild hyperplasia and a 15% reduction in body weight were observed. Chang and Bjeldanes (1985) have found that a significant increase in liver weight results from the feeding of 200 mg 5-vinyloxazolidine-2-thione kg^{-1} diet to rats for 14 days. An increase in the activity of intestinal and hepatic glutathione-S-tranferase and microsomal epoxide hydratase was also observed. Despite structural similarities to known teratogens, 5-vinyloxazolidine-2-thione has been found to be inactive in this respect (Khera, 1977). The major dietary significance of oxazolidine-2-thiones is their interference with the synthesis of thyroid hormones, which is a consequence of their thioamide structure. This antithyroid effect is not alleviated by iodide supplementation of the diet, in comparison with that of thiocyanate ion (see below). A significant increase in the thyroid weight of rats was observed following the daily administration of 40 μg 5-vinyloxazolidine-2-thione for 20 days (Langer and Michajlovskij, 1969). The same authors reported that the lowest single doses required to suppress radioiodine synthesis and radioiodine uptake were 20 and 80 μg respectively. Later Elfving (1980) found thyroxine synthesis in the rat to be inhibited by a daily dose of 1 μg kg^{-1} body weight while goitre could be clearly defined following the feeding of five times this level for 21 days.

In man a single dose of 50–200 mg 5-vinyloxaxolidine-2-thione caused inhibition of radioiodine uptake. According to Langer and Greer (1977), natural oxazolidine-2-thiones from rapeseed and *Crambe* possessed similar antithyroid activities, being

33% more active then propylthiouracil in man. A more recent investigation (McMillan *et al.*, 1986) found no change in sensitive indices of thyroid function (T_3, T_4 or TSH levels) in human volunteers during or following the daily ingestion of 100–150 g cooked Brussels sprouts for 28 days. This intake of the cooked vegetable corresponds to an equivalent of 46 mg 5-vinyloxazolidine-2-thione per day.

Enlargement of the thyroid gland is associated with the feeding of rapeseed meal to experimental animals and livestock. Unless the rations fed are marginal or deficient in iodine, when the effect of thiocyanate ion becomes significant, oxazolidine-2-thiones are considered to be the causative agents. These compounds are able to cross the placenta and are strong foetal goitrogens. They may accumulate in milk by diffusion processes, although the levels found and the claimed biological consequences are disputed. Kreula and Kiesvaara (1959) reported levels of 50–100 μg 5-vinyloxazolidine-2-thione l^{-1} in milk from areas of endemic goitre in Finland and, when such milk was fed to rats, clear evidence of thyroidal disturbance was noted. Recently Swiss workers (Bachmann *et al.*, 1985) have used HPLC techniques to measure the levels of this goitrogen in milk following the feeding of 4,19 and 39 g rapeseed meal kg^{-1} diet to cattle. Levels of 40–700 μg l^{-1} were measured, corresponding to a transfer of approximately 0.1% of the parent glucosinolate. Time course studies revealed that the level of this goitrogen dropped below the detection limit (7 μg l^{-1}) 12 h after the removal of the rapeseed from the diet. In view of the concerns expressed over the transfer of such goitrogens by indirect dietary exposure it is worth emphasizing that the potential levels of direct exposure due to cruciferous vegetable consumption in the UK are 50 mg (300 mg for the 95% percentile group).

Thiocyanate ion

This is now known to be the factor responsible for cabbage goitre. The origins of dietary thiocyanate include indole glucosinolates, isothiocyanates and nitriles. Since all three are found in brassicas, it is unsurprising that the feeding of seed meals, roots and leafy vegetable and forages has been shown to increase plasma thiocyanate levels and, when dietary iodine is limiting, to cause goitre. Increased thiocyanate ion concentrations have been found in the plasma and urine of pigs (Schöne and Paetzelt, 1985; Jahreis *et al.*, 1986), the milk of cattle (Laarveld *et al.*, 1981) and the urine of rats (Paik *et al.*, 1980) following the feeding of rapeseed meal. Astwood (1943) has reported that the feeding of potassium thiocyanate (500 mg daily) to patients suffering from hypertension increased plasma thiocyanate levels (500–100 mg l^{-1}) and inhibited radioiodine uptake. In contrast to the potency of oxazolidine-2-thiones, thiocyanate has much less activity, a single dose of 200–1000 mg being necessary to inhibit radioiodine uptake in man. In areas of Jugoslavia and Czechoslovakia, dietary brassicas contain sufficient thiocyanate ion and oxazolidine-2-thione precursors to increase significantly the incidence of iodine-deficient goitre. According to Matovinovic (1983), iodine deficiency was the main cause of goitre in only one of three goitrous communities in Sicily. In the other two the combination of thiocyanate (thought to originate from cabbage) and the lower than normal iodine levels was cited as the cause.

Recently Dahlberg *et al.* (1984) have examined the effect on thyroid function of prolonged exposure to thiocyanate ion (8 mg daily in milk for 84 days). No significant changes in thyroxine, triidothyronine or thyrotropic hormone levels were observed in subjects of normal iodine status. The results of a second study (Dahlberg *et al.*, 1985) revealed that no adverse effects on thyroid function in subjects from Western Sudan, where goitre is endemic, occurred following the daily administration of 4.75 mg thiocyanate ion for 28 days.

Nitriles

These products of glucosinolates are the most toxic of the common breakdown products, and as such have been studied in some detail (Tookey *et al.*, 1980). The feeding of 1-cyano-2-hydroxy-3,4-thiobutane (at up to 300 mg kg^{-1} diet) to rats for 90 days led to significant decreases in organ weights (Gould *et al.*, 1980). Dose-dependent lesions in the liver, pancreas and, especially, kidney were also noted, with renal lesions characterized by tubular epithelial karyomegaly. Further studies involved the feeding of the same compound to rats at 50 mg kg^{-1} body weight for up to 3 days (Gould *et al.*, 1985). The authors observed the result to be acute renal failure, typified by degeneration and necrosis of proximal tubular cells with preferential involvement of the pars recta.

In view of these detailed studies it is tempting to link glucosinolate-derived nitriles to the problem of liver haemorrhage which has been reported to occur in certain strains of laying hen fed rapeseed meal (Fenwick *et al.*, 1983). Recent studies are not completely in support of this, however.

While not teratogenic, 1-cyano-3,4-epithiobutane did cause decreased foetal weights and embryonal death in rats (Nishie and Daxenbichler, 1980). The lower homologue, 1-cyano-2,3-epithiopropane, proved to be a weak mutagen when screened against *S. typhimurium* TA100 and 1535 (Lüthy *et al.*, 1980).

Papas *et al.* (1979) have reported hydroxynitriles (2.9 mg l^{-1}) in milk following the feeding of low glucosinolate rapeseed meal (250 g kg^{-1} diet). In view of the toxicity of such compounds and the current use in the UK of rapeseed meal containing much higher levels of glucosinolates as a cattle feedstock, it would seem advantageous to re-examine this area. The acid conditions associated with certain food products, e.g. cole-slaw, pickled cabbage and sauerkraut, would be expected to facilitate nitrile, rather than oxazolidine-2-thione, production (Fenwick *et al.*, 1983). Such compounds may, however, undergo subsequent chemical reaction with the food matrix, or be broken down by the acid, hydrolytic conditions (section 10.2.4).

Indole products

Indole glucosinolates were amongst the most recent to be reported and their chemical and biological properties, and those of their products, have recently been reviewed (McDanell *et al.*, 1988). Products derived from indole glucosinolates have been shown to inhibit germination and recently Mithen *et al.* (1986) have found such products to have antifungal activity against the stem canker pathogen *Leptosphaeria maculans*. Indole glucosinolates are amongst the factors implicated in clubroot disease of brassicas. Butcher *et al.* (1984) have demonstrated that the development of 'clubbing' is associated with a large increase in both the synthesis and the degradation of indole glucosinolates. In agreement with this finding, Chong *et al.* (1985) have recently analysed 145 samples of cabbage and observed those possessing clubroot resistance to have low levels of indole glucosinolates.

A number of studies have demonstrated that cruciferous vegetables exert a protective effect against a variety of carcinogens. This has been attributed to the induction of glutathione-S-transferase, epoxide hydratase and a number of mono-oxygenases (e.g. aryl hydrocarbon hydroxylase). Indole glucosinolate products, notably indole-3-carbinol, indole-3-acetonitrile and diindolymethane, have been shown to possess such activity, although the presence of other active species cannot be discounted. Pantuck *et al.* (1979) demonstrated that consumption of Brussels sprouts or cabbage had a significant effect on the metabolism of drugs, such as antipyrin and phenacetin, in humans and suggested that diet could have an important effect on the variability of drug response. Wattenberg (1977) has shown that naturally occurring aromatic (e.g. benzyl, 2-phenylethyl) isothiocyanates possess a similar anticarcinogenic effect.

Recently Japanese workers (Wakabayashi *et al.*, 1985a) have reported that Chinese cabbage, after treatment with nitrite, shows significant direct-acting mutagenicity. Further work revealed the active principles to include stable indole nitrosamines (Wakabayashi *et al.*, 1985b,. 1986). A similar situation has been reported to result when nitrite and 5-vinyloxazolidine-2-thione are reacted under stomach conditions, the product, $N(1)$-nitroso-5-vinyloxazolidine-2-thione being a direct-acting mutagen with a specificity for base-pair substitution (Lüthy, 1984; Lüthy *et al.*, 1984). In both cases *in vivo* studies on the the extent of formation and metabolism of the products are necessary before a reliable risk assessment can be made.

10.3 ALLIIN, ALLICIN AND RELATED COMPOUNDS

10.3.1 Occurrence

Alliins (S-alk(en)ylcysteine sulphoxides) and their enzymic products, the allicins (alk(en)ylthiosulphinates), are found in the genus *Allium* (Liliaceae, subf. Allioidae). This comprises more than 600 different species found throughout North America, Europe, North Africa and Asia. Onion, garlic and leek are the most important cultivated alliums, others being listed in Table 5. In 1984 (Food and Agricultural Organisation, 1985) the production of onions was 23.1 Mt, the main production being centred in Asia (47%) and Europe, including the USSR (30%). Garlic production was about one-tenth of this figure (2.8 Mt), with Asia again predominating (65%). In addition to these cultivated species, there are a relatively small number of alliums which, according to various sources (Fenwick and Hanley, 1985a), have been used by man as foods. These are also listed in Table 5.

Most, but not all, *Allium* species when cut or bruised produce the pungent odours typical of onion and garlic, which are due to allicins and their sulphur-containing products (see below); hence this sensory characteristic has only limited usefulness for botanical classification. This is the more so since similar pungent odours have been reported from species of *Ipheion, Adenocalymma, Androstephium, Hesperocallis, Talbaghia, Nectarosiordum, Nilula* and possibly *Descurainia*. However, while in some cases the volatile products responsible for these odours have been identified and shown to be identical with those from certain alliums, their biosynthetic origin is largely unknown.

Table 5 — Allium species consumed by man

Cultivated

A.cepa L. var. *cepa*	Common onion
A.cepa L. var. *aggregatum*	Potato (multiplier) onion
A.cepa L. var. *ascalonium*	Shallot
A.cepa L. var. *proliferum*	Tree (Egyptian) onion
(syn. *A.cepa* L. var. *viviparum*)	
A. schoenoprasum L.	Chive
A. fistulosum L.	Japanese bunching (Welsh) onion
A. chinense G. Don (syn. *A. bakeri* Regel)	Rakkyo, Ch'iao t'ou
A. sativum L.	Garlic
A. ampeloprasum L. var. *porrum*	Leek
A. ampeloprasum L. var. *kurrat*	Kurrat
A. ampeloprasum L. var. *holmense*	Great headed garlic
A. tuberosum Rottl. ex. Spreng.	Chinese chive, Nira, Kui t'sai

Non-cultivated

A. akaka Gmel.	
A. angulosum L.	Mouse garlic
A. canadense L.	Wild garlic
A. cernuum Roth	Wild onion
A. grayi Regal	Nobiru
A. ledebourianum Schult.	
A. macleanii Baker	Royal salep
A. neopolitanum Cyr.	Daffodil garlic
A. nipponicum French and Sav.	
A. scorodoprasum L.	Rocambole, sand leek
A. senescens L.	
A. sphaerocephalum L.	Ballhead onion
A. splendens Willd.	
A. stellatum Fras.	
A. tricoccum Ait.	Wild leek
A. ursinum L.	Bear's garlic
A. victorialis L.	Long root onion
A. vineale L.	Crow garlic

10.3.2 Structure

It is no overstatement to describe the discovery of $(+)$-S-(prop-2-enyl)-L-cysteine sulphoxide (**XVa**, the original alliin isolated from garlic), its subsequent chemical characterization and synthesis, and the unravelling of its subsequent enzymically induced breakdown, as one of the triumphs of classical natural product chemistry (Stoll and Seebeck, 1947, 1948, 1949, 1951). When isolated and investigated, in the late 1940s, alliin was the first example of a natural product having optically active centres at C and S. Thereafter three other analogous compounds (**XVb–d**) were identified, their occurrence being predicted as a result of chromatographic analysis of the products of their reaction with thiamine. The alk(en)ylcysteine sulphoxides are the immediate precursors of the thiosulphinate allicins (**XVI**), regarded as the main biologically active sulphur compounds in onion, garlic, etc., and also important for their flavour characteristics. Since both symmetrical (**XVI**,R=R′) and asymmetrical (R≠R′) thiosulphinates may be formed from species possessing more than one substituted cysteine sulphoxide, only the general formula will be given here. Recently the presence of volatiles containing the $-S(CH_2)_4CH_3$ moiety has been reported in chives (Hashimoto *et al.*, 1983) and *A. fistulosum* (Kameoka *et al.*, 1984) and while other biosynthetic origins for this C_5 unit cannot be discounted it seems possible that hitherto unsuspected *S*-pentyl cysteine sulphoxide may occur in these species.

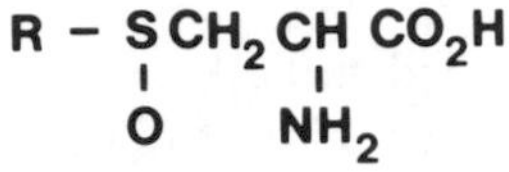

$$R - S\,CH_2\,CH\,CO_2H$$
$$\quad\;\; |\qquad\; |$$
$$\quad\;\; O\qquad NH_2$$

(**XV**)

$$R = CH_2 = CH\,CH_2-$$
$$R = CH_3 -$$
$$R = CH_3\,CH_2\,CH_2 -$$
$$R = CH_3\,CH = CH -$$

$$R - S - S - R'$$
$$\qquad\;\; |$$
$$\qquad\;\; O$$

(**XVI**)

10.3.3   Biosynthesis

The biosynthetic interconversions associated with the formation of the substituted cysteine sulphoxides are complex. Seven days after the injection of ^{35}S-sulphate into onion, 21 radioactive components were identified, while 46 days after the injection 18 compounds (differing both qualitiatively and quantitatively from those detected earlier) were observed (Ettala and Virtanen, 1962). In addition to cysteine, methionine and their derivatives (Table 6), the alliums are a rich source of γ-glutamyl

Table 6 — Involatile sulphur compounds of onion

L-cysteine
L-cystine
L-methionine
L-methionine sulphoxide
S-methyl-L-cysteine
S-propyl-L-cysteine
trans-*S*-(prop-1-enyl)-L-cysteine
S-(prop-2-enyl)-L-cysteine
S-(carboxymethyl)-L-cysteine
S-(2-carboxyethyl)-L-cysteine
S-(2-carboxyisopropyl)-L-cysteine

peptides (Table 7). These compounds, of which 14 are currently known, apparently have a storage reserve function since they are present in dormant seeds and resting bulbs. On seed germination or sprouting these compounds are readily broken down to the appropriate alk(en)ylcysteine sulphoxide via the action of γ-glutamyl peptidase or transpeptidase (Whitaker, 1976).

Table 7 — γ-glutamyl peptides isolated from alliums

	Source
γ-glutamyl	
-valine	Onion
-isoleucine	Onion, chive
-leucine	Onion
-methionine	Onion, chive
-phenylalanine	Onion
-tyrosine	Onion
-cysteine	Chive
-*S*-methylcysteine	Onion, chive
-*S*methylcysteine sulphoxide	Onion garlic?
-*S*-propylcysteine	Garlic, chive
-*S*-(prop-2-enyl)cysteine	Garlic, chive
-*S*-(prop-1-enyl)cysteine	Chive
-*S*-(prop-1-enyl)cysteine sulphoxide	Onion, chive
-*S*-(prop-1-enyl)cysteinyl-*S*-(1-propenyl)	
cysteine sulphoxide	Chive
-*S*-(2-carboxypropyl)cysteinylglycine	Onion, garlic
N,N'-bis (γ-glutamyl)cysteine	Chive
N,N'-bis(γ-glutamyl)-3,3'-(2-	
methylethylene-1,2-dithio)dialanine	
Glutathione	Onion
Glutathione cysteine disulphide	Onion
Glutathione-γ-glutamylcysteine disulphide	Onion
S-sulphoglutathione	Onion

It is important to realize that since alliinase (*S*-alk(en)yl-L-cysteine sulphoxide lyase, EC 4.4.1.4) has no effect on these γ-glutamyl derivatives, these represent 'potential' rather than 'actual' flavour. The importance of this for the formation of desirable organoleptic and biological characteristics is clearly seen when it is realized that 90% of the soluble organic-bound sulphur is present in this form (Virtanen, 1965).

Since there are significant differences between the biosynthetic origins of certain of the alk(en)yl cysteine (Granroth, 1970), these will be considered separately.

(+)-*S*-methyl-L-cysteine sulphoxide (**XVb**) has a rather complex origin. Under normal conditions the favoured pathway involves the transfer of a methyl group from methionine to cysteine. The occurrence of an alternative pathway involving transfer of –SCH$_3$ has also been demonstrated. Thus [^{35}S]methionine is readily metabolized into, amongst other products, S-methyl-L-cysteine and its sulphoxide. The mechanisms are indicated in Fig. 3. A variety of labelling studies using ^{14}C and ^{35}S precursors ruled out the possibility that the transfer of a methyl or thiomethyl grouping involved the intermediacy of methionine sulphoxide, S-methylmethionine sulphonium ion, S-adenosylmethionine or S-methylthioadenosine. In agreement with the latter mechanism, current research suggests that (+)-S-propyl- and (+)-S-(prop-2-enyl)-L-cysteine sulphoxides (**XVc,a**) are derived from serine and an appropriate S-alk(en)yl moiety. While the nature of the latter is unclear *in vivo*, the corresponding alk(en)yl mercaptans serve as excellent exogenous sources.

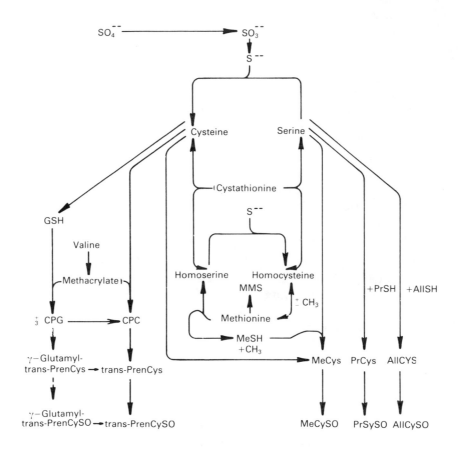

Fig. 3 — Biosynthesis of (+)-S-alk(en)yl-L-cysteine sulphoxide.

The biosynthesis of *trans*-(+)-S-(prop-1-enyl)-L-cysteine sulphoxide (**XVd**) has been examined in detail by Granroth (1970). The compound is considered to be derived from valine (Fig. 3) via oxidation, sulphur insertion, dehydration and reaction with cysteine to produce 2-carboxypropyl-L-cysteine (CPC), the absolute configuration $(-)$ of which has been determined (Parry and Naidu, 1983). To this point, the pathway is part of normal intermediary metabolism but the subsequent stages, involving decarboxylation and oxidation, are characteristic of members of the allium family. Whilst the nature of the C_1 fragment lost has not been unequivocally established it would seem probable that expulsion of CO_2 occurs, at least in part. Oxidation of S-alk(en)yl-L-cysteine occurs rapidly and is apparently reversible.

In conclusion it appears that the alk(en)ylcysteine sulphoxides of the allium family are metabolically inert, being biosynthetic end-products rather than intermediates. Thus the 2-propenyl and 1-propenyl compounds accumulate in garlic and onion at levels of 371 mg $(100 \text{ g})^{-1}$ (Stoll and Seebeck, 1951) and 200 mg $(100 \text{ g})^{-1}$ fresh weight (Matikkala and Virtanen, 1967) respectively.

10.3.4 Catabolism

As with the case of the glucosinolates, the sensory and biological properties of alliums are realized by the breakdown of involatile, inactive precursors.

A distinction will be made between methyl, 2-propenyl (or allyl) and propyl cysteine sulphoxides, which yield the corresponding thiosulphinates and, thereby, a plethora of sulphur-containing volatiles and 1-propenyl cysteine sulphoxide which is the precursor of the lachrymatory principle.

Before discussing the catabolism in detail, mention should be made of the enzyme (alliinase) which is responsible. Onion alliinase is a glycoprotein, of molecular weight 150 000 comprising three subunits of molecular weight 50 000. The carbohydrate content (5.8%) comprised simple sugars, hexosamines and methyl pentose. The pH optimum, which was buffer dependent, ranged from 7.6 (orthophosphate) to 8.0 (tricine). The amino acid content of the purified enzyme was rich in aspartic and glutamic acids, lysine and tryptophan (Tobkin and Mazelis, 1979). In contrast to earlier work indicating a requirement for pyridoxal phosphate, more recent studies have failed to find any such effect; rather it appears that one molecule of this salt may be strongly bound to each subunit, in support of which pyridoxal phosphate inhibitors were demonstrated to have a strong negative effect on the activity.

In comparison, much less is known about the corresponding enzyme from garlic which does not appear to be a glycoprotein. The pH optimum appears to be rather lower (pH 6.5) than that of the alliinase, but K_m values (mM) for methyl, ethyl, propyl and butyl cysteine sulphoxides are rather similar for onion and garlic, i.e. 16.6 and 15, 5.7 and 6, 3.8 and 3, 4.7 and 5 respectively. The K_m value for the natural substrate of garlic alliinase, 2-propenyl cysteine sulphoxide, was 6 mM.

The formation of the major sulphur-containing volatiles in the onion is shown in Fig. 4. The enzyme-induced breakdown of $(+)$-S-propyl-L-cysteine sulphoxide may be taken as typical of the $(+)$-S-methyl and $(+)$-S-(prop-2-enyl) compounds which are found in leek and garlic respectively. Following cellular disruption, the substrate and alliinase interact to produce, possibly via a transient sulphenic acid, the thiosulphinate and pyruvate. Since these products are formed stoichiometrically, measurement of pyruvate has commonly been used as an index of pungency. The thiosulphinates, which possess significant sensory and biological properties, are readily dissociated to yield disulphides and thiosulphonates, with the former undergoing disproportionation (to sulphides and trisulphides) and the latter eliminating

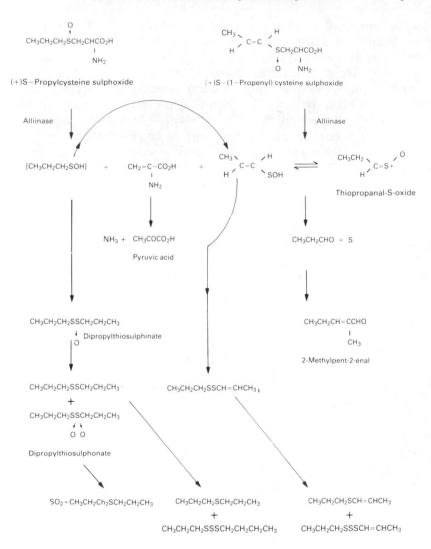

Fig 4 — Formation of major sulphur-containing volatiles.

sulphur dioxide to form sulphides. Further disproportionation can occur to form tri, tetra and higher sulphides which, however, become less volatile and develop characteristic 'rubbery' or 'sulphury' characteristics. More extensive reactions, including thiophene formation, may result from thermal processing or cooking and have been discussed elsewhere (Fenwick and Hanley, 1985b).

In many Allium species, more than one substituted cysteine sulphoxide occurs so that, after alliinase treatment, mixed thiosulphinates, sulphides, etc., can result via intermolecular processes. As an example of this, Iida *et al.* (1983) have identified the following major components in the volatile oil from steam distilled Chinese chives: dimethyl sulphide (32% total), dimethyl trisulphide (16%), methyl (prop-2-enyl) sulphide (8%) and methyl (2-propenyl) trisulphide (10%). More generally, quantitative and qualitative differences in the composition of the volatiles obtained from different *Allium* species have been used by Bernhard (1970) and Freeman and Whenham (1975) to classify members of this genus on a chemotaxonomic basis (Table 8).

Table 8 — Relative proportions of alk(en)yl radicals in the volatile fraction of *Allium* species[a]

| Species | Relative proportions of alk(en)yl radicals | | | | |
	Propyl	Methyl	prop-1-enyl	prop-2-enyl	Lachrymatory score
A cepa L.	88	6	6	—	s
A cepa L. var. *proliferum*	70	22	9	—	m
A cepa L. var. *ascalonicum*	62	30	8	—	s
A. chinense G. Don	35	24	10	—	s
A. ampeloprasum L. var. *porrum*	62	29	8	—	m
A. schoenoprasum L.	64	29	7	—	m
A. sativum L.	2	13	—	85	—
A. tuberosum Rottl. ex Sprengel	1	39	—	60	—
A. ursinum L.	5	32	—	64	—
A. karavatiense Regel	5	95	Trace	—	—
A siculum Ucria	8	93	—	—	—

[a]Extracted from Freeman and Whenham (1975).
s=strong; m=medium.

The characteristic pungent flavour of alliums is thus dependent on the relative and absolute amounts of these individual thiosulphinates, sulphides, disulphides, etc. The typical flavour of onion is due largely to the presence of propyl compounds; garlic contains mainly 2-propenyl (allyl) compounds while leek contains both propyl and methyl compounds. However, analysis of such products does not explain the origins of the lachrymatory characteristic which is such an important feature of onions. Examination of Table 8 suggests that this may be related to the presence of products possessing 1-propenyl moieties which in turn means the presence in the intact allium of (+)-S-(prop-1-enyl)-L-cysteine sulphoxide. However, the relative instability of the lachrymatory principle made its isolation and characterization difficult.

Spåre and Virtanen (1963) originally suggested this compound to be a sulphenic acid. However, later work by Brodnitz and Pascale (1971) proved the compound to be thiopropanal-S-oxide (**XVII**). Elegant studies by Block and coworkers (1979, 1980) showed this compound to possess the Z configuration and to undergo facile self-condensation. The S-oxides derived from thioacetone, thiobutanal and thiohexanal also exhibited lachrymatory properties (Brodnitz *et al.*, 1971), as did the products of the alliinase-mediated decomposition of synthetic 5-vinyl- and S-(but-1-enyl)-L-cysteine sulphoxide (Muller and Virtanan, 1965).

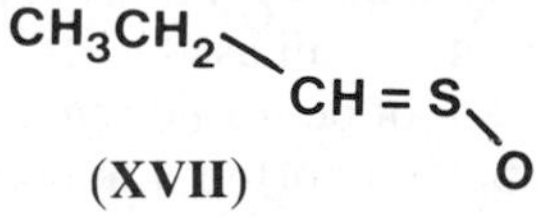

$$CH_3CH_2 \diagdown \atop CH = S \diagdown \atop O$$

(**XVII**)

As indicated above, thiopropanal-S-oxide is relatively unstable and, in addition to forming the dimer, may decompose to form propionaldehyde. This may undergo self-condensation or react with other aldehydes (for example pyruvate-derived acetaldehyde) to produce a range of oxygen-containing volatiles which are important components of the headspace above cut onions (Fig. 5). While there is no firm evidence for the occurrence of di(prop-1-enyl)thiosulphinate and its sulphides in onion, mixed thiosulphinates and the sulphide products do occur.

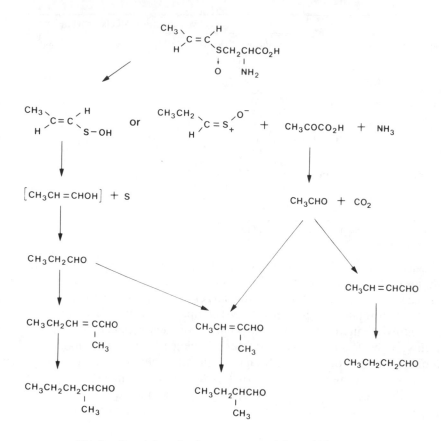

Fig. 5 — Formation of major oxygen-containing volatiles.

It is clear from a variety of studies that the rate of development of volatile aroma products depends on the degree of tissue disruption. Total production of volatile aroma compounds increased as tissue breakdown became increasingly greater following cubing, slicing or homogenizing.

(+)-S-(prop-1-enyl)-L-cysteine sulphoxide may undergo breakdown under alkaline conditions to produce cycloalliin, 3-methyl-1,4-thiazane-5-carboxylic acid -S-oxide (**XVIII**), which is resistant to alliinase (Virtanen and Matikkala, 1956). It is not certain that this compound cannot be formed *in vivo*; levels of 140–320 mg (100 g)$^{-1}$ fresh weight of onion have been reported, but may have arisen as an artefact of isolation. Cycloalliin possesses biological activity (for example, decreasing fibrinoly-tic activity in man) but none of the undesirable sensory characteristics of allicins or sulphur-containing volatiles.

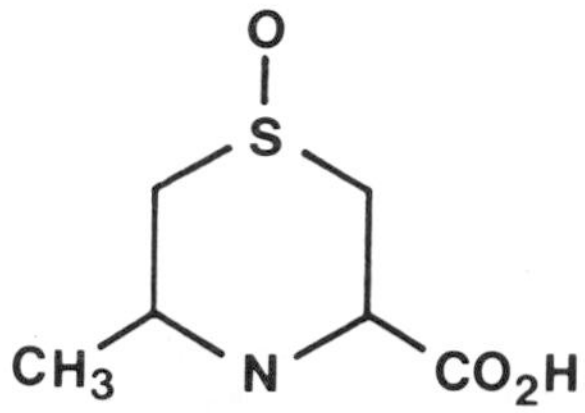

(**XVIII**)

10.3.5 Metabolism

Since the metabolism of sulphoxides, disulphides and related compounds has been dealt with elsewhere in this volume (Part A, chapters 6 and 7), mention will only be made here of a study into the metabolism of ^{35}S-labelled di(prop-2-enyl) disulphide in female Swiss mice. Maximal uptake by the liver occurred 90 min after dosage, with 70% of the label being present in the liver cyctosol 30 min later, mostly in the form of sulphates (Pushpendran *et al.*, 1980).

10.3.6 Biological effects

Onion and garlic, their oils, and expressed juices and extracts possess a wide range of biological activities, which are generally interpreted as being a result of the presence of sulphur-containing compounds (Fenwick and Hanley, 1985c). The present section will thus consider the various classes of sulphur-containing compounds in detail before dealing with the effects of onion and garlic oils and extracts.

10.3.6.1 *S-substituted cysteine sulphoxides*

The antilipidemic effects of *S*-methyl- and *S*-(2-propenyl)-L-cysteine sulphoxide have been demonstrated by Itokawa *et al.* (1973); both compounds were much more active than *S*-methyl-L-cysteine. *S*-(prop-2-enyl)-L-cysteine sulphoxide has been shown to be an effective antiplatelet aggregation factor *in vitro*, being active against collagen-induced platelet aggregation at a concentration of 0.01 mM (Liakopolou-Kyriakides *et al.*, 1985). Recently Hikino *et al.* (1986) have shown *S*-(prop-2-enyl)-L-cysteine sulphoxide to possess marked antihepatotoxic activity *in vivo* and *in vitro*, although this was much less than that associated with the volatile oil. Whereas the latter produced a significant inhibition of free radical and lipid peroxidation, no such effects were associated with the sulphoxide. In support of the general argument that *S*-alk(en)yl cysteine sulphoxides are the precursors of the biologically active princi-ples in onion and garlic, they proved to be without antitumour activity (see below).

10.3.6.2 Thiosulphinates

The wide-ranging biological effects of these compounds (allicins) is generally interpreted as a result of their facile reaction with biological –SH groups (Cavallito, 1946), interfering with disulphydryl reductase and intracellular reductase functions of –SH moieties. Thus di(prop-2-enyl) thiosulphinate has broad spectrum antibacterial activity and possesses activity at a dilution of $1/10^5$. The antimicrobial, antifungal and antibacterial effects of thiosulphinates have been discussed by Fenwick and Hanley (1985c). Small *et al.* (1947) found diethyl thiosulphinate and thiosulphonate to possess similar activity against a range of organisms and while *S. aureus* 209P is known to develop resistance to penicillin and streptomycin the same is not true for di(2-propenyl) thiosulphinate (Klimek *et al.*, 1948).

Recent work (Mirelman, 1987) has shown di(prop-2-enyl) thiosulphinate to be as effective as metronidazole in controlling the parasite *Entamoeba histolytica*, which causes amoebic dysentery. A common observation of thiosulphinate-related activity is its loss with storage. Thus, at room temperature, anti-amoebic activity is lost within 48 h. Storage in alcoholic solutions (4°C) or in solid form (-20°C) was without effect on activity.

Since the antibacterial activity of isolated di(prop-2-enyl) thiosulphinate is less than that of expressed garlic juice, it has been suggested that the activity of the latter is potentiated by other garlic components or that the active principles are chemically protected. Thus reduction of di(prop-2-enyl) thiosulphinate to the disulphide with cysteine or dithiothreitol reduced the activity, but this was regained on oxidation with hydrogen peroxide. In a series of papers, Mizutani and coworkers (Minura *et al.*, 1976, 1979; Tahara *et al.*, 1977; Tahara and Mizutani, 1979) have described the synthesis and antibacterial properties of a number of *S*-alkylthiomethylhydantoin-*S*-oxides, which were considered to generate the active dialkylthiosulphinates by *in situ* β-elimination. Although the authors suggested that these compounds could be useful as convenient, readily prepared (and presumably odour-free) antibacterial precursors, there appears to have been no clinical follow-up.

Di(prop-2-enyl) thiosulphinate has been shown to be a potent inhibitor of platelet aggregation *in vitro* (Mohammed and Woodward, 1986). The same compound (100 mg kg^{-1} body weight) reduced liver cholesterol and serum and liver lipids in 4-month old rats and, after 15 days of treatment, activities of α-glucan phosphorylase, glucose-6-phosphatase and lipase were increased (Augusti and Mathew, 1975). At a dose of 250 mg kg^{-1}, di(2-propenyl) thiosulphinate was found to possess 60% of the activity of tolbutamide when tested against mildly diabetic rats, but neither compound was effective in cases of severe alloxan diabetes (Mathew and Augusti, 1973), a finding interpreted as indicating the dependence of the hypoglycaemic effect of the thiosulphinate on endogenous insulin. Studies by Augusti (1975) have shown the blood-sugar-lowering effect of thiosulphinate to be dose related. Garlic and onion have been shown to produce contact dermatitis in vegetable handlers (Table 9) and Papageorgiu *et al.* (1983) have recently shown di(prop-2-enyl) thiosulphinate to be amongst the allergens present in the former.

A number of thiosulphinates have shown activity as antitumour agents. In early studies, Weisberger and Pensky (1957, 1958) demonstrated the diethyl compound (not naturally occurring) to inhibit the uptake of sulphur amino acid by leukaemic leucocytes *in vitro*. *In vivo* and *in vitro* studies demonstrated the effectiveness of

Table 9 — Incidence of positive patch tests with various vegetables (Pasricha and Guru, 1979)

Vegetable	Number of patients (53 total) showing positive patch tests
Garlic	44
Onion	30
Tomato	10
Carrot	9
Radish	6
Cauliflower	5

Patch tests performed using 50 μl sample.

diethyl thiosulphinate against sarcoma 180 ascites cells and, later, Ehrlich ascites tumour cells (Fujiwara and Natata, 1967). DiPaolo and Carruthers (1960) have found di(prop-2-enyl) thiosulphinate to inhibit solid and ascites tumour cell growth completely following pre-incubation. A marked reduction in tumour development was noted following intratumoral injections of di(prop-2-enyl) thiosulphinate into mice pre-treated with sarcoma 180 tumour cells (Cheng and Tung, 1981). Amongst various synthetic thiosulphinates which have been tested against Ehrlich ascites tumour development, Kametani *et al.* (1959) found those possessing iso-butyl, *t*-butyl and amyl substituents to be most active. Thiosulphinates, in part at least, are considered responsible for the inhibition of tumour promotion associated with onion and garlic oils (Belman, 1983).

According to Foushee *et al.* (1982) dietary di(prop-2-enyl) thiosulphinate is the probable cause of diarrhoea in people not used to consuming large quantities of garlic.

While a number of sulphur compounds, particularly disulphides, have been identified amongst the compounds eliciting egg-laying responses in the onion maggot (*H. antiqua*), it appears that such behaviour in the leek moth (*A. assectella*) is stimulated to a greater degree by thiosulphinates than disulphides (Auger and Thibout, 1979, 1981).

10.3.6.3 *Cycloalliin*

A single dose of cycloalliin (125 mg) was found to increase fibrinolysis in man, albeit by less than dipropyl disulphide. In a subsequent trial, comprising both healthy subjects and those having a history of myocardial infarction, a 250 mg dose was shown to increase the fibrinolytic activity of venous blood to a highly significant degree (Agarwal *et al.*, 1977). The dose, equivalent to approximately 1.2 kg onions, was well tolerated, produced no harmful effects (at least in the short term) but, unlike other S products, was without effect on platelet aggregation.

10.3.6.4 Sulphides, disulphides, polysulphides

While sulphides and polysulphides obtained from alliums exhibit little or no antimicrobial activity, disulphides do possess significant activity, although much less than that of the above thiosulphinates. Thus dipropyl and di(prop-2-enyl)disulphides, the most active of a number of such compounds examined, were bacteriostatic against various organisms at the 0.1% level. Murthy and Amonkar (1974) found di(2-propenyl) disulphide to be active against ten plant pathogenic fungi. Complete inhibition of mycelial growth occurred at concentrations between 0.1% and 0.01%, with lower concentrations (0.01–0.0025%) inhibiting spore growth. In contrast, this compound had no effect on *Candida albicans* or *C. monosa*, the activity of garlic oil against the former being due to the presence of the corresponding thiosulphinate (Petricic *et al.*, 1978). Adetumbi *et al.* (1986) have found lipid synthesis in *C. albicans* to be completely inhibited by aqueous extract of garlic and suggest this, at least in part, as an explanation for the anticandidal activity.

Di(prop-2-enyl)disulphide and trisulphide proved toxic to mosquito larvae (Amonkar and Banerji, 1971), the effect being due to inhibition of normal protein synthesis. Such compounds are undoubtedly the basis of the wide-ranging insecticidal and nematicidal activity of garlic. Venugopal and Narayan (1981) have reported the successful use of 'allitin', a mixture of synthetic di(prop-2-enyl) disulphide and trisulphide, to control the green peach aphid (*Myzus persicae*). No phytotoxic symptoms were observed on plants sprayed with this mixture.

There have been many examples of domestic and farm animals suffering severe anaemia, often with fatal consequences, following the consumption of wild or decaying onions, cull onions or onion waste (Fenwick and Hanley, 1985c). Cattle are most susceptible, dogs and horses being intermediate while sheep and goats are more resistant. The toxic principle has been identified as dipropyl disulphide which alters glucose-6-phosphate dehydrogenase within the erythrocytes, thus causing denaturing and precipitation of haemoglobin. Such damaged erythrocytes are removed from circulation by the spleen and, if sufficiently large numbers are removed, haemolytic anaemia results. A similar condition (kale anaemia) is known to occur in ruminants consuming kale which is rich in *S*-methyl cysteine sulphoxide, this being converted *in vivo* to the active haemolysin dimethyl disulphide (Smith, 1980). Breeding programmes aimed at reducing sulphoxide levels have been developed as a consequence. It should be noted that such disulphides are not produced from the cysteine sulphoxides in man, so that concerns about such possible anaemic responses following brassica or allium consumption are misplaced.

Sulphur-containing volatiles have been shown to be important oviposition regulators for the onion maggot (*H. antiqua*). Pierce *et al.* (1978) showed methyl propyl disulphide and *cis* and *trans* prop-1-enyl propyl disulphides to be ovipositing stimuli while dimethyl disulphide and methyl *cis* (or *trans*) prop-1-enyl were inactive. This, and other, work led to the suggestion that a thiopropyl grouping was necessary for egg laying stimuli in *H. antiqua*. Traps containing dipropyl disulphide have been used in the field to catch male and female onion flies. However, the finding that males were more attracted than females to sulphur-containing volatiles in the field (Vernon *et al.*, 1981) and the observation that multicomponent mixtures of volatiles were more attractive than individual, purified components (Dindonis and Miller, 1981) indicates the complex nature of this phenomenon.

Sulphur-containing compounds from onion and garlic have been shown to possess a range of beneficial properties against the various manifestations of ischaemic heart disease. Ariga and coworkers (Ariga *et al.*, 1981; Oshiba *et al.*, 1981) have found various disulphide and trisulphide components of garlic oil to inhibit platelet aggregation *in vitro*, with most the potent compound, methyl(prop-2-enyl) trisulphide, being active at a concentration of <10 μmol 1^{-1} human platelet-rich plasma. A large-scale, clinical study of this compound in man is in progreess in Japan. Subsequently particular attention has been directed towards the platelet aggregation properties of the sulphoxide E,Z-ajoene (see below).

Many of the effects on fibrinolytic activity, blood and tissue lipids which are evident following the feeding of garlic and, to a lesser extent, onion oil are presumably due to the component di- and other sulphides, but in general few studies with these isolated products have been conducted. One such study has recently shown that propyl(prop-2-enyl) disulphide (100 g kg^{-1}, daily), administered intragastrically in rats for 45 days produces a significant lowering of plasma lipid levels (Bobboi *et al.*, 1984). Adoga and Osuji (1986) observed that the hypolipidaemic effects of garlic oil extract were related to a lowering of the tissue levels of alkaline phosphatase and alcohol dehydrogenase and further work showed the same extract to reduce high levels of tissue glutathione reductase to near-normal values in rats fed high sucrose and alcohol diets (Adoga, 1986). The author suggested that this reduction is brought about by di(prop-2-enyl) disulphide which interacts with the –SH groups of the enzyme and its substrate. In view of the reactivity of the thiosulphinate precursor it is likely that this compound also has a part to play in reducing the levels of the above enzymes(s).

Wargovitch and coworkers (Wargovitch and Goldberg, 1985; Wargovitch, 1987a) have reported di(prop-2-enyl) sulphide to inhibit carcinogen-induced nuclear damage in colon epithelial cells and shown an oral dose of the same compound (200 mg kg^{-1}, administered weekly 3 h prior to an injection of 1,2-dimethylhydrazine in C57BL/6J mice, to offer significant protection against resultant development of colorectal adenocarcinoma in mice exposed to 1,2-dimethylhydrazine (Wargovitch, 1987b). Other findings (Wargovitch, 1987a) indicate that the same compound is capable of total suppression of oesophageal cancer in rats. Recently Hayes *et al.* (1987) have conducted experiments, the results of which suggest that di(prop-2-enyl)-sulphide may prevent macromolecular alkylation by dimethylhydrazine, and that its ability to reduce the promotion of post-necrotic regeneration mechanisms is a more important factor than, for example, any preventative effect on turnover initiation. The authors suggest that this may explain the antipromoting effects of onion and garlic oils, for example, as studied by Belman (1983).

The allergenicity of garlic has been known for many years and recent data suggest that prop-2-enyl thiol, (prop-2-enyl) propyl disulphide and di(prop-2-enyl) disulphide were active allergens thereof. In contrast, dimethyl disulphide was without effect (Papageorgiu *et al.*, 1983).

10.3.6.5 E,Z-4,5,9-trithiadodeca-1,6,11-triene-9-oxide (E,Z-ajoene)
This compound (**XIX**) has recently been identified in fresh, but not processed, garlic and is derived from di(prop-2-enyl) thiosulphinate (Block *et al.*, 1984). The considerable interest in its antithrombotic properties justifies its being considered

separately in this section. The compound produced complete inhibition of collagen-induced platelet aggregation for 24 h after the feeding of 20 mg to rabbits. Aggregation caused by all known inducing agents was inhibited as was rabbit granulocyte aggregation. Various related compounds (both synthetic and natural) were also tested for antithrombotic activity but all proved markedly less active (Block et al., 1986). It is suggested that the enhanced activity of E,Z-ajoene is due to a combination of disulphide exchange reactions with the –SH groups of the blood platelet membrane and coordination of calcium ions to the sulphoxide group, a process which facilitates binding to the platelet membranes.

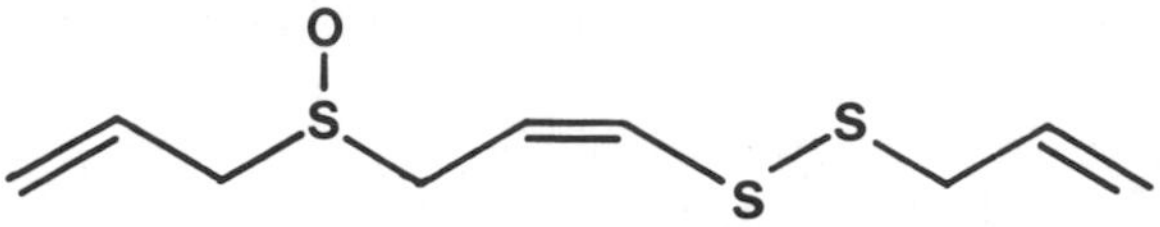

(XIX) (Z – ISOMER)

Currently patent applications are being filed for this compound which, by analogy with the thiosulphinates and disulphides, possesses antimicrobial activity.

10.3.6.6 Garlic and onion oils

There have been many hundreds of reports about the biological activities of onion and garlic oils as well as the raw or processed vegetable. These have been reviewed by Petkov (1966, 1979), Ikram (1971, 1972), Ernst (1981), Lau et al. (1983), Reuter (1983), Fenwick and Hanley (1985c) and Sprecher (1986). It is probable that most of the observed effects are due to the content and complement of S-containing compounds as described above. In general garlic and its products has been found to be more active than onion, a finding generally attributable to the greater volatile oil content of the former and the greater reactivity of prop-2-enyl as opposed to 2-propyl sulphur compounds.

There is little evidence to suggest that any additional benefit can be gained from administration of garlic capsules of health supplements over and above that available from the fresh or cooked plants. Indeed, it is probable that in some capsules and tablets there is very little of the active ingredient and that prolonged storage may further reduce the levels.

A particularly detailed epidemiological study has been carried out by Sainani et al. (1979a,b). This involved three populations from the Jain community, similar in regard to intake of calories, fat and carbohydrate and differing in their consumption of onions and garlic. The results are shown in Table 10 and clearly demonstrate the protective effects of these alliums.

Amongst the most recent researches on onion and garlic oils are the findings that dietary garlic oil reduced blood sugar, serum and liver cholesterol, serum and liver triglycerides, liver lipids and proteins associated with streptozotocin-diabetic rats (Farva et al., 1986) and that odour-modified garlic extracts may be used in

Table 10 — Effect of dietary onion and garlic on three groups selected from the Jain community (Sainani et al., 1979a,b)

Group-	Plasma fibrogen (mass%)	Clotting time (s)	Prothrimbin time (s)	Euglobin clot lysis time (s)
L	482	220	16	51
M	364	254	16	65
H	327	240	16	87

	Fatty cholesterol levle (mg(100 ml)$^{-1}$)	Serum triglycerides (mg (100 ml)$^{-1}$)	Serum lipoproteins (%)	Phospholipids (%)
L	208	109	71	18
M	172	75	63	6
H	159	53	70	6

L, group consuming no garlic and onion; M, group consuming <200 g onion, <10 g garlic per week; H, group consuming >600 g onion, >50 g garlic per week.

conjunction with dietary modification for control of hyperlipidaemia in man (Lau *et al.*, 1987). However, the recent report of Zelikoff *et al.* (1986) that onion and garlic oils shorten the cell doubling time and stimulate growth and proliferation of NIH-3T3 cells in a concentration-dependent manner clearly is deserving of further investigation.

10.4 CYCLIC DISULPHIDES AND RELATED COMPOUNDS

10.4.1 Occurrence

The presence of sulphur-containing compounds in asparagus, based largely on the volatile products obtained during heating or cooking, has long been recognized. It is almost a century since Nencki (1891) described investigations into the nature and origins of the sulphury odour which is frequently detected in the urine of certain individuals following the consumption of asparagus. This phenomenon continues to attract interest and is discussed in Volume 2, chapter 3.

10.4.2 Structure

A novel sulphur-containing compound, 3,3′-dimercaptoisobutyric acid (**XXa**) was isolated from asparagus juice by Jansen (1948) and the same compound, its *S*-acetyl derivative (**XXb**) and cyclic, reduced form (1,2-dithiolane-4-carboxylic acid, asparagusic acid, **XXIa**) were later isolated and identified in asparagus shoots (Yanagawa *et al.*, 1972). A number of other related compounds (**XXIb,c, XXII, XXIII, XXIV, XXV, XXVI**) were identified only partially by Tressl *et al.* (1977b) although some of these, notably the esters, may be artefacts of isolation. A cyclic trisulphide, 1,2,3-trithiane-5-carboxylic acid (**XXVI**), was subsequently isolated from raw asparagus shoots (Kasai and Sakamura, 1982). Despite previous claims to the contrary (Tressl *et al.*, 1977a), this appears to be an authentic natural product rather than being exclusively derived from asparagusic acid during thermal processing or cooking. Two structurally related disulphides (**XXc,d**) have also been reported, together with *S*-(2-carboxylpropyl)- and *S*-(1,2-dicarboxyethyl)-L-cysteines (**XXVIIa,b**) (Kasai *et al.*, 1981). *S*-methylmethionine salts have also been found in asparagus and are considered to be the precursor of the dimethylsulphide which is produced on cooking.

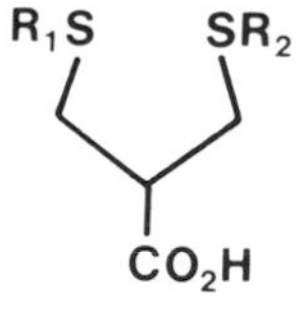

$R_1S \quad SR_2$

(structure with CO_2H)

(**XX**)

a) $R_1 = R_2 = H$

b) $R_1 = -COCH_3$; $R_2 = H$

c) $R_1 = HO_2C\,CHCH_2-$; $R_2 = -SCH_2CHCH_3$
        $\overset{|}{NH_2}$ $\overset{|}{CO_2H}$

d) $R = HO_2C\,CH-CH_2-$; $R_2 = -SCH_2CH-CH_2\,SCH_2CHCH_3$
        $\overset{|}{NH_2}$ $\overset{|}{CO_2H}$ $\overset{|}{CO_2H}$

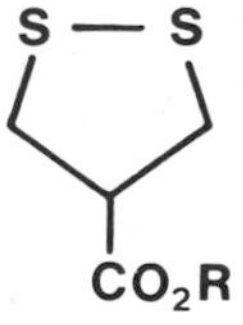

a) R = H
b) R = −CH$_3$
c) R = −CH$_2$CH$_3$

(XXI)

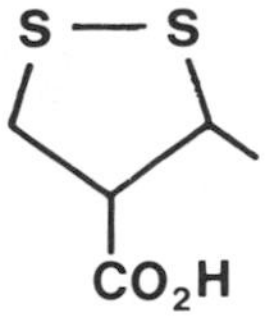

(XXII)

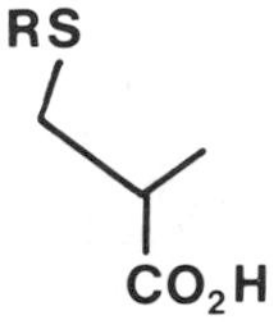

a) R = H
b) R = −CH$_3$
c) R = −SCH$_2$CHCH$_3$
              |
             CO$_2$H

(XXIII)

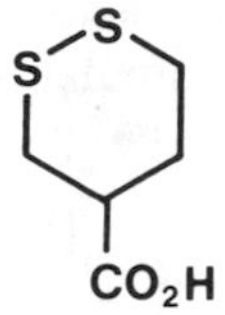

(XXIV) 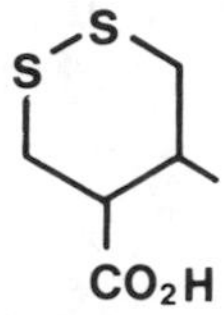   (XXV) 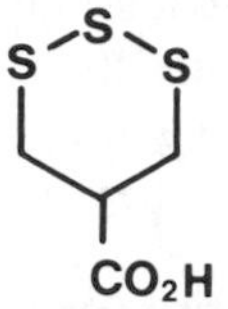   (XXVI)

R−SCH$_2$CHCO$_2$H
        |
       NH$_2$

(XXVII)

a)   R = −CH$_2$CHCH$_3$
                  |
                 CO$_2$H

b)   R = −CHCH$_2$CO$_2$H
           |
          CO$_2$H

10.4.3 Biosynthesis

Despite their both possessing a 1,2-dithiolane ring, asparagusic acid and lipoic acid are biosynthetically unrelated. Biosynthetic studies on the former using a variety of labelled precursors have shown methacrylate to be a key biosynthetic intermediate (Tressl *et al.*, 1977a; Parry *et al.*, 1982). The overall process, shown in Fig. 6, is analogous to the formation of *S*-(prop-1-enyl)-L-cysteine sulphoxide in onion.

10.4.4 Biological effects

Kitahara *et al.* (1972) and Yanagawa *et al.* (1972, 1973) have shown asparagusic acid and its derivatives to possess growth inhibitory properties when tested against a range of plant seedlings. The activity which was evident against lettuce seedlings at a concentration of 6.67×10^{-7}–6.67×10^{-4} M, is thus comparable with that of abscisic

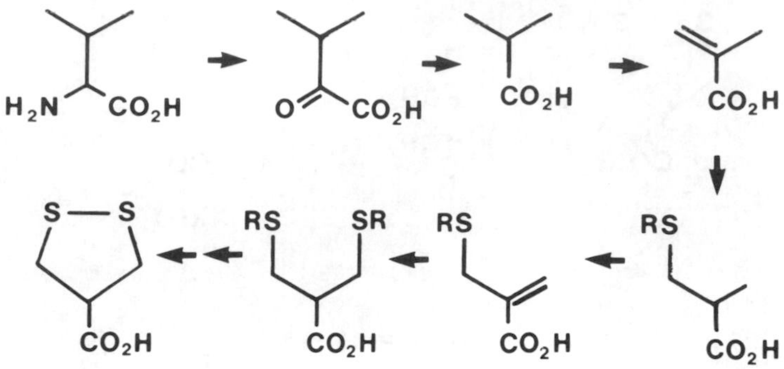

Fig. 6 — Biosynthesis of asparagusic acid.

acid. The same compounds also stimulated growth and pyruvate oxidation in *Streptococcus faecalis* 10 Cl and asparagus mitochondria. It has been suggested that these compounds may act as cofactors in the latter since lipoic acid had no effect when tested in such a system, in marked contrast to its stimulatory action in *S. faecalis* 10 Cl.

ACKNOWLEDGEMENT

The authors are grateful to Mrs J. Wilson for her skill and patience in typing this manuscript.

REFERENCES

Adetumbi, M., Javor, G. T., and Lau, B. H. S. (1986). *Allium sativum* (garlic) inhibits lipid synthesis by *Candida albicans. Antimicrob. Agents Chemother.,* **30**, 499–501.

Adoga, G. I. (1986). Effect of garlic oil extract on glutathione reductase levels in rats fed on high sucrose and alcohol diets, a possible mechanism of the activity of the oil. *Biosci. Rep.,* **6**, 909–912.

Adoga, G. I. and Osuji, J. (1986). Effect of garlic oil extract on serum, liver and kidney enzymes of rats fed on high sucrose and alcohol diets. *Biochem. Int.,* **13**, 615–624.

Agarwal, R. K., Dewar, H. A., Newell, D. J., and Das, B. (1977). Controlled trial of the effect of cycloalliin on the fibrinolytic activity of venous blood. *Atherosclerosis.* **27**, 347–352.

Al-Shehbaz, I. A. (1985). The genera of Thelypodieae (Cruciferae: Brassicaceae) in the South Eastern United States. *J. Arnold Arboretum,* **66**, 95–111.

Amonkar, S. V. and Banerji, A. (1971). Isolation and characterization of larvicidal principle in garlic. *Science,* **174**, 1343–1344.

Ariga, T., Oshiba, S., and Tamada, T. (1981). Platelet aggregation inhibitor in garlic. *Lancet,* **8212**, 150–151.

Astwood, E. B. (1943). The chemical nature of compounds which inhibit the function of the thyroid gland. *J. Pharm. Exp. Ther.,* **78**, 79–89.

Auger, J. and Thibout, E. (1979). Action des substances soufrees volatiles du poireau (*Allium porrum*) sur la ponte d'*Acrolepiopsis assectella* (Lepidoptera: Hyponomentoidea): preponderance des thiosulphinates. *Can. J. Zool.,* **57**, 2223–2229.

Auger, J. and Thibout, E. (1981). Emission por le poireau *Allium porrum* de thiosulphinates actif sur la teigne de Poireau, *Acrolepiopsis assectella* (Lepidoptera). *C.R. Acad. Sci. Paris, Ser. III,* **292**, 217–220.

Augusti, K. T. (1975). Studies on the effect of allicin (diallyl dusulphide oxide) on alloxan diabetes. *Experientia,* **31**, 1263–1265.

Augusti, K. T. and Mathew, P. T. (1975). Effect of allicin on certain enzymes of liver after a short-term feeding to normal rats. *Experientia,* **31**, 148–149.

Bachmann, M., Theus, R., Lüthy, J., and Schlatter, C. L. (1985). Vorkommen von Goitrogen Stoffen in Milch. *Z. Lebensm. Unters. Forsch.,* **181**, 375–379.

Belman, S. (1983). Onion and garlic oils inhibit tumor promotion. *Carcinogenesis,* **4**, 1063–1065.

Bernhard, R. A. (1970). Chemotaxonomy: distribution studies of sulphur compounds in *Allium. Phytochemistry,* **9**, 2019–2027.

Bille, N., Eggum, B. O., Jacobsen, I., Olsen, O., and Sørensen, H. (1983). Antinutritional and toxic effects in rats of individual glucosinolates ($\pm$myrosinase) added to a standard diet. *Z. Tierphysiol., Tiernahr. Futtermittelk.,* **49**, 195–210.

Block, E., Penn, P. E., and Revelle, R. K. (1979). Structure and origin of the onion lachrymatory factor, a microwave study. *J. Am. Chem. Soc.,* **101**, 2200–2201.

Block, E., Reville, L. K., and Buzzi, A. A. (1980). The lachrymatory factor of the onion — an nmr study. *Tet. Lett.,* **21**, 1277–1280.

Block, E., Ahmed, S., Jain, M. K., Crecely, R. W., Apitz-Castro, R., and M. R. (1984). (E,Z)-Ajoene: a potent anti-thrombotic agent from garlic. *J. Am. Chem. Soc.,* **106**, 8295–8296.

Block, E., Ahmad, S., Catalfamo, I. L., Jain, M. K., and Apitz-Castro, R. (1986). Antithrombotic organosulfur compounds from garlic: structural, mechanistic and synthetic studies. *J. Am. Chem. Soc.,* **108**, 7045–7055.

Bobboi, A., Augusti, K. T. and Joseph, P. K. (1984). Hypolipidaemic effects of onion iol and garlic oil in ethanol-fed rats. *Ind. J. Biochem. Biophys.,* **21**, 211–213.

Brocker, E. R. and Benn, M. H. (1983). The intramolecular formation of epithioalkanenitriles from alkenylglucosinolates by *Crambe abyssinica* seed flour. *Phytochemistry,* **22**, 770–772.

Brodnitz, M. H. and Pascale, J. V. (1971). Thiopropanal-S-oxide, a lachrymatory factor in onions. *J. Agric. Food Chem.,* **19**, 269–272.

Brodnitz, M. H., Matawan, P., and Pascale, J. V. (1971). Neue Schwefelverbindungen mit Zwiebel-oder Lauch-geschmack. *German Patent 210,8,727.*

Brüsewitz, G., Cameron, B. D., Chausseaud, L. F., Görler, K., Hawkins, D. R., Koch, H., and Mennicke, W. H. (1977). The metabolism of benzyl isothiocyanate and its cysteine conjugate. *Biochem. J.,* **162**, 99–107.

Butcher, D. N., Chamberlain, K., Rausch, T., and Searle, L. M. (1984). Changes in indole metabolism during the development of clubroot symptoms in brassicas. *Br. Plant Growth Regulator Group, Monogr.,* **11**, 91–101.

Butler, E. J. and Fenwick, G. R. (1984). Trimethylamine and fishy taint in eggs. *World's Poultry Sci. J.,* **40**, 38–51.

Carlson, D. G., Daxenbichler, M. E., VanEtten, C. H., Tookey, H. L., and Williams, P. H. (1981). Glucosinolates in crucifer crops: turnips and rutabagas. *J. Agric. Food Chem.,* **29**, 1235–1239.

Carlson, D. G., Daxenbichler, M. E., VanEtten, C. H., Hill, C. B., and Williams, P. H. (1985). Glucosinolates in radish cultivars. *J. Am. Hort. Sci.,* **110**, 634–638.

Cavallito, C. J. (1946). Relationship of thiol structures to reaction with antibiotics. *J. Biol. Chem.,* **164**, 29–34.

Chang, Y. and Bjeldanes, L. F. (1985). Effects of dietary R-goitrin on hepatic and intestinal glutathione S-transferase, microsomal epoxide hydratase and ethoxy-coumarin O-deethylase activities in the rat. *Food Cosmet. Toxicol.,* **23**, 905–909.

Cheng, H. H. and Tung, T. C. (1981). Effect of allithiamine on sarcoma-180 tumour growth in mice. *Chem. Abstracts,* **95**, 197366q.

Chong, C., Chiang, S. M., and Crete, R. (1985). Studies on glucosinolates in clubroot resistant selections and susceptible commercial cultivars of cabbages. *Euphytica,* **34**, 65–73.

Dahlberg, P. A., Bergmark, A., Björck, L., Bruce, A., Hambraeus, L., and Claesson, O. (1984). Intake of thiocyanate by way of milk and its possible effect on thyroid function. *Am J. Clin. Nutr.,* **39**, 416–420.

Dahlberg, P. A., Bergmark, A., Eltom, M., Björck, L., and Claesson, O. (1985). Effect of thiocyanate levels in milk on thryroid function in iodine deficient subjects. *Am. J. Clin. Nutr.,* **41**, 1010–1014.

Daun, J. K. (1986). Glucosinolate analysis in rapeseed and canola — an update. *Yakugaku,* **35**, 426–434.

Daxenbichler, M. E., VanEtten, C. H., and Williams, D. H. (1979). Glucosinolates and derived products in cruciferous vegetables. Analysis of fourteen varieties of Chinese cabbage. *J. Agric. Food Chem.,* **27**, 34–37.

Dindonis, L. L. and Miller, J. R. (1981). Onion fly and little house fly host finding selectively mediated by decomposing onion and microbial volatiles. *J. Chem. Ecol.,* **7**, 419–426.

DiPaulo, J. A. and Carruthers, C. (1960). The effect of allicin from garlic on tumour growth. *Cancer Res.,* **20**, 431–437.

Eder, E., Neudecker, T., Lutz, D., and Henschler, D. (1980) Mutagenic potential of allyl and allylogenic compounds. Structure–activity relationships as determined by alkylating and direct *in vitro* mutagenic properties. *Biochem. Pharmacol.,* **29**, 993–998.

Elfving, S. (1980). Studies on the naturally-occurring goitrogen, 5-vinyl-2-thiooxa-zolidone. *Ann. Clin. Res.,* **28**, 1–47.

Ernst, E. (1981). Therapie mit Knoblauch-Theorien uber ein volkstumliches Heil-prinzip. *Münch. Med. Wschr.,* **123**, 1537–1538.

Ettala, T. and Virtanen, A. I. (1962). Labelling of sulphur-containing amino acids and gamma-glutamyl peptides after injection of labelled sulphate into an onion. *Acta Chem. Scand.*, **16**, 2061–2072.

Ettlinger, M. G. and Lundeen, A. J. (1956). The structure of sinigrin and sinalbin: an enzymatic rearrangement. *J. Am. Chem. Soc.*, **78**, 4172–4173.

Farva, D., Goji, I. A., Joseph, P. K., and Augusti, K. T. (1986). Effects of garlic oil in streptozotocin-diabetic rats maintained on normal and high fat diets. *Ind. J. Biochem. Biophys.*, **23**, 24–27.

Fenwick, G. R. and Hanley, A. B. (1985a). The genus *Allium* — part 1. *CRC Crit. Rev. Food Sci. Nutr.*, **22**, 199–271.

Fenwick, G. R. and Hanley, A. B. (1985b). The genus *Allium* — part 2. *CRC Crit. Rev. Food Sci. Nutr.*, **22**, 273–376.

Fenwick, G. R. and Hanley, A. B. (1985c). The genus *Allium* — part 3. *CRC Crit. Rev. Food Sci. Nutr.*, **23**, 1–73.

Fenwick, G. R., Heaney, R. K., and Mullin, W. J. (1983). Glucosinolates and their breakdown products in foods and food plants. *CRC Crit. Rev. Food Sci. Nutr.*, **18**, 123–201.

Fenwick, G. R., Heaney, R. K., and Mawson, R. (1989). Glucosinolates. In P. R. Cheeke (ed.), *Natural Toxicants*, Vol. II, *Glucosides*, CRC Press, Boca Raton.

Food and Agriculture Organisation (1985). *FAO Production Yearbook*, Vol. 38, Rome.

Foushee, D. B., Ruffin, J., and Banerji, U. (1982). Garlic as a natural agent for the treatment of hypertension — a preliminary report. *Cytobios*, **34**, 145–152.

Freeman, G. G. and Whenham, R. J. (1975). A survey of the volatile components of some *Allium* species in terms of S-alk(en)yl-L-cysteine sulphoxides present as flavour precursors. *J. Sci. Food Agric.*, **26**, 1869–1886.

Fukiwara, M. and Natata, T. (1967). Induction of tumour immunity with tumour cells treated with extracts of garlic (*Allium sativum*). *Nature*, **216**, 83–84.

Goetz, J. K. and Schraudolf, H. (1983). Two natural indole glucosinolates from Brassicaceae. *Phytochemistry*, **22**, 905–907.

Görler, K., Krumbiegel, G., Mennicke, W. H., and Siehl, H. U,. (1982a). The metabolism of benzyl isothiocyanate and its cysteine conjugate in guinea-pigs and rabbits. *Xenobiotica*, **12**, 535–542.

Görler, K., Krumbiegel, G., Lorenz, D., and Mennicke, W. H. (1982b). Untersuchungen zur Verstoffwechsung von Benzylisothiocyanat beim Menschen. *Planta Med.*, **45**, 160–166.

Gould, D. H., Gumbmann, M. R., and Daxenbichler, M. E. (1980). Pathological changes in rats fed the *Crambe* meal–glucosinolate hydrolytic products — 2S-1-cyano-2-hydroxy-3,4-epithiobutanes (*erythro* and *threo*) — for 90 days. *Food Cosmet. Toxicol.*, **20**, 279–287.

Gould, D. H., Fettman, M. J., Daxenbichler, M. E., and Bartuslea, B. M. (1985). Functional and structural alterations of the rat kidney induced by the naturally-occurring organonitrile, 2S-1-cyano-2-hydroxy-3,4-epithiobutane. *Toxicol. Appl. Pharmacol.*, **78**, 190–201.

Gramberg, L. G., Rijk, M. A. H., Schouten, A., and deVos, R. H. (1986). Glucosinolates in Dutch cole crops. *Proc. Eurofoodtox II, Zurich*, pp. 279–284.

Granroth, B. (1970). Biosynthesis and decomposition of cysteine derivatives in onion and other allium species. *Ann. Acad. Sci. Fenn.*, **154, 411**, 9–71.

Hälvä, S., Hirvi, T., Mäkinen, S., and Honkanen, E. (1986). Yield and glucosino-lates in mustard seeds and volatile oil in caraway seeds and coriander fruit, I. *J. Agric. Sci., Finland*, **58**, 157–160.

Hashimoto, S., Miyazawa, M., and Kameoka, H. (1983). Volatile flavour com-pounds of chive (*Allium schoenoprasum* L.). *J. Food Sci.*, **48**, 1858–1859.

Hayes, M. A., Rushmore, T. H., and Goldberg, M. T. (1987). Inhibition of hepatocarcinogenic responses in 1,2-dimethylhydrazine by diallyl sulphide, a component of garlic oil. *Carcinogenesis*, **8**, 1155–1157.

Heaney, R. K. and Fenwick, G. R. (1980). Glucosinolates in *Brassica* vegetables. Analysis of twenty-two varieties of Brussels sprout (*Brassica oleracea* L. var. *gemmifera*). *J. Sci. Food Agric.*, **31**, 785–793.

Heaney, R. K. and Fenwick, G. R. (1987). Identifying toxins and their effects: glucosinolates. In D. H. Watson (ed.), *Natural Toxicants in Food*, Ellis Horwood, Chichester, pp. 76–109.

Hikino, H., Tohkin, M., Kiso, Y., Namiki, T., Nishimura, S., and Takeyama, K. (1986). Antihepatotoxic actions of *Allium sativum* bulbs. *Planta Med.*, 163–168.

Iida, H., Hashimoto, S., Miyazawa, M., and Kameoka, H. (1983). Volatile flavour components of nira (*Allium tuberosum* Rottl.). *J. Food Sci.*, **48**, 660–661.

Ikram, M. (1971). A review on chemical and medicinal aspects of *Allium cepa*. *Pak. J. Sci. Ind. Res.*, **14**, 395–398.

Ikram, M. (1972). A review on chemical and medicinal aspects of *Allium sativum*. *Pak. J. Sci. Ind. Res.*, **15**, 81–86.

Itokawa, Y., Inoue, K., Sasagawa, S., and Fujiwara, M. (1973). Effect of S-methylcysteine sulphoxide, S-allylcysteine sulphoxide and related sulphur-containing amino acids on lipid metabolism of experimental hypercholesterolae-mic rats. *J. Nutr.*, **103**, 88–92.

Jahreis, G., Hesse, V., Schöne, F., Lüdke, W., Hennig, A., and Mehnert, E. (1986). Einfluss von Nitrat and pflanzlichen Goitrogenen auf die Schilddrüsenhormone, den Somatonedinstatus und das Wachtum beim Schwein. *Mh. Vet.-Med.*, **41**, 528–533.

Jansen, E. F. (1948). The isolation and identification of 2,2'-dithiolisobutyric acid from asparagus. *J. Biol. Chem.*, **176**, 657–664.

Kameoka, H., Iida, H., Hashimoto, S., and Miyazawa, M. (1984). Sulphides and furanones from steam volatile oils of *Allium fistulosum* and *A. chinense*. *Phyrochemistry*, **23**, 155–158.

Kametani, T., Fukumoto, K., and Umezawa, O. (1959). Studies on anticancer agents. I. Synthesis of various alkyl thiosulphinates and their tumour-inhibiting effects. *Yakugaku Kenkyu*, **31**, 60–64.

Kasai, T. and Sakamura, S. (1982). 1,2,3-trithiene-S-carboxylic acid in raw aspara-gus shoots. *Agric. Biol. Chem.*, **46**, 821–822.

Kasai, T., Hirakuri, Y., and Sakamura, S. (1981). Two cysteine derivatives in asparagus shoots. *Phytochemistry*, **20**, 2209–2211.

Khera, K. S. (1977). Non-teratogenicity of D- and L-goitrin in the rat. *Food Cosmet. Toxicol.*, **15**, 61–62.

Kitahara, Y., Yanagawa, H., Kato, T., and Takahashi, N. (1972). Asparagusic acid, a new plant growth inhibitor in etiolated young asparagus shoots. *Plant Cell Physiol.*, **13**, 923–925.

Kjaer, A. (1974). The natural distribution of glucosinolates: a uniform class of sulphur-containing glucosides. In G. Bendz and J. Santesson (eds.), *Chemistry in Botanical Classification*, Academic Press, London, pp. 229–257.

Kjaer, A. and Skrydstrup, T. (1987). Seleno-glucosinolates: synthesis and enzymatic hydrolysis. *Acta Chem. Scand.,* **B41**, 29–33.

Klimek, J. W., Cavallito, R. J., and Bailey, C. H. (1948). Induced resistance to *Staphylococcus aureus* to various antibiotics. *J. Bacteriol.,* **55**, 139–145.

Kojima, M., Hamada, H. and Toshimitsu, N. (1986). Simultaneous quantitative determination of allyl isothiocyanate and 2-phenylethyl isothiocyanate by gas chromatography equipped with FPD. *Nippon Shokuhin Kogyo Gakkaishi,* **33**, 155–159.

Kreula, M. and Kiesvaara, M. (1959). Determination of L-5-vinyl-2-thiooxazolidone from plant material and milk. *Acta Chem. Scand.,* **13**, 1375–1382.

Laarveld, B., Brockman, R. P., and Christensen, D. A. (1981). The effects of Tower and Midas rapeseed meals on milk production and concentrations of goitrogens and iodine in milk. *Can. J. Anim. Sc.,* **61**, 131–139.

Langer, P. and Greer, M. A. (1977). *Antithyroid Substances and Naturally Occurring Goitrogens*, Karger, Basle.

Langer, P. and Michajlovskij, N. (1969). Studies on the antithyroid activity of naturally-occuring L-5-vinyl-2-thiooxaxolidone and its urinary metabolite in rats. *Acta Endocrinol. (Copenhagen),* **62**, 21–30.

Lanzani, A., Piana, G., Piva, G., Cardillo, M., and Rastelli, A. (1974). Changes in *Brassica napus* progoitrin induced by sheep rumen fluid. *J. Am. Oil Chem. Soc.,* **51**, 517–518.

Lau, B. H. S., Adetumbi, M. A., and Sanchez, A. (1983). *Allium sativum* (garlic) and atherosclerosis: a review. *Nutr. Res.,* **3**, 119–128.

Lau, B. H. S., Lam, F. and Wang-Cheng, R. (1987). Effect of an odour-modified garlic preparation in blood lipids. *Nutr. Res.,* **7**, 139–149.

Lewis, J. A. and Fenwick, G. R. (1987). Glucosinolate content of *Brassica* vegetables. Analysis of twenty four cultivars of calabrese (green sprouting broccoli, *Brassica oleracea* L. var. *botrytis* subvar. *cymosa* Lam.). *Food Chem.,* **25**, 259–268.

Lewis, J. A. and Fenwick, G. R. (1988). Glucosinolate content of *Brassica* vegetables, Chinese cabbage (*Brassica campestris* ssp. *pekinensis*, pe-tsai and *B. campestris* spp. *chinensis*, pak-choi). *J. Sci. Food Agric.,* **45**, 379–386.

Liakopolou-Kyriakides, M., Sinakos, Z., and Kyriakides, D. A. (1985). Identification of alliin, a constituent of *Allium cepa* with an inhibitory effect on platelet aggregation. *Phytochemistry,* **24**, 600–601.

Lichtenstein, E. P., Morgan, D. G., and Mueller, C. H. (1964). Naturally-occurring insecticides in cruciferous crops. *J. Agric. Food Chem.,* **12**, 158–161.

Linscheid, M., Wendisch, D., and Strack, D. (1980). The structures of sinapic acid esters and their metabolism in cotyledons of *Raphanus sativus*. *Z. Naturforsch.,* **35C**, 907–914.

Lüthy, J. (1984). Identizifierung und Mutagenität des Reaktionsproduktes von Goitrin und Nitrit. *Mitt. Geb. Lebensm., Hyg.,* **75**, 101–102.

Lüthy, J., von Daniken, A., Friedrich, U., Manthey, B., Zweifel, U., Schlatter, C.

L., and Benn, M. H. (1980). Synthesis and toxicology of three natural cyanoe-pithiolkanes. *Int. J. Vit. Nutr. Res.*, **50**, 423–424.

Lüthy, J., Carden, B., Friedrich, U., and Bachmann, M. (1984). Goitrin — a nitrosatable constituent of plant foodstuffs. *Experientia*, **40**, 452–453.

Macholz, R., Ackermann, H., Diedrich, M., Henschel, K. P., Kujawa, M., Lewerenz, H. J., Przybilski, H., Schnaak, E., Schulze, J., and Woggon, H. (1986). Studies on the degradation of glucotropaeolin and progoitrin, toxicity and reactivity of splitting products. *Proc. Eurofoodtox, II, Zürich*, pp. 40–45.

MacLeod, A. J. (1976). Volatile flavour compounds of the Cruciferae. In J. G. Vaughan, A. J. MacLeod, and B. M. G. Jones (eds.), *The Biology and Chemistry of the Cruciferae*, Applied Science, London, pp. 307–330.

Mathew, P. T. and Augusti, K. T. (1973). Studies on the effect of allicin (diallyl disulphide oxide) on alloxan diabetes. I. Hypoglycaemic action and enhance-ment of serum insulin effect and glycogen synthesis. *J. Ind. Biochem. Biophys.*, **10**, 209–212.

Matikkala, E. J. and Virtanen, A. I. (1967). On the quantitative determination of the amino acids and gamma-glutamyl peptides of onions. *Acta Chem. Scand.*, **21**, 2891–2893.

Matovinovic, J. (1983). Endemic goitre and cretinism at the dawn of the third millenium. *Ann. Rev. Nutr.*, **3**, 341–412.

McDanell, McLean, A. E. M., Hanley, A. B., Heaney, R. K., and Fenwick, G. R. (1988). Chemical and biological properties of indole glucosinolates (glucobrassi-cins) — a review. *Food Chem. Toxicol.*, **26,** 59–70.

McGregor, D. I., Mullin, W. J., and Fenwick, G. R. (1983). Analytical methodology for determining glucosinolate composition and content. *J. Assoc. Off. Anal. Chem.*, **66**, 825–849.

McMillan, M., Spinks, E. A., and Fenwick, G. R. (1986). Preliminary observations on the effect of dietary Brussels sprouts on thyroid function. *Human Toxicol.*, **5**, 15–19.

Mennicke, W. H., Görler, K., and Krumbiegel, G. (1983). Metabolism of some naturally-occurring isothiocyanates in the rat. *Xenobiotica*, **13**, 203–207.

Mirelman, D. (1987). Garlic the talisman tackles dysentery, *The Times (London) Science Report*, April 29.

Mitchell, J. C. (1974). Contact dermatitis from plants of the caper family, Cappari-daceae. *Br. J. Dermatol.*, **91**, 13–20.

Mitchell, J. C. and Jordan, W. P. (1974). Allergic contact dermatitis from radish. *Raphanus sativa. J. Dermatol.*, **91**, 183–185.

Mithen, R. F., Lewis, B. G., and Fenwick, G. R. (1986). The *in vitro* activity of glucosinolates and their products against *Leptosphaeria maculans. Trans. Br. Mycol. Soc.*, **87**, 433–440.

Miura, Y., Hohara, S., Tahara, S., and Mizutani, J. (1976). Antibacterial activity of L-5-alkylthiomethylhydantoin-S-oxides. *Agric. Biol. Chem.*, **40**, 1907–1908.

Miura, Y., Tahara, S., and Mizutani, J. (1979). The structure activity relationships of antibacterial L-5-alkylthiomethylhydantoin S-oxides and related compounds. *J. Pesticide Chem.*, **4**, 25–29.

Mohammed, S. F. and Woodward, S. G. (1986). Characterisation of a potential inhibitor of platelet aggregation and release reaction isolated from *Allium sativum* (garlic). *Thrombosis Res.,* **44**, 793–806.

Muller, A. L. and Virtanen, A. I. (1965). On the biosynthesis of cycloalliin. *Acta Chem Scand.,* **19**, 2257–2258.

Mullin, W. J. and Sahasrabudhe, M. R. (1978). An estimate of the average daily intake of glucosinolates via cruciferous vegetables. *Nutr. Rep. Int.,* **18**, 273–278.

Mullin, W. J., Proudfoot, K. G., and Collins, M. J. (1980). Glucosinolate content and clubroot of rutabaga and turnip. *Can. J. Plant Sci.,* **60**, 605–6512.

Murthy, N. B. K. and Amonkar, S. V. (1974). Effect of a natural insecticide from garlic (*Allium sativum* L.) and its synthetic form (diallyl disulphide) on a plant pathogenic fungi. *Ind. J. Exp. Biol.,* **12**, 208–209.

Nencki, M. (1891). Uber das Vorkommen von Methylmercaptan un wenschlicken Harn nach Spargelgenuss. *Arch. Experim. Pathol. Pharmacol.,* **28**, 206–209.

Neudecker, T. and Henschler, D. (1985). Allyl isothiocyanate is mutagenic in *Salmonella typhimurium. Mutat. Res.,* **156**, 33–37.

Nishie, K. and Daxenbichler, M. E. (1980). Toxicity of glucosinolates, related compounds (nitriles, R-goitrin and isothiocyanates) and vitamin K found in Cruciferae. *Food Cosmet. Toxicol.,* **18**, 159–172.

Nugen-Boudon, L. and Szylit, O. (1987). Unpublished observation.

Oshiba, S., Ariga, T., Sawai, H., Imai, H., and Endoh, H. (1981). The inhibitory effect of garlic oil on platelet aggregation, II. *J. Physiol. Soc. Japan,* **43**, 407–412.

Piak, I. K., Robblee, A. R., and Clandinin, D. R. (1980). The effect of sodium thiosulphate and hydroxo-cobalamin in rats fed nitrile-rich or goitrin-rich rapeseed meals. *Can. J. Anim. Sci.,* **60**, 1003–1013.

Pantuck, E. J., Pantuck, C. B., Garland, W. A., Mins, B., Wattenberg, L. W., Anderson, K. E., Kappas, A., and Conney, A. H. (1979). Stimulatory effect of Brussels sprouts and cabbage on human drug metabolism. *Clin. Pharm. Ther.,* **25**, 88–95.

Papageorgiu, C., Corbet, J.-P., Menezes-Brandao, F., Pecegueiro, M., and Benezra, C. (1983). Allergic contact dermatitis to garlic (*Allium sativum* L.). Identification of the allergens; the role of mono-, di- and trisulphides present in garlic. A comparative study in man and animal guniea pig. *Arch. Dermat. Res.,* **275**, 229–234.

Papas, A., Ingalls, J. R., and Campbell, L. D. (1979). Studies on the effect of rapeseed meal on thyroid status of cattle. Glucosinolate and iodine content of milk and other parameters. *J. Nutr.,* **109**, 1129–1139.

Parry, R. J. and Naidu, M. V. (1983). Determination of the absolute configuration of (−)-*S*-(2-carboxypropyl)-L-cysteine. *Tet. Lett.,* **24**, 1133–1134.

Parry, R, J., Mizusawa, A. E., and Ricciardone, M. (1982). Biosynthesis of sulphur compounds. Investigations of the biosynthesis of asparagusic acid. *J. Am. Chem. Soc.,* **104**, 1442–1448.

Pasricha, J. S. and Guru, B . (1979). Preparation of an appropriate antigen extract for patch tests with garlic. *Arch. Dermatol.,* **115**, 230–235.

Petkov, V. (1966). Pharmakologische under Klinische Untersuchungen des Knoblauchs. *Deutsche Apoth. Z.,* **106**, 1861–1867.

Petkov, V. (1979). Plants with hypotensive, antiatheromatous and coronaradilating action. *Am. J. Chinese Med.,* **7**, 197–236.

Petricic, J., Kupinic, M., and Lulic, B. (1978). Garlic. (*Allium sativum* L.) antifungal effects of some components of volatile oil. *Acta Pharm. Jugosl.*, **28**, 41–48.

Petroski, R. J. and Kwolek, W. F. (1985). Interactions of a fungal thioglucoside glucohydrolase and cruciferous plant epithiospecifier protein to form 1-cyanoepithioalkalanes: implications of an allosteric mechanism. *Phytochemistry*, **24**, 213–216.

Pierce, Jr. H. D., Vernon, R. S., Borden, V. H., and Oelschlager, A. C. (1978). Host selection by *Hylemya antiqua* (Meigen): identification of three new attractants and oviposition stimulants. *J. Chem. Ecol.*, **4**, 65–72.

Pushpendran, C. K., Devasagayam, T. P. A., Chintalwar, G. J., Banerji, A. and Eapen, J. (1980). The metabolic fate of (^{35}S) diallyl disulphide in mice. *Experientia*, **36**, 1000–1001.

Reuter, H. D. (1983). Antiarteriosklerotische Wirkung von Knoblauchinhaltstoffen. *Therapie Woche*, **33**, 2474–2487.

Röbbelen, G. and Thies, W. (1980). Variations in rapeseed glucosinolates and breeding for improved meal quality. In A. Tsunoda, K. Hinata and G. Gomez-Campos (eds.), *Brassica Crops and Wild Allies, Biology and Breeding*, Japanese Scientific Societies Press, Tokyo, pp. 285–299.

Rodman, J. E. (1978). Glucosinolates, methods of analysis and some chemosystematic problems. *Phytochem. Bull.*, **11**, 6–31.

Rodman, J. E. (1981). Divergence, convergence and parallelism in phytochemical characters: the glucosinolate–myrosinase system. In D. A. Young and D. A. Seigler, (eds.), *Phytochemistry and Angiosperm Phylogeny*, Praeger, New York, pp. 43–79.

Sainani, G. S., Desai, D. B., Natu, S. M., Pise, D. V., and Sainani, P. G. (1979a). Dietary garlic, onion and some coagulation parameters in the Jain community. *J. Assoc. Phys. Ind.*, **27**, 707–716.

Sainani, G. S., Desai, D. B., Gorhe, N. H., Natu, S. M., Pise, D. V., and Sainani, P. G. (1979b). Effect of dietary garlic and onion on serum lipid profile in the Jain community. *Ind. J. Med. Res.*, **69**, 776–780.

Schöne, F. and Paetzelt, H. (1985). Excretion of thiocyanate in urine of growing pigs after rapeseed meal feeding. *Die Nahrung*, **29**, 541–543.

Small, D. L., Bailey, J. H., and Cavallito, R. J. (1947). Alkylthiosulphinates. *J. Am. Chem. Soc.*, **69**, 1710–1713.

Smith, R. H. (1980). Kale poisoning: the brassica anaemia factor. *Vet. Record*, **107**, 12–15.

Sones, K., Heaney, R. K., and Fenwick G. R. (1984a). An estimate of the mean daily intake of glucosinolates from cruciferous vegetables in the UK. *J. Sci. Food Agric.*, **35**, 712–720.

Sones, K., Heaney, R. K., and Fenwick, G. R. (1984b). The glucosinolate content of UK vegetables — cabbage (*Brassica oleracea*), swede (*B. napus*) and turnip (*B. campestris*). *Food Addit. Contamin.*, **1**, 289–296.

Sones, K., Heaney, R. K., and Fenwick, G. R. (1984c). Glucosinates in *Brassica* vegetables. Analysis of twenty-seven cauliflower cultivars (*Brassica oleracea* L. var. *botrytis* subvar. *cauliflora*, DC). *J. Sci. Food Agric.*, **35**, 762–766.

Spåre, C.-G. and Virtanen, A. I. (1963). On the lachrymatory factor in onion (*Allium cepa*) vapours and its precursors. *Acta Chem. Scand.,* **17**, 641–650.

Sprecher, E. (1986). *Allium sativum* L. — Wundermittel oder Arnzneipflanze? *Pharmazeut. Z.,* **50**, 3161–3168.

Stoll, A. and Seebeck, E. (1947). Alliin, the pure mother substance of garlic oil. *Experientia, 3,* 114–115.

Stoll, A. and Seebeck, E. (1948). Allium compounds, I. Alliin, the true mother compounds of garlic oil. *Helv. Chim. Acta,* **31**, 189–210.

Stoll, A. and Seebeck, E. (1949). Uber den enzymatischen Abbaü der Alliins und die Eigenschaften der Alliinase. *Helv. Chim. Acta, 32,* 197–205.

Stoll, A. and Seebeck, E. (1951). Chemical investigations on alliin, the specific principle of garlic. *Adv. Enzymol.,* **11**, 377–399.

Tahara, S. and Mizutani, J. (1979). L-5-alk(en)ylthiomethylhydantoin-($\pm$)-S-oxides: non-enzymatical precursors of fresh flavours of *Allium* plants. *Agric. Biol. Chem.,* **43**, 2021–2028.

Tahara, S., Miura, Y., and Mizutani, J. (1977). Alkylthiosulphinates, antimicrobial principles of S-alkylthiomethylhydantoin-S-oxides. *Agric. Biol. Chem.,* **41**, 221–222.

Tobkin, H. E. and Mazelis, M. (1979). Alliin lyase: preparation and characterisation of the homogeneous enzyme from onion bulbs. *Arch. Biochem. Biophys.,* **193**, 150–157.

Tookey, H. L., VanEtten, C. H. and Daxenbichler, M. E. (1980). Glucosinolates. In I. E. Liener (ed.), *Toxic Constituents of Food Crops,* 2nd edn., Academic Press, New York, pp. 103–142.

Tressl, R., Holzer, M., and Apetz, M. (1977a). Formation of flavour components in asparagus, I. Biosynthesis of sulphur-containing acids in asparagus. *J. Agric. Food Chem.,* **25**, 455–459.

Tressl, R., Bahri, D., Holzer, M., and Kossa, T. (1977b). Formation of flavour components in asparagus, II. Formation of flavour components in cooked asparagus. *J. Agric. Food Chem.,* **25**, 459–463.

Uda, Y., Kurata, T., and Arakawa, N. (1986a). Effects of pH and ferrous ion on the degradation of glucosinolates by myrosinase. *Agric. Biol. Chem.,* **50**, 2735–2740.

Uda, Y., Kurata, T., and Arakawa, N. (1986b). Effects of thiol compounds on the formation of nitriles from glucosinolates. *Agric. Biol. Chem.,* **50**, 2741–2746.

Underhill, L. R. (1980). Glucosinolates. In E. A. Bell and B. V. Charlwood (eds.), *Encyclopaedia of Plant Physiology*, Vol. 8, Springer, Heidelberg, pp. 493–511.

Underhill, L. R. and Wetter, E. W. (1973). Biosynthesis of glucosinolates. *Biochem. Soc. Symp.,* **38**, 303–326.

VanEtten, C. H., Gagne, W. E., Robbins, D. J., Booth, A. N., Daxenbichler, M. E., and Wolff, I. A. (1969). Biological evaluation of *Crambe* seed meals and derived products in rat feeding. *Cereal Chem.,* **46**, 145–155.

VanEtten, C. H., Daxenbichler, M. E., Tookey, H. L., Kwolek, W. F., Williams, P. H., and Yoder, O. C. (1980). Glucosinolates: potential toxicants in cabbage cultivars. *J. Am. Soc. Hort. Sci.,* **105**, 710–714.

Venugopal, M. S. and Narayan, V. (1981). Effects of allitin on the green peach aphid (*Myzus persicae* Sutzer). *Int. Pest Control,* 130–131.

Vermorel, M., Heaney, R. K., and Fenwick, G. R. (1986). Nutritive value of rapeseed meal: effects of individual glucosinolates. *J. Sci. Food Agric.*, **37**, 1197–1202.

Vermorel, M., Heaney, R. K., and Fenwick, G. R. (1988). Antinutritional effects of the rapeseed meals, Darmor and Jet Neuf and the glucosinolate progoitrin, together with myrosinase in the growing rat. *J. Sci. Food Agric.*, **45**, 321–334.

Vernon, R. S., Judd, G. J. R., Bordon, J. H., Pierce Jr., H. D., and Oelschlager, A. G. (1981). Attraction of *Hylemya antiqua* Meigen (Diptera: Anthomyiidae) in the field to host-produced oviposition stimulants and their new host analogues. *Can. J. Zool.*, **59**., 872–881.

Virtanen, A. I. (1965). Studies on organic sulphur compounds and other labile substances in plants — a review. *Phytochemistry*, **4**, 207–228.

Virtanen, A. I. and Matikkala, E. J. (1956). A new sulphur-containing amino acid in the onion. *Suom. Kemi.*, **B29**, 134–135.

Wakabayashi, K., Nagao, M., Ochiai, M., Tahira, T., Zamaizumi, Z., and Sugimura, T. (1985a). A mutagen precursor in Chinese cabbage indole-3-acetonitrile which becomes mutagenic on nitrite treatment. *Mutat. Res.*, **143**, 17–21.

Wakabayashi, K., Nagao,, M., Tahira, T., Saito, H., Katayama, M., Maramo, S., and Sugimura, T. (1985b). 1-nitrosoindole-3-acetonitrile, a mutagen produced by nitrite treatment of indole-3-acetonitrile. *Proc. J. Japan*, **618**, 190–192.

Wakabayashi, K., Nagao, M., Tahira, T., Yamaizumi, Z., Katayama, M., Marumo, S., and Sugimura, T. (1986). 4-methoxyindole derivatives as nitrosable precursors of mutagens in Chinese cabbage. *Mutagenesis*, **1**, 423–426.

Wargovitch, M. J. (1987a). Diallyl sulphide, a flavour component of garlic (*Allium sativum*), inhibits dimethylhydrazine-induced colon cancer. *Carcinogenesis*, **8**, 487–489.

Wargovitch, M. J. (1987b). Personal communication.

Wargovitch, M. J. and Goldberg, M. T. (1985). Diallysulphide, a naturally-occurring thioether inhibits carcinogen-induced nuclear damage to colon epithelial cells *in vivo*. *Mutat. Res.*, **143**, 127–129.

Wattenberg, L. W. (1977). Inhibition of carcinogenic effects of polycyclic hydrocarbons by benzyl isothiocyanate and related compounds. *J. Natl. Cancer Inst.*, **58**, 395–403.

Weisberger, A. J. and Pensky, J. (1957). Tumour-inhibiting effects derived from an active principle of garlic (*Allium sativum*). *Science*, **126**, 1112–1114.

Weisberger, A. J. and Pensky, J. (1958). Tumour inactivation by a sulphydryl-blocking agent related to an active principle of garlic (*Allium sativum*). *Cancer Res.*, **18**, 1301–1308.

Whitaker, J. R. (1976). Development of flavour, odour and pungency in onion and garlic. *Adv. Food Res.*, **22**, 73–133.

Yamaguchi, T. (1980). Mutagenicity of isothiocyanates, thiocyanates and thioureas on *Salmonella typhimurium*. *Agric. Biol. Chem.*, **44**, 3017–3018.

Yanagawa, H., Kato, T., Kitahara, Y., Takahashi, N., and Kato, Y. (1972). Asparagusic acid, dihydroasparagusic acid and S-acetyldihydroasparagusic acid, a new plant growth inhibitors in etiolated young *Asparagus officinalis*. *Tet. Lett.*, 2549–2552.

Yanagawa, H., Kato, T., and Kitahara, Y. (1973). Stimulation of pyruvate oxidation in asparagus mitochondria by asparaguric acids. *Plant Cell Physiol.,* **14**, 1213–1216.

Zelikoff, J. T., Atkins Jr., N. M., and Belman, S. (1986). Stimulation of cell growth and proliferation of NIH-3T3 cells by onion and garlic oils. *Cell Biol. Toxicol.,* **2**., 369–378.

Index